ELEMENTS OF
STATISTICAL COMPUTING

Numerical computation

ELEMENTS OF STATISTICAL COMPUTING

Numerical computation

Ronald A. Thisted
The University of Chicago

New York   London
CHAPMAN AND HALL

To Linda
Without whose encouragement and good cheer
this book would never have been completed

First published in 1988 by
Chapman and Hall
29 West 35th Street, New York NY 10001
Published in the UK by
Chapman and Hall Ltd
11 New Fetter Lane, London EC4P 4EE

© 1988 R. A. Thisted

Printed in the United States of America

ISBN 0 412 01371 1

Library of Congress Cataloging in Publication Data

Thisted, Ronald A. (Ronald Aaron), 1951–
    Elements of statistical computing.

    Bibliography: p.
    Includes index.
    1. Mathematical statistics——Data processing
I. Title
QA276.4.T47   1988   519.5'028'5   86-23227
ISBN 0–412–01371–1

British Library Cataloguing in Publication Data

Thisted, Ronald A.
    Elements of statistical computing.
    1. Mathematical statistics——Data processing
    I. Title
    519.5'028'5   QA276.4
    ISBN 0–412–01371–1

CONTENTS

PREFACE

The purpose of this book is to describe the current state of the art in a number of branches of statistical computing. The very term "statistical computing" is ill-defined, and one aspect of this book is to illustrate the pervasive nature of computation in statistics at all levels, from the most mundane and straightforward applications to the most esoteric and formidable theoretical formulations. What is more, these computational aspects are often at the frontier of knowledge, so that statistical computing now contains a rich variety of research topics in almost every branch of statistical inquiry. A second purpose of this book, then, is to demonstrate the breadth of study and the great room for advances that are possible in areas that simultaneously involve computational and statistical ideas.

Graduate students in Statistics or Computer Science particularly may find this book helpful in suggesting topics worthy of research that would otherwise not be apparent. Researchers in statistical theory and methods may also find this book of value, both for pointing out areas of inquiry that are little known and for the suggestions provided for further reading in each of the areas. In this latter respect, my intent is to provide a source book that can be used as a starting point from which readers can locate results of current research on a wide spectrum of topics in computational statistics. Finally, I hope that this book just makes plain good reading for those who want to stay abreast of advances in this field.

This book makes some assumptions about the background of its reader. The general prerequisite is that attained in the typical first year of graduate study in Statistics, together with a first-year graduate knowledge of computer programming, data structures, and algorithmic analysis. That "typical" first year in Statistics is assumed to provide familiarity both with the basic theoretical ideas and frameworks used in statistics, as well as some experience with fundamental statistical methods and their applications to real problems, that is, to designing real experiments or to analyzing real data. Although this is a *general* level of necessary background, it is not always either necessary or sufficient. Some of the chapters will require deeper understanding of or fuller experience in statistical theory, computer science, or data analysis, while other chapters will require far less.

In particular Chapter 1, which is an overview of the field of statistical computing, is purposely written with a much lower level of background in mind. This chapter assumes the smallest possible exposure to statistical methods and to computation. It can serve as the point of departure for a one-quarter undergraduate course that makes particular use of the experience and background of its students through judicious selection of material. The first chapter also identifies the various threads running through the cloth of the book, and consequently, is the one chapter that serves as introduction to each of the others.

The remaining chapters are meant to stand more or less on their own. Chapter 3, for instance, could serve as the basis for a one semester course in computational aspects of linear models without relying heavily on other chapters. To a lesser extent, I have tried to make the major sections within chapters as independent of one another as possible so as to enhance their value as a reference. Some topics require more space than would be appropriate for a single chapter, and some topics are inherently closely related. I have tried to make these connections natural and obvious.

The chapters of this book could be organized along several different lines. The most obvious choices are discussed at some length in the first chapter, namely, organizing according to branches of statistics (applications, methods, theory, meta-theory), according to branches of computer science (numerical analysis, graphics, software engineering, symbolic computation, artificial intelligence), or according to style of computation (numerical, seminumerical, graphical, symbolic, environmental). Except for Chapter 1, the chapters are ordered more or less according to this last organizing principle, as it seems to me the best way to demonstrate that seemingly disparate topics in statistical computing are really related to one another and lie on a spectrum. In order to present an overview for a statistical audience, however, and to emphasize the pervasiveness of computing in statistics, the first chapter concentrates on illustrating the role of computation in various branches of statistical inquiry.

This is not the first book written on statistical computing. Unlike its predecessors, however, it is not primarily a book of algorithms, nor is it a collection of sources for algorithms. The field has now grown sufficiently that it is now possible to step back somewhat and to attempt an overview which takes account of some of the philosophical issues as well as the purely methodological ones. One fact which makes such an attempt possible is the existence of an excellent work by Kennedy and Gentle (1980) which is a superb compendium of statistical computing methods, emphasizing numerical algorithms. References to Kennedy and Gentle abound here, and with good reason. An earlier book, Chambers (1977), is also notable in this area. Chambers's book is a somewhat broader collection of computational methods relevant to data analysis. At the same time, it is considerably more compact. Although it has been overtaken somewhat by the rapid developments both in computation and in statistics in the last nine years, Chambers's book was the first attempt at a systematic introduction to statistical computing, and it remains valuable.

The emphasis in this book is on computational methods which are of general importance in statistics, either as fundamental building blocks on which other computations are based, or as algorithms which provide insight into the statistical structure of computational problems or the computational structure of statistical problems. There are many computational methods which are highly specialized or highly complex, which do not fall into this category despite their prominence in statistical data analysis (and computing centers' income streams). Examples include computation for the general linear model and for factor analysis. One could easily devote an entire book to the computational issues in either of these areas, indeed, people have done so. What distinguishes these particular techniques is that they are not really building blocks for other statistical computations; rather, they constitute an end in themselves. Many of these computations represent special-purpose elaboration on the more fundamental algorithms and data structures discussed here. Because such excellent programs as SAS and SPSS exist and are generally available, it is unlikely that many statisticians will find the need to write a general analysis of variance program, for example, either for data analysis or for use in their research. Much the same can be said about factor analysis. For this reason, little space is devoted here to the computational details of these procedures. Rather, we have focused on the nature of these computations, and have provided pointers to the literature for the interested reader.

To make this work more suitable for self-study, most sections conclude with a series of exercises based on the material in the section. Answers, hints, or suggestions are provided for many of these exercises. Most of the answers are in outline form only, and require details to be filled in by the reader. It is hoped that the comments are sufficiently detailed to make them useful to the solitary reader while maintaining their utility for classroom use.

Each exercise has been assigned a grade roughly indicating its difficulty. The grading scheme is a Richter-like scale devised by Donald Knuth, in which the rating is approximately logarithmic in the difficulty. Readers who find problems that are either much easier or much more difficult than the difficulty rating suggests are invited to correspond with the author. The following description of the rating system is quoted from Knuth (1968).

Rating Interpretation

00 An extremely easy exercise which can be answered immediately if the material of the text has been understood, and which can almost always be worked "in your head."

10 A simple problem, which makes a person think over the material just read, but which is by no means difficult. It should be possible to do this in one minute at most; pencil and paper may be useful in obtaining the solution.

20 An average problem which tests basic understanding of the text material but which may take about fifteen to twenty minutes to answer completely.

30 A problem of moderate difficulty and/or complexity which may involve over two hours' work to solve satisfactorily.

40 Quite a difficult or lengthy problem which is perhaps suitable for a term project in classroom situations. It is expected that a student will be able to solve the problem in a reasonable amount of time, but the solution is not trivial.

50 A research problem which (to the author's knowledge at the time of writing) has not yet been solved satisfactorily. If the reader has found an answer to this problem, he is urged to write it up for publication; furthermore, the author of this book would appreciate hearing about the solution as soon as possible (provided it is correct)!*

This volume contains the first six chapters of what I once envisioned as a single-volume survey of statistical computing. The outline below of the *Elements of Statistical Computing* contains tentative chapter titles for this lengthier work. It is likely that a second volume containing Chapters 7 through 11 will appear, and occasional references to these chapters crop up in this volume. The field of computing is changing so rapidly, however, that the remainder of the outline may be obsolete by the time the earlier chapters have been finished. For this reason, readers should consider this outline to be the expression of my fond hope that someday these topics will receive an appropriate treatment.

Part I. Introduction and History

1. Introduction to Statistical Computing

Part II. Numerical Computation

2. Basic Numerical Methods

---

* Donald Knuth, *The Art of Computer Programming*, ©1981, Addison-Wesley, Reading Massachusetts. Reprinted with permission.

This book was typeset using TEX, the mathematical typesetting system designed and implemented by Donald Knuth (1984). That means that even typogryphical errors are the author's fault. I hope that readers will be kind enough to inform me of any errors they discover, so that they may be expunged from future editions. [Of course, I am in favor of expunging errors, not readers!]

It is a pleasure to acknowledge the support over many years of the National Science Foundation, most recently under Grant DMS-8412233.

Computing equipment used in the preparation of this work was supported in part by NSF Grant MCS-8404941 to the Department of Statistics at the University of Chicago. The manuscript was (nearly) completed while I was on leave at the Department of Statistics at Stanford University. Stanford's hospitality contributed greatly to the completion of this work.

I am indebted to many colleagues whose comments and suggestions upon reading early versions of this material have led to marked improvements and, on occasion, to salvation from committing major blunders. I am particularly grateful in this regard for comments received from Tony Cooper, Sir David Cox, Peter McCullagh, Robert Tibshirani, David Wallace, Sanford Weisberg, and Wing Wong. A number of people responded to a query on the modern history of statistical computation, and I thank them for the recollections and materials that they provided: John Chambers, Sir David Cox, Wilfrid Dixon, James Gentle, John Gower, Peter Huber, and William Kennedy. Any errors of omission or commission are, of course, the author's responsibility.

To my many teachers I owe a great intellectual debt. Although I have received benefit from each of them, I am particularly grateful to four without whom this book arguably would never have come to pass: to Bill Boyd, who first taught me how to write; to Larry Shaw, who taught me how to conduct an experiment; to Isabelle Walker, who introduced me both to statistics and to computation; and to Frederick Sontag, who taught me how to ask questions. Finally, I thank my first and best teachers, my parents.

Ronald A. Thisted
Department of Statistics
The University of Chicago

Chicago, Illinois
September, 1987

1 INTRODUCTION TO STATISTICAL COMPUTING

It is common today for statistical computing to be considered as a special subdiscipline of statistics. However such a view is far too narrow to capture the range of ideas and methods being developed, and the range of problems awaiting solution. Statistical computing touches on almost every aspect of statistical theory and practice, and at the same time nearly every aspect of computer science comes into play. The purpose of this book is to describe some of the more interesting and promising areas of statistical computation, and to illustrate the breadth that is possible in the area. Statistical computing is truly an area which is on the boundary between disciplines, and the two disciplines themselves are increasingly finding themselves in demand by other areas of science. This fact is really unremarkable, as statistics and computer science provide complementary tools for those exploring other areas of science. What is remarkable, and perhaps not obvious at first sight, is the universality of those tools. Statistics deals with how information accumulates, how information is optimally extracted from data, how data can be collected to maximize information content, and how inferences can be made from data to extend knowledge. Much knowledge involves processing or combining data in various ways, both numerically and symbolically, and computer science deals with how these computations (or manipulations) can optimally be done, measuring the inherent cost of processing information, studying how information or knowledge can usefully be represented, and understanding the limits of what can be computed. Both of these disciplines raise fundamental philosophical issues, which we shall sometimes have occasion to discuss in this book.

These are exciting aspects of both statistics and computer science, not often recognized by the lay public, or even by other scientists. This is partly because statistics and computer science — at least those portions which will be of interest to us — are not so much scientific as they are fundamental to all scientific enterprise. It is perhaps unfortunate that little of this exciting flavor pervades the first course in statistical methods, or the first course in structured programming. The techniques and approaches taught in these courses are fundamental, but there is typically such a volume of material to cover, and the basic ideas are so new to the students, that it is difficult to show the exciting ideas on which these methods are based.

1.1 Early, classical, and modern concerns

I asserted above that statistical computing spans all of statistics and much of computer science. Most people — even knowledgeable statisticians and computer scientists — might well disagree. One aim of this book is to demonstrate that successful work in statistical computation broadly understood requires a broad background both in classical and modern statistics and in classical and modern computer science.

We are using the terms "classical" and "modern" here with tongue slightly in cheek. The discipline of statistics, by any sensible reckoning, is less than a century old, and computer science less than half that. Still, there are trends and approaches in each which have characterized early development, and there are much more recent developments more or less orthogonal to the early ones. Curiously, these "modern" developments are more systematic and more formal approaches to the "primordial" concerns which led to the establishment of the field in the first place, and which were supplanted by the "classical" formulations.

In statistics, which grew out of the early description of collections of data and associations between different measurable quantities, the classical work established a mathematical framework, couched in terms of random variables whose mathematical properties could be described. Ronald A. Fisher set much of the context for future development, inventing the current notion of testing a null hypothesis against data, of modeling random variation using parameterized families, of estimating parameters so as to maximize the amount of information extracted from the data, and of summarizing the precision of these estimates with reference to the information content of the estimator. Much of this work was made more formal and more mathematical by Jerzy Neyman, Egon S. Pearson, and Abraham Wald, among others, leading to the familiar analysis of statistical hypothesis tests, interval estimation, and statistical decision theory.

More recently, there has been a "modern" resurgence of interest in describing sets of data and using data sets to suggest new scientific information (as opposed to merely testing prespecified scientific hypotheses). Indeed, one might say that the emphasis is on analyzing data rather than on analyzing procedures for analyzing data.

In computer science, too, there were primordial urges — to develop a thinking machine. Early computers were rather disappointing in this regard, although they quickly came to the point of being faster and more reliable multipliers than human arithmeticians. Much early work, then, centered on making it possible to write programs for these numerical computations (development of programming languages such as FORTRAN), and on maximizing the efficiency and precision of these computations necessitated by the peculiarities of machine arithmetic (numerical analysis). By the 1960s, computers were seen as number crunchers, and most of computer science dealt with this model of what computers do. More recently, computer science has come again to view computing machines as general processors of symbols, and as machines which operate most fruitfully for humans through interaction with human users (collaborators?). The fields of software engineering and of artificial intelligence are two "modern" outgrowths of this view.

To see how a broad understanding of statistics and a broad understanding of computer science is helpful to work in almost any area of statistical computing — and to see why there is such interest in this area — it is helpful to examine how and where computation enters into statistical theory and practice, how different aspects of computer science are relevant to these fields, and also how the field of statistics treats certain areas of broad interest in computer science. With this in hand it will be easier to see that computer science and statistics are, or could well be, closely intertwined at their common boundary.

1.2 Computation in different areas of Statistics

It is difficult to divide the field of statistics into subfields, and any particular scheme for doing so is somewhat arbitrary and typically unsatisfactory. Regardless of how divided, however, computation enters into every aspect of the discipline. At the heart of statistics, in some sense, is a collection of *methods* for designing experiments and analyzing data. A body of knowledge has developed concerning the mathematical properties of these methods in certain contexts, and we might term this area *theory of statistical methods.*The theoretical aspects of statistics also involve generally applicable abstract mathematical structures, properties, and principles, which we might term *statistical meta-theory.* Moving in the other direction from the basic methods of statistics, we include *applications* of statistical ideas and methods to particular scientific questions.

1.2.1 Applications

Computation has always had an intimate connection with statistical applications, and often the available computing power has been a limiting factor in statistical practice. The applicable methods have been the currently computable ones. What is more, some statistical methods have been invented or popularized primarily to circumvent then-current computational limitations. For example, the centroid method of factor analysis was invented because the computations involved in more realistic factor models were prohibitive (Harman, 1967). The centroid method was supplanted by principal factor methods once it became feasible to extract principal components of covariance matrices numerically. These in turn have been partially replaced in common practice by normal-theory maximum likelihood computations which were prohibitively expensive (and numerically unstable) only a decade ago.

Another example, in which case statistical methods were advanced to circumvent computation limitations, is the invention of Wilcoxon's test (1945), a substitute for the standard t-test based only on the ranks of the data rather than the numerical values of the data themselves. Wilcoxon invented the procedure as a quick approximate calculation easier to do by hand than the more complicated arithmetic involved in Student's procedure. It was later discovered that in the case of data with a Gaussian (normal) distribution — where the t-test is optimal — Wilcoxon's procedure loses very little efficiency, whereas in other (non-Gaussian) situations, the rank-based method is superior to the t-test. This observation paved the way for further work in this new area of "nonparametric" statistics. In this case, computational considerations led to the founding of a new branch of statistical inquiry, and a new collection of statistical methods.

COMMENT. Kruskal (1957) notes that, as is often the case in scientific inquiry, the two-sample version of Wilcoxon's procedure was anticipated by another investigator, in this case Gustav Adolf Deuchler (1883–1955), a German psychologist. Deuchler published in 1914; others working independently around 1945 also discovered the same basic procedure.

There are ironies in the story of nonparametric statistics. Wilcoxon's idea was based on the assumption that it is easier for a human to write down, say thirty, data values in order (sorting) than it would be to compute the sum of their squares and later to take a square root of an intermediate result. Even with most hand calculators this remains true, although the calculator can't help with this sorting process. It turns out, however, that even using the most efficient sorting algorithms, it is almost always faster *on a computer* to compute the sums of squares and to take square roots than it is to sort the data first. In other words, on today's computers it is almost always more efficient to compute Student's t than it is to compute Wilcoxon's statistic! It also turns out that most other nonparametric methods require much more computation (if done on a computer, anyway) than do their Gaussian-theory counterparts.

A third example is that of multiple linear regression analysis, possibly the most widely used single statistical method. Except for the simplest problems the calculations are complicated and time-consuming to carry out by hand. Using the generally available desk calculators of twenty years ago, for instance, I personally spent two full days computing three regression equations and checking the results. Even in the early sixties, regression analyses involving more than two variables were hardly routine. A short ten years later, statistical computer packages (such as SPSS, BMD, and SAS) were widely available in which multiple linear regression was an easily (and inexpensively) exercised option. By the late seventies, these packages even had the capability of computing all 2^p regression models possible to form based on p candidate independent variables, and screening out the best fitting of these.

COMMENT. Actually, this last assertion is over-stated. Of the 2^p possible models, many of these are markedly inferior to the others. To screen this many regression models efficiently, algorithms have been developed which avoid computing most of these inferior models. Still, as the number of predictors p increases, the amount of computation increases at an exponential rate, even using the best known algorithms. It is not known whether less computation would suffice. We shall return to this problem in Chapter 22.

Thus, as computing power has increased, our notion of what a "routine regression analysis" is has changed from a single regression equation involving a single predictor variable, to screening possibly a million possible regression equations for the most promising (or useful) combinations of a large set of predictors.

The statistical methods that people use are typically those whose computation is straightforward and affordable. As computer resources have fallen dramatically in price even as they have accelerated in speed, statistical practice has become more heavily computational than at any time in the past. In fact, it is more common than not that computers play an important role in courses on statistical methods from the very start, and many students' first introduction to computing at the university is through a statistical computer package such as Minitab or SPSS, and not through a course in computer science. Few users of statistics any longer expect to do much data analysis without the assistance of computers.

COMMENT. Personal computers are now quite powerful and inexpensive, so that many students coming to the university have already had experience computing, perhaps also programming, without any exposure to data analysis or computer science. As of this writing, software for statistical computing or for data analysis is primitive. It is likely, however, that personal computers will greatly influence the direction that statistics — and computer science — will take in the next two decades. Thisted (1981) discusses ways in which personal computers can be used as the basis for new approaches to data analysis.

1.2.2 Theory of statistical methods

Postponing for the moment a discussion of statistical methods themselves, let us turn to theoretical investigation of properties of statistical procedures. Classically, this has involved studying the mathematical behavior of procedures within carefully defined mathematical contexts.

> COMMENT. The conditions under which a procedure is optimal, or has certain specified properties, are often referred to in common parlance as the "assumptions" under which the procedure is "valid," or simply, the "assumptions of the procedure." This is really a misnomer, in that it is often entirely permissible—even preferable—to use a procedure even if these conditions are known not to hold. The procedure may well be a valid one to apply in the particular context. To say that a procedure is not valid unless its assumptions hold is an awkward and misleading shorthand for saying that the standard properties which we attribute to the procedure do not hold exactly unless the assumptions (that is, the conditions) of certain theorems are satisfied. These properties may still hold approximately even if the standard assumptions are violated, or the properties we lose may not be important ones.
>
> Sometimes mathematical theory can indicate the extent to which good properties are affected or degraded when the usual assumptions fail to hold, and often we can use the data at hand to assist us in determining whether the data are consistent with these assumptions. Looking at the data in this way is said to be "checking the assumptions," and is generally a good thing to do.

Often these contexts in which properties of statistical procedures are derived are limited to those which admit of exact mathematical analysis, even though they may not reflect very well the situations in the real world in which the statistical procedures under study are likely to be used. Nonetheless, these investigations often lead to deep understanding of how, when, and why particular procedures work well, and sometimes to circumstances in which they cannot be expected to behave adequately. An example may help to clarify this. Consider the problem of testing whether the midpoint of a population (say, its median or its mean) is zero. In many ways, this is the simplest of all statistical problems. The simplest theory for this problem is obtained if we assume that we have n independent observations from this population, which we shall denote by F. If F is symmetric, as we shall assume here, its mean and its median coincide. A common notation is to write

$$X_1, X_2, \ldots, X_n \sim \text{iid } F,$$

where "iid" stands for independent and identically distributed, and "$\sim$" means, "has the distribution of." If we assume that F is a Gaussian distribution with mean (and median) μ and variance σ^2, we can test whether $\mu = 0$ using the t-statistic

$$t = \frac{\sum_{i=1}^{n} X_i}{n} \bigg/ \frac{s}{\sqrt{n}} = \frac{\overline{X}}{s/\sqrt{n}},$$

where

$$s = \sqrt{\sum (X_i - \overline{X})^2/(n-1)}.$$

Within this mathematical structure it can be shown that the t procedure is the best possible, in the sense that it makes optimal use of the information in the data about the question of interest, that is, whether $\mu = 0$. Moreover, the theory tells us exactly what the distribution of the t-statistic is under this hypothesis (namely, Student's t on $n - 1$ degrees of freedom). If, however, the population F from which we are sampling is not Gaussian, then the theory mentioned so far does not directly apply. In fact, the t-statistic is no longer optimal, and even under the null hypothesis, no longer has a Student distribution. (More sophisticated theory, however, tells us that under fairly unrestrictive assumptions about F the t procedure will do quite well, and its null distribution will be close to that in the Gaussian case.)

The Wilcoxon procedure mentioned above in effect replaces the data values X_i themselves by the corresponding "signed rank." That is, the smallest observation in absolute value is replaced by $+1$ (if it is positive) or -1 (if it is negative), the next smallest observation is replaced by ± 2 (again, depending on the sign of the corresponding X_i), and so on, up to the largest observation, which is replaced by $\pm n$. The test statistic here is essentially the sum of these signed ranks. Theory tells us that under the null hypothesis of a zero median, this test statistic has a given distribution *independent of the underlying population F*. This means that the properties of the Wilcoxon procedure are the same, regardless of whether we are sampling from a Gaussian population, or a Cauchy population, or one with some completely arbitrary and mathematically ugly distribution. It does not matter whether the variance of the population is known or not; in fact the population variance need not even exist.

The Wilcoxon procedure may, on this account, look very attractive, even preferable to the t-procedure. But it is not assumption-free. The assumptions under which the results quoted above apply, both for the t-test and for the Wilcoxon test, include three quite restrictive assumptions. First, all of the observations must have been sampled from the same population F. This implies, among other things, that each of the observations must have the same variance (provided the variance exists). Second, the observations must be statistically independent of one another. Often this assumption fails to hold, particularly when the observations are taken sequentially in time. Third, F must be symmetric. What happens if we enlarge the mathematical context to allow non-identical distributions, or to allow observations with some dependence, or to allow non-symmetric distributions? Immediately, the mathematics becomes very difficult, perhaps impossible, even in this simplest of statistical settings.

This is one situation, however, in which computation can come to the rescue, to help answer theoretical questions about statistical procedures. One of the basic tools of the theoretical statistician is *simulation*, or *Monte Carlo experimentation*. Consider the specific problem of testing whether $\mu = 0$ discussed above, provided that the observations are no longer independent, but that observations taken near to one another in time are more highly correlated than observations separated by long time intervals. If the statistician can specify a particular mathematical structure having these properties, and if he can generate samples from the population specified in this structure, then he can 1) generate many thousands of such samples, 2) for each such sample, construct the Wilcoxon statistic and the t-statistic, and 3) over these thousands of instances, compare the average performance of these two methods. Using the Monte Carlo method to address theoretical questions depends upon three factors: the ability to define an appropriate structure for the problem, the ability to use the computer to generate (that is, to simulate) samples from this structure, and the ability to design experiments tailored to answer efficiently the theoretical questions of interest.

There is another area in which computation can be of great service to the theoretical statistician, even when the investigations are entirely mathematical. Much statistical theory involves approximations of the classical type of mathematical analysis, approximating non-linear functions using Taylor expansions, for instance, and approximating non-Gaussian distributions by Edgeworth expansions (Kendall and Stuart, 1977, 169ff), which are similar in spirit. Distributions are also described or approximated in terms of their moments or cumulants. Each of these often requires extensive algebraic computation and laborious computation of derivatives of functions with many terms to obtain the coefficients in the expansions. It is not uncommon to have to produce ten pages of intermediate calculations simply to obtain a single coefficient of interest. Not surprisingly, it is easy to make mistakes and hard to see patterns. The sort of mechanical rule-following required in differentiation and algebraic manipulation is just the sort of thing that we might expect computers to be good at, even though these computations are computations on symbols rather than the numerical sort of computations we have come to associate with Statistics. Algebraic systems such as REDUCE and MACSYMA which can do these and other computations will be discussed in Chapter 16.

A third area in which computation figures in theoretical statistics brings us back to the numerical. Much statistical theory deals with average behavior of estimators or tests, and averages, of course, are sums. If we are averaging over infinitely many possibilities, these sums become integrals, so it is easy to see how complicated integrals might arise in theoretical investigations. Often we wish to demonstrate that one such integral is everywhere less than another as some parameter of interest varies, and sometimes the only way to do so is to compute the numerical value of these integrals over a grid of parameter values. Sometimes, too, we are interested in computing the probability that a random variable will exceed a given value, and this is equivalent to computing the area under the probability density to the right of the given point. In each case, we are led to numerical quadrature to obtain the desired solution.

1.2.3 Numerical statistical methods

Until relatively recently, computation has entered into statistical methods primarily at the level of implementation, that is to say, at the level of algorithms for computing values for statistical formulae. Statistics has always involved considerable arithmetical work. Computers have been helpful primarily in automating these calculations, and the concerns of statistical computing as a discipline in this area has largely been to develop efficient, accurate, and numerically stable algorithms implementing the statistical formulæ.

As an example of this kind of concern, consider the problem of computing the sample variance, which is defined to be

$$s^2 = \sum_{i=1}^{n} (X_i - \overline{X})^2 / (n-1). \tag{1.2.1}$$

For purposes of this discussion, let us consider how to compute the numerator of (1.2.1). There are many alternative ways in which s^2 can be computed on a digital computer, and the answers produced may be very different. Let us consider a few. An obvious approach is to take expression (1.2.1) literally, which requires that the sample mean $\overline{X}$ be computed before computing the sum of squared deviations, giving us

Algorithm A: $sum := 0$
$sq\_sum := 0$
for $i := 1$ **to** n **do** $sum := sum + X_i$
$\overline{X} := sum/n$
for $i := 1$ **to** n **do** $sq\_sum := sq\_sum + (X_i - \overline{X})^2$
$s^2 := sq\_sum/(n-1).$

This method is a "two-pass" algorithm, and requires either that the data all be in memory simultaneously, or that the data must be read twice through from a file. If the number of observations n is large (say on the order of a million data points), these restrictions could become substantial limitations. Moreover, algorithm A requires $3n - 2$ additions and subtractions, and $n + 1$ multiplications and divisions to obtain the final result (omitting the final division, since we are only dealing with the numerator).

An alternative is to use the algebraically equivalent formulation

$$s^2 = \frac{\sum_{i=1}^{n} X_i^2 - \frac{1}{n}(\sum_{j=1}^{n} X_j)^2}{n - 1}. \qquad (1.2.2)$$

This formulation, used in the obvious way, yields a "one-pass" algorithm (Algorithm B), since both the sum of the observations and the sum of their squares can be accumulated at the same time as each observation is read.

> **Algorithm B.** $\quad sum := 0$
> $\qquad\qquad\qquad sq\_sum := 0$
> $\qquad\qquad\qquad$ **for** $i := 1$ **to** n **do**
> $\qquad\qquad\qquad\quad$ **begin**
> $\qquad\qquad\qquad\qquad sum := sum + X_i$
> $\qquad\qquad\qquad\qquad sq\_sum := sq\_sum + X_i^2$
> $\qquad\qquad\qquad\quad$ **end.**

This algorithm requires very little intermediate storage. This algorithm requires only $2n - 1$ additions and subtractions, and only $n + 2$ multiplications and divisions, so that Algorithm B is roughly fifty percent more efficient than Algorithm A in terms of the computations required.

A third equivalent alternative is to use the algebraically equivalent formulation given in Algorithm C. In this algorithm, the ith largest value for X is denoted by $X_{(i)}$.

> **Algorithm C.** $\quad Sort(X)$
> $\qquad\qquad\qquad sum := 0$
> $\qquad\qquad\qquad$ **for** $i := 1$ **to** n **do** $sum := sum + X_{(i)}$
> $\qquad\qquad\qquad \overline{Y} := sum/n$
> $\qquad\qquad\qquad$ **for** $i := 1$ **to** n **do** $Z_i := (X_{(i)} - \overline{Y})^2$
> $\qquad\qquad\qquad Sort(Z)$
> $\qquad\qquad\qquad sq\_sum := 0$
> $\qquad\qquad\qquad$ **for** $i = 1$ **to** n **do** $sq\_sum := sq\_sum + Z_{(i)}$
> $\qquad\qquad\qquad s2 := sq\_sum/(n - 1).$

This bizarre looking algorithm simply rearranges the X's before computing the mean, and then rearranges the squared deviations before adding them up. Ignoring the sorting steps, this algorithm is at least a three-pass algorithm, and in addition to the costs of algorithm A, this algorithm also requires some multiple of $n \log n$ more operations in order to perform the pair of sorting steps.

The most *efficient* of the three algorithms is clearly method B, which requires less space and fewer basic operations than either A or C. However, method A is much more *stable* numerically, since it avoids subtracting two large numbers, many of whose significant digits are likely to cancel. It is even possible for *all* of the significant digits to cancel in Algorithm B, leading to results that are less than useful. (Old computer programs for data analysis sometimes produced negative estimates for variances, simply due to using algorithms such as B.) The accuracy of the result produced by Algorithm B depends a great deal on the relative sizes of the mean and the standard deviation of the data, the ratio of which is called the *coefficient of variation, CV* $= \sigma/\mu$. The larger the mean relative to the standard deviation (the smaller the CV), the worse results algorithm B will produce. Algorithm A, on the other hand, will produce results of the same quality regardless of the size of the CV; hence, algorithm A is said to be numerically stable.

Is there any redeeming virtue to algorithm C? In fact, there is. If there is a very large number of observations, the accumulated sums may become very "full," and additional observations may be so small as to represent less than one digit in the last significant place of the accumulating sum. For instance, suppose that the particular computer we are using carries only three significant (decimal) digits. And suppose that our data set consisted of $X_1 = 1000$, and X_2 through $X_{101} = 1$. Then the sum of these 101 numbers is $\sum X_i = 1.10 \times 10^3$. If we add up the observations in the order given using our mythical 3-digit computer, however, adding 1 to 1.00×10^3 gives 1.001×10^3, which rounds to 1.00×10^3— since this computer retains only three digits. Repeating this process 99 more times gives a final result of 1.00×10^3, which is in error by 10%. This problem is avoided by adding the observations from smallest to largest, however, and this is precisely what algorithm C does wherever a sum is involved. Thus, Algorithm C is the most *accurate* of the three. It is also the most costly in every respect: storage requirements, execution speed, and complexity of program. The phenomenon described here (accumulation error) can occur on any computer using floating-point fixed-precision arithmetic.

Much of the early literature in statistical computing dealt directly with these issues of efficiency, accuracy, and numerical stability, particularly within the context of regression and linear models, where the details of the algorithms for the linear algebraic computations can greatly affect the precision and the stability of the results. Longley (1967) is now a classic paper in this area. Longley compared several widely available computer programs which performed regression analysis in terms of the results they produced for a single data set consisting of economic variables. This data set was ill-conditioned, which is akin to saying that the CV is small in the variance example discussed above. The various packages produced wildly different results from one another, which in turn were wildly different from the exact computations. Largely on the strength of this single paper, much more attention was directed to the numerical quality of the algorithms employed in statistical software. For the most part, however, statistics and statistical data analysis do not require super-accurate algorithms (such as algorithm C in our example), since most of the quantities we compute are estimates for parameters whose standard errors are far larger than possible numerical resolution of the computation. What we do require are relatively efficient numerically stable algorithms (such as algorithm A). Beaton, Rubin, and Barone (1976) give a balanced discussion of the relative roles of numerical and statistical accuracy. By these standards, most generally available statistical packages are quite good for most of the standard statistical computations.

Statistical computing, then, has been part of statistical methods largely through algorithms implementing the computation of standard quantities (such as the variance in our discussion above). And up until, say, the late 1970's, the important concerns in statistical computing were focussed on those algorithms and their efficiency, accuracy, and stability. The primary areas in which statistical computing played a role in the statistical methods people used were in the areas of numerical linear algebra (basically, in regression and analysis of variance), and in computing tail areas of probability functions (replacing standard tables such as the Gaussian, the Student, and chi-squared distributions).

Over the past ten years, however, concerns have shifted, as it has become increasingly apparent that other aspects of statistical methodology could also be affected (positively) through attention to computational matters.

1.2.4 Graphical statistical methods

Pictorial methods of displaying data and of summarizing their contents predate many of the numerical descriptive statistical methods. As with the latter, we seem now to be coming full circle with a restored interest in some of the earliest statistical ideas. Graphical displays, once out of favor as being approximate, inexact or subjective, are now the subject of a resurgence of interest. This renewed interest is due to at least two factors. First, it is once again considered respectable to explore data. It is again legitimate to use data as a vehicle for discovery and not just as a test-bed against which preformulated hypotheses are to be tested. In searching for patterns — especially unexpected ones — the eye assisted by a suitably designed pictorial display is often superior to an array of numerical summaries. The renewal of interest in graphical methods, then, is a side effect of renewed interest in data analysis in general. Data analytic methods are naturally graphical and pictorial, whereas formal statistical procedures such as hypothesis tests do not easily lend themselves to graphical depiction. Second, and by no means less important, the rapid decline in computing costs has made it feasible to automate the process of drawing statistical pictures. Since it is now cheap, fast, and easy to construct residual plots from regression analyses, for instance, and since these plots can be constructed at the same time the other basic regression computations are being done, they are now taught as part of the standard regression methodology and are performed automatically (either by the computer program or by the conscientious student of data analysis). Other plots whose value has long been recognized, such as normal probability plots, used to require laborious table lookups or special paper in order to construct, and it required some effort to produce these plots on a routine basis. With the computing resources now generally available, however, the corresponding entries in the table can be recomputed as needed (to an adequate approximation), and the plot can again be produced automatically. Indeed, now that it is so easy to compute the raw ingredients for these and other plots, it is natural to take one more step and to have the statistical program automatically draw the picture for the user.

Because cheap computing has contributed substantially to the current interest in statistical graphics, much recent work has centered on using the computer to generate the graphical displays. The emphasis has shifted almost entirely away from pencil-and-paper graphics towards computer graphics, so much so that in some circles, statistical graphics is nearly synonymous with statistical computer graphics. Recent sessions at the annual meetings of the American Statistical Association (ASA) sponsored by the ASA's Committee on Statistical Graphics, for example, have largely dealt with work on computer-generated graphical methods. As computing equipment for graphics has become less expensive and of higher quality, it has become possible to construct very high-resolution displays in vivid color at low cost. Vendors of statistical software have made it possible to access these facilities when they are available, but their efforts have simply made it possible to draw the standard graphical displays with the embellishments that color, high resolution, and texture affords. These efforts have to date had no impact whatsoever on the practice of data analysis.

It is becoming fairly common to distinguish between two types of statistical displays: those used for exploration and those used for presentation. The distinction is a useful one. Presentation graphics are those whose purpose is to present or to display some aspect of the data. The purpose is to convey information extracted from the data set. Tufte (1983) is concerned with such visual displays, and with how such displays can be designed "to give visual access to the subtle and the difficult" — in short, to tell a story honestly and clearly. The efforts of statistical computer package developers have been almost exclusively in the realm of presentation graphics, and by some accounts what is now available represents a step backward in displaying information clearly and effectively (see Tufte, 1983). Unfortunately, the recent statistical graphics incorporated in commercial packages have made it possible to construct pie charts and bar charts of the highest quality. But if something is not worth doing, it is not worth doing well.

Although presentation graphics are of some interest to practicing statisticians and data analysts, their day-to-day work uses graphical displays to augment the process of building models, testing hypotheses, and searching for structure. In a sense, the work that must be done is not so much to tell a story, but to discover a plausible story to be told, and a set of graphical tools have been developed which makes this discovery process easier. These exploratory uses of graphics can more appropriately be called graphical *methods*. Developing useful graphical methods is even more difficult than developing numerical statistical methods, since there is a rich body of theory to be drawn on in the latter case. Nonetheless, the field of statistics has built a small but impressive collection of highly useful tools. Among these are the histogram (and relatives such as the rootogram and stem-and-leaf display), scatter plots, residual plots, and probability plots.

These traditional methods, designed for pencil and paper use, are restricted to displaying at most two-dimensional data. The scatter plot, for instance, can only show the relationship between two variables at a time. In a data set with, say, ten variables, there are 45 bivariate scatter plots which could be drawn, and even looking carefully at all of these does not guarantee that some three- or four-way relationship could be detected. This is not to denigrate the bivariate scatter plot — probably the single most useful graphical method that we have. Rather, it is easy to see just how limited even our best methods are, and how much room for improvement there is. In fact, much effort has been directed at displaying three or four aspects of the data in a single display. Many such efforts have involved embellishing the plotted points in the scatter plot by using color, or size of points, or by adding "weather vanes" to each point whose length and direction represent other variables.

How might computers be brought to bear on this problem? There are at least two ways. First, the speed of computation and of computer-driven plotters and graphics terminals makes it possible to construct a large number of pictures, even if each one of them is very complex or requires substantial computation. This makes it possible to consider graphical displays which would take a day or two to construct by hand as plausible candidates for routine use in data analysis (since the computer and the graphics output device could create the same display in minutes or seconds). Thus, for instance, Chernoff's technique of representing highly multivariate data by cartoon faces in which each feature is controlled by the magnitude of one of the variables in the data set (Chernoff, 1973) would not have been possible without the speed and precision of computer plotting equipment.

Second, it is possible to do things with computer graphics display devices that are simply not possible using pencil and paper. For instance, it is possible to show a three-dimensional scatter plot through animation in which the three-dimensional cloud of points appears to rotate on the display screen. This effect is achieved by successively plotting the two-dimensional projections of the three-dimensional point cloud that would be seen by "walking around" the three-dimensional display. An early effort in projection and rotation for data analysis was PRIM-9 (Fisherkeller, Friedman, and Tukey, 1974), which will be discussed at greater length in Chapter 12. A fast computer and display device can produce these successive views at a rate in excess of ten frames per second, which is adequate for the human eye to perceive smooth motion. These "motion graphics," or "kinematic displays," are the basis of some exciting new tools which are now becoming widely available on inexpensive personal computers. Because it is now possible to generate displays so rapidly, it has become possible for the data analyst to interact with an evolving picture on the screen, directing and controlling the evolution of the graphical display. This potential for true interaction, sometimes called *real-time graphics*, suggests that the data analyst may soon be able to bring new skills to bear on old problems of data analysis. All of these methods, of course, are intimately tied to computers, and there are many algorithmic and methodological issues to be addressed in this area, one of the most exciting branches of statistical computing.

1.2.5 Meta-methods: strategies for data analysis

As computers have become faster and cheaper, and as software for data analysis has become easier to use and more widely available, it has become possible for data analysts to look at more aspects of their data sets. For some time — at least twenty years — it has been possible to obtain twenty to thirty pages of computer output pertaining to a particular statistical analysis, much of which is potentially useful. For the past ten years at least, it has been possible to perform data analyses interactively, with each step of the exploration informed by the earlier steps taken. As computers, telecommunications links, and terminals have all become much faster, a requested analysis or graphical display can be produced almost instantaneously. Indeed, as mentioned in the previous section, it is now possible to produce a continuous stream of animated graphics — in effect, graphical statistical analyses are purposely being generated and displayed faster than they can be requested! Our access as data analysts to a panoply of statistical methods has expanded enormously, so that today every beginning student of statistics has a full range of stock methods such as descriptive statistics, regression, correlation, scatter plots, probability plots, analysis of variance, random number generation, basic significance tests, time-series plots and model-building methods (literally) at his or her fingertips.

While it is certainly still appropriate to refer to the computation of, say, a correlation coefficient as a statistical method, actually doing this computation is now a triviality. In Minitab, for instance, one need only type `CORR C1,C2` to obtain the message that "`The correlation is -0.509`"; obtaining a correlation matrix for the first ten variables in a data set requires typing only `CORR C1-C10`. Even two decades ago, a full regression analysis involving just a single predictor variable could easily absorb the better part of a day; now one can do a better and more complete job, and also produce a set of pictures helpful in explaining the analysis, in less than an hour. In that hour, one would typically draw several scatter plots, including plots of the raw data and residuals from the regression fit, examine diagnostic statistics, perhaps look at a probability plot of the residuals, and perhaps examine alternative models involving functions of the original variables. In effect, what we would do today would be more than simply the regression computation: we would combine that computation with several other basic statistical methods, both numerical and graphical.

We might well say that basic statistical methods are building blocks for *strategies of statistical data analysis.* That is, we no longer focus on (or teach) statistics as simply a collection of unrelated techniques, or as a collection of techniques unified only through being maximum likelihood solutions in the normal-theory case. Because computer software and hardware make these basic techniques instantly, cheaply, and painlessly available to us, we can now begin to focus on approaches to data analysis which put these tools together into "analysis strategies" appropriate to different kinds of data. An analogy might be helpful. Orthopædic surgeons use the same basic tools in several different operations; what changes is the order and manner in which each of the tools is used. Still, surgeons in training study and observe the different operations; they do not study the individual tools.

In data analysis, because we are now at the stage where the basic techniques can be easily combined (thanks to advances in computing), we are just now able to focus on more global approaches to data analysis. These approaches or strategies are techniques for appropriately and effectively combining other more basic techniques, and as such we might call them *meta-methods* of data analysis. One implication of such an emphasis is that the statistical software used in data analysis must be increasingly interwoven with statistical methodology. Three widely distributed interactive software products are Minitab, IDA, and SCSS. An approach that is easy and natural using Minitab may well be cumbersome using SCSS or impossible using IDA. [The same can be said for each of the other five permutations.] The statistical computing challenge is twofold, and both aspects have to do with how statistical software systems are designed. First, we must consider ways in which basic techniques can usefully be combined, and then develop methods for structuring those computations. This aspect of the problem has to do with concerns internal to the software system. Second, and possibly more important, is an examination of the software *environment* for data analysis. This has to do with the ways in which the user of the software, that is, the data analyst, interacts with the program in order to actually accomplish his work. Roughly speaking, this aspect is concerned with how to make it easy and comfortable for the user to get things done. This has less to do with "user friendliness" than it has to do with "goodness of fit" between program and user — a package that makes it possible to implement data-analytic strategies should fit like a glove. It should feel like a natural adjunct to the process of analysis.

To date, most software vendors have spent their efforts on making their statistical systems non-intimidating, flexible, and easy to learn. Some quite recent research has begun to consider both the human element in software design and the natural proclivities of experienced data analysts, and to reflect these aspects in the data analysis system itself. Just as the particular software (statistical package) used by a data analyst increasingly affects that analyst's style, so too will particular choices of computing hardware determine approaches to data analysis in the future.

COMMENT. This fact raises serious problems for statistical education, and even for research in statistical methodology. As particular (expensive) pieces of hardware replace pencil and paper as the medium on which statistical results are computed and displayed, it becomes less and less feasible for everyone to learn the same approaches to data analysis simply because not everyone will have access to the particular equipment needed to employ certain new techniques. In addition, if a particular computer and graphics display device is required in order to implement and demonstrate a new method, how is this contribution to the field to be judged? What will be the equivalent of publication? How will this work receive the scrutiny of colleagues comparable to that of a paper in a refereed journal? How will the results of this research be adequately disseminated? Who should bear the cost of dissemination? These questions are by no means trivial, nor are the answers to them at all obvious.

In this regard, however, there is some room for hope. Useful advances of the research community in computer science ultimately reach public availability, even though the delay may be ten or fifteen years. It may be that statistics will become less like mathematics and more like computer science, in which the entire community does not share immediately in each new advance, but rather there are pockets of investigators who enjoy access to the most recent work, and this circle gradually expands as experience is gained and computing costs come down.

The challenge in the arena of meta-methods, then, is to design software which helps to make data analysis more of an integrated activity, and to discover and incorporate ways of structuring both data and computations to facilitate such an approach. A key to advancement in this area is a thorough understanding of data analysis, so that greater emphasis can be placed on strategies for analysis and less on the individual techniques. To accomplish this, the software that is produced must somehow take account of strategies which are generally useful, so as to make such strategies easily followed by users of the system. Which strategies are in fact "generally useful," and even what counts for a strategy in the first place, must arise from the experience and knowledge of seasoned data analysts themselves. In some sense, the best computer systems for data analysis in the future will incorporate knowledge about the process of data analysis itself. This leads statistical computing as a field to the doorstep of artificial intelligence, particularly work in expert systems. This topic will be addressed in the context of statistical data analysis in Chapter 20.

1.2.6 Statistical meta-theory

In the preceding sections we have argued that until quite recently, the role of computing in statistics has been to speed up and to automate some of the basic statistical methods. In the early eighties, though, we are starting to use computers in statistics as fundamental tools in their own right, tools on which new statistical methods can be based. Instead of being viewed as a more or less automatic (and more ecologically sound) substitute for pencil and paper, the computer is now being viewed as a new medium: sometimes superior to pencil and paper, sometimes inferior, and often just fundamentally different.

With this in mind, theoretical statisticians are beginning, in Bradley Efron's words, to "think the unthinkable," and to study the theoretical properties of statistical procedures which require so much computation to carry out that they never would have been thought of seriously before the advent of the computer. Paths which several years ago were dismissed as dead ends due to computational considerations, and along which no one has since thought to travel, are now being reconsidered, often with surprising results. In this way, then, advances in computing in general, and in statistical computing in particular, have changed the shape of statistical theory, in at least two ways. First, there has been a shift in those properties of estimators that are now of interest. For example, the small-sample adequacy of large-sample approximations is of increasing interest, because the requisite calculations are becoming more feasible, and even where this is not the case, much usable information can be collected through simulation of the procedure under study. As with statistical methods, the theoretical issues that we address are those for which we can carry out the calculations. What is different is the view that it may be legitimate to employ computing machines in performing those computations relevant to theoretical questions.

Second, there has been a shift in the kind of statistical procedure worth considering. Two examples may be helpful. Projection pursuit regression (Friedman and Stuetzle, 1981) is a technique for what might be called "multiple nonlinear regression," that is, for constructing a model for a response variable as a nonlinear function of a collection of predictors. The procedure requires so much computation that even ten years ago its routine use would have been hideously expensive. That is no longer the case. Now that projection pursuit regression has been used both for data analysis and for several test cases, empirical questions have been raised which are being addressed theoretically. One such question is this: which multivariable functions can be approximated arbitrarily well by sums of smooth univariate functions whose arguments are projections of the underlying variables (Diaconis and Shahshahani, 1984)? Roughly speaking, the answer is that most nice functions can be so represented. A second example is the "bootstrap" method for obtaining standard errors and confidence intervals for parameters (Efron, 1982). Actual computation of bootstrap estimates often require extensive simulation, which also would have been prohibitively expensive until recently. The bootstrap technique is really a theoretical formulation of a general problem in statistical inference which most of the time can be implemented in specific situations only through the use of substantial computational resources. Because these computations were so horrendous, the theoretical formulation was "unthinkable"—or perhaps totally uninteresting—until the cost of computation fell to current levels. The bootstrap, which raises important and fundamental theoretical questions concerning the foundations of inference, is an example where the availability of computing resources affected one of the basic frameworks of statistical theory.

In our discussion of data analysis environments (encompassing networks of data analytic techniques), we termed these higher-level considerations meta-methodological. Similarly, it is appropriate to think of computational issues which affect the entire approach to statistical theory as meta-theoretical. Although not directly concerns of theoretical statistics, they influence and even help determine what those concerns ought to be. These meta-theoretical issues arise from time to time throughout the field of statistical computing. They are largely philosophical issues to which we shall refer occasionally in the rest of this book.

1.3 Different kinds of computation in Statistics

So far, we have indicated the many different areas of statistics in which computing plays a role—often a pivotal one. While such a picture may be helpful in obtaining an overview of statistical computing, it is hard to organize a systematic investigation of statistical computing as a field of study in its own right along similar lines. For the remainder of this book, we shall consider different kinds or styles of computation—different computational concerns, if you will—and the aspects of statistical computing related to them. This will make it possible to proceed so as to make best use of the similarity of one computational problem to another. In the rest of the book we shall divide the kinds of computational concerns into six overlapping categories: numerical, seminumerical, graphical, symbolic, environmental, and theoretical. These kinds of computation lie more or less on a continuum. Although the topics to be discussed can easily be arranged as points in two dimensions, we do not do much violence to them by forcing them to lie along a single dimension; perhaps the effective dimension is something like 1.2. The remainder of this section will elaborate somewhat on what is contained in these six categories, and will outline topics to be covered under each of them.

COMMENT. It should be noted that these categories refer, for the most part, to kinds of algorithms and have very little to say about data structures, the ways in which data are organized for computation. This is not because data structures are unimportant. Indeed, the choice of a suitable data structure may often be critical to obtain adequate performance. Wherever it is important to do so, we shall point out the role data structures play in each of the areas of computation as they arise in different contexts.

1.3.1 Numerical computation

Numerical calculations are the ones that most people ordinarily associate with statistics: grinding out numbers obtained by applying (complicated) formulæto the numerical values in a data set. These are the sorts of things that students are required to do in courses on statistical methods, sometimes with the aid of a computer but most frequently—particularly at exam time—with recourse to nothing more than a pocket calculator. The computational concerns here chiefly involve numerical analysis (that is, concerns of precision and accuracy), convergence properties of iterative algorithms, and algorithmic efficiency, both in terms of time and space required to carry out the computation.

The computations we are concerned with in the numerical category are those which involve *floating-point numbers,* that is, non-integer numbers represented internally using scientific notation. Chambers (1977, chapter 4), Kennedy and Gentle (1980, chapter 2), and most programming textbooks discuss the fundamentals of floating-point arithmetic, a basic understanding of which is assumed. The majority of statistical methods incorporated in commercial statistical packages such as SPSS, BMDP, and Minitab deal largely with floating-point computation.

Many specific topics fall under the rubric of numerical statistics. *Numerical linear algebra* arises in regression, the analysis of variance, and factor analysis (among other places), through the numerical solution of matrix expressions. These matrix formulæ themselves arise directly from the mathematical formulation of these problems, and they require computing such things as correlation and covariance matrices, solutions to systems of linear equations, matrix inverses, and eigenvalues and eigenvectors of matrices. Two problems which often arise in practice are computing *tail areas* of probability distributions ("p-values"), and the inverse problem of computing *percentage points* of distributions. These problems often require such numerical techniques as approximation by rational functions and computation of repeated fractions.

Numerical quadrature, the numerical evaluation of integrals, arises frequently in statistics, both in statistical theory (for instance, computing asymptotic relative efficiencies, or mean squared errors of estimators) and in data analysis (for instance, in computing the posterior moments of a parameter's distribution). *Optimization* is the process of finding the parameter values (or other conditions) under which a function of the parameters is maximized (or minimized). Optimization problems also arise frequently; for instance, finding maximum likelihood estimates for parameters often involves numerical maximization of the likelihood function, and computing least-absolute-deviations regression estimates also requires numerical optimization. Numerical optimization can also arise in some theoretical calculations, such as in finding an optimal hypothesis test of a given size, or a confidence interval of minimum length.

The problem of estimating a smooth function also arises frequently. Estimating a probability density from observed data, and estimating the mean of a response variable as a nonlinear function of a predictor are two examples. In both cases, techniques for numerical *smoothing* enter the picture. Some of these smoothing algorithms are easily understood and straightforward; others, designed to be useful in a wide variety of circumstances, can be enormously complicated and expensive to compute. Methods such as projection pursuit regression mentioned above, and more general *projection pursuit* methods, are based on a special combination of smoothing and optimization.

Other numerical procedures involve those for estimating *missing data,* including the EM algorithm, for numerical computation of *Fourier transforms,* for *root-finding,* and for estimating variability of estimates using techniques such as the *jackknife* and other methods of *cross-validation*.

1.3.2 Seminumerical computation

Here we have borrowed a term from Knuth (1981), who uses the term "seminumerical" to refer to algorithms requiring attention both to the numerical characteristics of the algorithm proper and to characteristics of the actual machine for which the algorithm is designed. We are using the term in a related, but somewhat different way, to refer primarily to computations whose basic components are manipulations of integers or strings of bits and not of floating-point "real" numbers. Moreover, when these methods are used in data analysis, the actual (floating-point) values of the data themselves are peripheral to the main computation. For example, if we wish to compute the *order statistic* for a random sample, a natural approach is to sort the data. The sorting algorithm in effect provides the permutation of the original data vector which puts the vector in ascending order. The actual data values enter only through comparison to one another (or rather, comparison of their internal bit representation), and not through any arithmetic operation on them. On this view, then, sorting problems are "seminumerical."

There are other statistical problems which also fall into this category. Along with all sorts of sorting tasks, statistical computations sometimes involve *searching* through explicit or implicit tree structures for specific data values or for optima of objective functions. An example of the latter is the computation of all possible subsets regression, in which the goal is to find, for a given number of predictors, the regression equation with maximum F statistic. The number of equations which are candidates for the "best subset" grow exponentially fast in the number of predictors from which to choose. Doing all of these regressions rapidly becomes infeasible. However, clever algorithms can reduce the number of regressions which must be computed to a fraction of the number of subsets by employing a more systematic search.

Monte Carlo methods are used in theoretical work to determine through simulation quantities which would be difficult or impossible to evaluate analytically. A typical simulation study generates thousands of realizations from the theoretical structure of interest, simulating the random variables which enter into the computation. Statistics based on these simulated data can then be computed, and their behavior learned by watching what happens in thousands of these repeated samples. For instance, the mean value of the statistic can be approximated by the average of those values actually observed in the thousands of simulated instances. Similarly, such quantities as mean squared errors, coverage probabilities of confidence sets, operating characteristics of hypothesis tests, and percentiles of distributions can all be approximated through simulation. Two issues arise. First, how should such studies, which are really experiments in theoretical statistics, be designed? Second, how can a computer be used to simulate random variables? The literature on the first topic is scant, but there is a large literature on the topic of *pseudorandom numbers,* so called because the sequence of realizations of "random" variables is in fact algorithmically determined—and so not random at all. Curiously, most of the results on generating random numbers depend heavily on results from number theory.

Simulation methodology also arises in data analysis, in particular in problems employing *bootstrap* techniques. In the simplest of these problems, many samples, perhaps thousands, are drawn directly from the empirical distribution of the observed data, and computations based on these drawings are used as the basis for inference about the estimated parameters in the model.

> COMMENT. One might argue whether the bootstrap is properly classified as seminumerical, as it involves substantial arithmetic computation, and the raw ingredients for that computation come from the raw data. At the same time, however, much of the technique depends upon simulation methodology and selection of random subsets of the data. Perhaps these methods should be classified as hemiseminumerical.

Related to the topic of designing computer simulation experiments is the more general question of designing experiments in general. The theory of "experimental design" has led to a number of theoretical combinatorial problems, with which the computer has been enormously helpful. In actually carrying out experiments it is often necessary to make random assignments to experimental conditions (treatments), and often these assignments must be made subject to certain constraints which preserve balance or symmetries in certain respects. The necessary computations are often feasible only with computer assistance.

Another sort of combinatorial problem which arises in statistics is computing the *exact distribution of test statistics,* often in discrete data situations. Examples include computing the null distribution of nonparametric procedures, of permutation tests, and computing p-values for tests of no association in contingency tables with fixed marginals. Methods based on generating functions or discrete Fourier transforms (empirical characteristic functions) are often quite helpful in computing distributions; network algorithms often are the basis for computing probabilities associated with contingency tables.

Occasionally, problems arise in which the estimation problem itself requires a network algorithm for its solution. An example is computing *isotonic modal regression* estimators (Sager and Thisted, 1982).

1.3.3 Graphical computation

Graphical displays often involve the pictorial or visual representation of numerical evidence, and thus involve at once both the numerical and the symbolic. It is tempting to resort to neologism once again to describe this class of computations as "semisymbolic," although in this instance we shall generally stick with the less unwieldy term "graphical." The feature that distinguishes these computations from the ones we have already discussed is that the goal of these computations is the construction of a symbol or picture representing some aspect of a problem. In theoretical situations, the symbol is typically the graph of a function, or several such functions for comparison. In data analytic situations, the picture is typically one of our arsenal of graphical methods. I call these computations semisymbolic, because they are not computational manipulation of symbols (hence, not symbolic), but the computations are all directed at constructing an artifact which has symbolic and not numerical significance to its ultimate user (and hence, they are not numerical or seminumerical).

The most obvious examples of this kind of computation are *standard graphics* of all sorts, including graphs of functions, histograms, probability plots, stem-and-leaf displays, contour plots, and scatter plots. Efficient and effective construction of even these most straightforward displays depends on good algorithms tailored to the characteristics of the display device.

Algorithms for computing *minimal spanning trees* of the points in a multidimensional data set have important applications in multivariate data analysis. Similarly, algorithms for computing *convex hulls* are used in multivariate outlier detection methods.

An emerging area of graphical computation discussed earlier is the area of *kinematic graphics* for data analysis, in which animated displays are used to simulate motion, which in turn is used to obtain one or more additional dimensions that can be displayed simultaneously. The output from such a computation can best be described as an "interactive movie."

Multidimensional scaling is a collection of procedures for data-analytic dimension reduction, in which points in a high-dimensional space are forced into spaces of successively lower dimension so as to minimize a measure of stress. The output from this procedure is a map of points in the lower dimensional representations which as nearly as possible preserve the relationships between points in the original data set.

As far as statistical computing is concerned, all of the examples discussed in this section involve graphical displays as the result of a computation. Issues which urgently need addressing, but on which little energy has been spent to date, include graphical displays (and other information-laden symbols) as basic data types, which in turn are objects for further computation.

1.3.4 Symbolic computation

At a very basic level, computers are high-speed manipulators of symbols. In the eye of the general public, computers perform arithmetic calculations. With the advent of personal computers people now more generally think of word processing as another sort of computation that computers do. At the lowest level, numerical computations involve electronic manipulation of bits whose patterns correspond to numerical values. The rules according to which, say, two numbers are multiplied are well determined mathematically, and are easily implemented in hardware. This sort of quantitative symbol manipulation is so commonplace, and has been in the repertoire of general-purpose computers for so long, that it is unremarkable. Word processing is somewhat more interesting, since it involves little overt numerical computation. Rather, it involves manipulation of strings of typographical symbols, combined by the human user into meaningful compound symbols: words. Despite the name "word processing," such programs perform no computations based on the *values*, or meanings, of the words themselves. Instead, they make it easy for a human user to combine words and phrases. Word processors do not write (or edit) essays or poems, and they are not capable of correcting flaws in syntax or in logical development. (They are by now fairly adept at correcting speling errors, however.)

Symbolic computation takes that next step forward, operating on the values of symbols. That is, the meaning of the symbol is a vital determinant of how the symbol enters into the computation. There are few instances today of statistical software in which any sort of symbolic computation takes place in this sense, although we shall argue both that there is much room for symbolic computation in statistics and data analysis, and that important strides in data analysis software will, of necessity, depend in part on symbol manipulation.

Statistical theory has, however, been the beneficiary of some advances in symbolic computation: the algebraic computation systems such as MAC-SYMA or REDUCE. These programs, which are important products of research into artificial intelligence, perform computations on algebraic and other symbolic mathematical expressions. An example will make this clearer. Consider the solution of the quadratic equation

$$x^2 + bx + c = 0.$$

If asked to solve this expression for a solution x using the command

```
solve(x**2 + b*x + c, x);
```

MACSYMA would recognize that x is the indeterminate and that b and c

are constants. MACSYMA would then return the expression

```
                                    2
                           SQRT(B   - 4 C) + B
(E1)                   X = - -------------------
                                    2
                                    2
                           B - SQRT(B   - 4 C)
(E2)                   X = - -------------------
                                    2
(D2)                       [E1, E2]
```

as the solution. Thus, MACSYMA is a program which computes general solutions to algebraic problems, in algebraic terms. Contrast this to a FORTRAN program which could compute a numerical solution for each given set of constants. MACSYMA can also perform integration and differentiation problems that are far too difficult for most calculus students, compute with continued fractions, sum infinite series, and the like. In each case, both the input and the output are symbolic expressions, not numerical values. Such programs have been used by theoretical statisticians to perform (or to check) laborious algebraic or combinatorial calculations which figure in distribution theory and elsewhere.

We shall explore aspects of symbolic computation with an eye toward its applicability in data analysis software. To this end, we shall examine some of the basic ideas of artificial intelligence, discuss the notion of heuristics, develop suitable data-analytic heuristics, and consider approaches toward representing knowledge about data analysis in computer programs.

1.3.5 Computing environments

Roughly speaking, a computing environment is a collection of integrated hardware and software tools designed for a specific range of computational activity. An environment is designed and implemented in order to increase both the productivity of the user and the quality of his or her work. A good computing environment should be comfortable and should make it feel natural to do those things which are naturally part of the task to be accomplished. A prime example of a *software development environment* is the UNIX environment developed by Ritchie and Thompson at what is now AT&T Bell Laboratories. UNIX is, among other things, an operating system that combines editors, compilers, data management, mail and communications, and text processors in such a fashion that the process of writing, debugging, and documenting large software projects is greatly enhanced.

By comparison to UNIX, an environment for developing software, statistical data analysis environments are meager. Typically, they consist of a single computer program that interacts badly, if at all, with other available programs. This is not entirely due to neglect of the subject by statisticians. UNIX was designed and implemented in the early seventies, and experience with software development environments has accumulated into the eighties. By contrast, most of the widely used statistical programs are those whose design was essentially fixed by 1975, and some are still based on essentially the same design as their first appearance in the early sixties. During this time, the statistical community has treated statistical software as essentially given, and has been content to evaluate the software products offered to it. It is instructive to consider the problems of *evaluation of statistical software,* as much of the earliest organized effort in statistical computing was directed toward that end. Thisted (1984b) examines methods for comprehensive software evaluation and the difficulties inherent in on-going monitoring of software quality.

Turning from the merely critical to the constructive, it is important to go from software evaluation to *statistical software design.* The issues addressed in this realm must include not only those of sound data analytic practice, but also the concerns of what has been called *human engineering:* conscious design of the product with the abilities and limitations of the human user fully taken into account.

The natural next step is to consider what an ideal *environment for data analysis* would look like, and to take some first steps in designing such a system. Thisted (1986a) examines the relationship data analysts have with the hardware and software systems they currently use, and how such systems could be extended to make data analysis easier, more natural, and more productive. Thisted (1986b) describes inexpensive hardware and software components which can be integrated to achieve some of the goals of an integrated environment for data analysis. More sophisticated systems would incorporate substantial knowledge about both the language in which data-analytic computation is expressed and the process of data analysis itself. Such environments would borrow heavily from the literature on *expert systems,* programs which embody and make use of expert knowledge about a subject area, in this case, data analysis. We shall take a dual view in our exploration of these topics, examining aspects of expert system design particularly well suited to data analysis, as well as studying ways in which such expert systems could be integrated into a comprehensive environment for data analysis.

1.3.6 Theoretical computation

A final kind of computation is the (mainly) mathematical work in theoretical computer science. The formulations of computer science theory have largely to do with abstract computation on abstract machines, and the inherent difficulty of certain kinds of computation. In a sense, a main concern of theoretical computer science is the question, "Which questions have answers that can be computed effectively by a machine?" A consequence of this theory is that some kinds of computation are inherently difficult, in the sense that as the size of the problem grows, the amount of computation required to solve the problem grows extraordinarily fast. Other problems are very well behaved.

Some problems arise in statistics, such as that of computing the median of a list of n numbers, naturally give rise to the question, "how efficient can an algorithm be for solving this problem?" This is a question of *computational complexity*, and the same question arises in a variety of statistical contexts, such as computing all-subsets regression, and computing the sampling distribution of certain rank-based statistics. There are several open questions about statistical computation that fall squarely in the realm of theoretical computer science.

1.4 Statistics in different areas of Computer Science

This section will be very brief, not so much because statistical ideas and statistical methods do not arise in computer science, but rather, the flavor they take on when they do arise is that of standard application of standard methods, and hence, does not have the flavor of statistical computing as we have been discussing it. There are, however, two areas in which the field of statistics has a potential contribution to make to computer science, *per se*.

First, there are important notions such as the average performance of algorithms, or the performance of algorithms when the inputs are randomly distributed. In fact, much of computational complexity theory is based either on the "worst-case" analysis or the "average performance" analysis, the latter typically based on the assumption of a uniform random input. There are often intermediate cases which would be of interest, and statistical theory can help in both formulation and solution of these problems.

Second, the field of statistics, broadly construed, is concerned with extracting pattern against a background of variability, and of constructing maximally informative summaries based on observed (noisy) data. Statistical principles (although not always standard statistical methods) are applicable in certain areas of artificial intelligence. Statistical methods and formulations have been given exceedingly short shrift in the artificial intelligence literature, largely because attention has been focussed exclusively on statistical decision theory, which emphasizes optimal procedures developed within narrowly-defined mathematical contexts. We shall argue in Chapter

20 that a broader view of statistics can be beneficial in constructing certain kinds of expert systems.

1.5 Some notes on the history of statistical computing

The preceding sections have been devoted to the breadth of statistical computing. Curiously, the history of organized activity in the field has been rather narrow, at least since the advent of programmable digital computers. In this section I trace some history of the interactions between data analysts and digital computers. Strictly speaking, of course, the history of statistical computing is coextensive with that of statistics itself. The focus here is on those developments which have most influenced the current state of computational statistics based upon general-purpose machines programmable in higher-level languages. The view presented here is somewhat idiosyncratic; for balance, I recommend the accounts by Chambers (1968), Mann (1978), Nelder (1984), Sint (1984), Griffiths and Hill (1985), and Eddy and Gentle (1985). An account of the Chilton meeting of 1966 (discussed below) appeared in Volume 16, Number 2, of *Applied Statistics.*

Griffiths and Hill mark the arrival of a prototype Elliott-NRDC 401 computer at Rothamsted Experimental Station in 1954 as the beginning of the modern era of statistical computing. In the early days of general access to computers by scientific investigators, it was essential to transfer advanced statistical methods from the hand-crank desk calculators to the newer and faster computing machines. Such tasks were by no means trivial; the Elliott 401, for instance had no compilers, a random-access memory of five 32-bit words, and a "main memory" of 2944 words, which resided on a disk. Floating-point arithmetic operations had to be coded by hand. Gower (1985) reports that visual inspection of an attached oscilloscope was used as a convergence test for some algorithms. The Elliott 401 was retired at Rothamsted in 1965.

The early (pre-1965) experience at Rothamsted is typical of that worldwide. For each computing task an individual program was written, often in the machine language of a particular computer. With the advent of FORTRAN in the late 1950s and ALGOL in the early 1960s, it became possible to write programs that were both more general and that were more unified, in the sense that they could draw on libraries of subroutines for common calculations rather than having to program everything *de novo* each time. At the same time, it became easier for working statisticians to write programs for themselves.

In the early and middle 1960s work had begun both in England and in the United States to construct general-purpose programs with some degree of portability using the newly available languages, these replacing the individual machine-language programs of the previous half-decade. These programs generally settled on a rectangular data matrix as the basic structure for input data, and a partially mnemonic control language for specifying analyses to be done. Genstat was developed in England and Australia starting in 1965; it has evolved considerably since then at Rothamsted Experimental Station and survives today. In the United States, the National Institutes of Health began funding projects with the same idea in mind—to produce programs for general-purpose computers implementing the now feasible statistical methods, so as to make them accessible tools for life scientists. These projects produced two of the most widely used and respected statistical software programs in the world. The Bimed (for Biomedical computer programs) project—later, BMD and BMDP—from UCLA began in 1960. The first BMD programs appeared in 1962, and by mid-decade there were over fifty separate programs for statistical analysis in the series. SAS (Statistical Analysis System, from North Carolina State) began in 1966 and was also supported initially by NIH. At roughly the same time (1965–1966), two graduate students in political science at Stanford saw a similar need to make the statistical methods that they were using in their research more accessible to their colleagues. In mid-1967 they began implementing their ideas on the newly-installed IBM 360 computer at Stanford. Their efforts led to the SPSS program (Statistical Package for the Social Sciences).

Within the statistics profession in the United States, this process of making it possible to do statistical computations on the new computers was viewed as being a necessary task for *someone* to do, but rather a mundane one; all that was really necessary was to appropriate the standard formulæ for, say, correlation coefficients, and translate them into a computer programming language such as FORTRAN, itself appropriately mnemonic (for *Formula Translator*). This process was widely thought to have little or no statistical content, and perhaps no intellectual content either. This is not surprising given the capabilities of computers at that time, the emphasis on mathematical statistics in the 1950s and 1960s, and the generally poor understanding throughout the scientific community as a whole of the subtleties of what is now called computer science.

Professional recognition of statistical computation and its role in statistical practice took rather different forms in England and the United States. On the 15th of December, 1966, John Nelder and Brian Cooper organized a meeting on the topic of "Statistical Programming" at the Atlas Computer Laboratory in Chilton, England. From this symposium arose the Working Party on Statistical Computing, consisting initially of John Chambers, J. M. Craddock, E. S. Page, J. C. Gower, and M. J. R. Healy, in addition to Nelder and Cooper, organized to consider the problem of standardizing data structures. It later became an official working party of the Royal Statistical Society, and in that capacity was largely responsible for instituting the algorithms section of *Applied Statistics*, which was instituted in 1968. One later project of the RSS Working Party was the Generalised Linear Model Project which in 1973 completed the first version of GLIM. Thus in England the emphasis was on standardization and on statistical software.

Another significant meeting with international participation took place in April, 1969, in Madison, Wisconsin. The proceedings of that conference were published the same year (Milton and Nelder, 1969).

In contrast to the direction taken in England, early organizational efforts in the United States were aimed at exploring common interests of statisticians and computer scientists. The annual Symposium on the Interface between computer science and statistics, which now draws several hundred people each year, began as a one-day symposium held in Los Angeles on February 1, 1967. The first conference was organized by a committee led by Arnold Goodman and was chaired by Nancy Mann. It consisted of four sessions on the topics of simulation, applications, computational linguistics, and artificial intelligence. Following the initial meeting, the symposium has been held annually (except in 1980, in which the ill-fated thirteenth symposium was postponed for a year). The first four meetings were all held in Los Angeles, and were carried out as a joint project of the local chapters of the American Statistical Association (ASA) and the Association for Computing Machinery (ACM). During these years particularly, the Bimed group at UCLA was active in Interface activities. Later both sponsorship and participation became broader, and the Interface was established as a self-sustaining national activity. Proceedings of these meetings have been published since the fourth meeting, in 1970.

The first active participation by a U. S. statistical society comparable to the RSS Working Party came in the area of *software evaluation*. Efforts to evaluate statistical programs arose with the realization that, by the middle 1970s, the most used programs at many university computing centers were statistical packages such as SPSS, BMD, and SAS. This fact implied that most statistical calculations were being performed by nonstatisticians, most of whom had little or no statistical training. Many felt a professional responsibility to assess the quality of these software products from the standpoint of sound statistical practice. In 1973, the American Statistical Association appointed a Committee on Statistical Program Packages, which launched an ambitious program of software review.

Software evaluation as a statistical computing activity is essentially backward looking. It can only examine and evaluate existing programs, and perhaps lament that those currently available are deficient in one way or another. Still, this was a valuable first step, for several reasons. First, in order to do a good job of assessing the quality of a statistical computer program, it became necessary to think hard about just what makes a program good. That turned out to involve far more than a faithful translation of statistical formulae into FORTRAN code, and that led to increased concern about algorithms for statistical computations. Second, it became clear that objective evaluation criteria were simply not adequate to tell the whole story, but that subjective criteria such as "ease of use," "ease of learning," and "flexibility" came into play. This realization led to improvements in statistical programs, as well as an increased understanding that there was something more to data analysis than simply computing the numbers correctly. Third, much early software was rotten by today's standards: algorithms were numerically unstable, or slow, or failed to converge; programs were inflexible in allowing the output of one procedure to be used as input to another; programs were difficult to learn and to use effectively. As a result of evaluation efforts, these problems came to the fore and were quickly addressed by package producers. Indeed in the United States, developers of statistical packages have, for better or for worse, been a driving force behind professional recognition of statistical computing.

John Chambers and Paul Meier were both visitors in London during 1966–67 and participants at the Chilton meeting, and Meier was Vice-President of the American Statistical Association during 1965–67. They wrote a joint memorandum to the ASA Board of Directors, urging them to take steps similar to those being taken in Europe. The ASA established a Committee on Computers in Statistics in 1968, which became an official section of the ASA in 1972. Since 1975, the Statistical Computing Section has published Proceedings containing papers presented under its auspices at the annual meeting of the ASA. The size of the section has increased tremendously, and a perusal of the topics of papers presented indicates that current interest and activity spans the entire range of topics outlined in this chapter.

The International Statistical Institute recognized statistical computing as a separate area of activity for the first time at its Vienna meetings in 1973. The beginning of regular organized activity in Europe outside Great Britain is marked by the first COMPSTAT conference held in Vienna, Austria, in 1974. The COMPSTAT meetings have been held biennially thereafter. The International Association for Statistical Computing was established in 1977 as a section of the International Statistical Institute. Since its inception, the IASC has been one of the sponsors of the COMPSTAT meetings.

Computer graphics is the fastest growing area in statistical computing. In 1985 the American Statistical Association once again expanded its list of sections, adding a Statistical Graphics Section. This section succeeds the Committee on Statistical Graphics of the ASA which, among its other activities, has acted as co-sponsor of sessions at the annual meetings of the National Computer Graphics Association.

2 BASIC NUMERICAL METHODS

This chapter is concerned with the basic ideas from numerical analysis that enter into statistical computation. Broadly speaking, numerical analysis is concerned with the properties of numerical algorithms, particularly their convergence, their stability, and the size of possible errors in the result. These questions arise because computations on a digital computer are generally not exact, and the inexactness can produce side effects in an otherwise well-behaved context. One view, then, is that numerical analysis is concerned with understanding these side effects, and constructing algorithms which minimize them. We begin with a review of floating-point numbers and arithmetic operations on them on a digital computer.

2.1 Floating-point arithmetic

It is sometimes surprising for people to discover that computers do not "always compute the right answer." One reason why computers fail to do so is that the numbers on which computers perform their computations are rarely the "right numbers" to begin with. Typically, real numbers are *represented* in a computer using a fixed number of bits, 32, 36, and 60 being the most common choices. On a computer using 32 bits for real arithmetic, say, there are only 2^{32} different real numbers that can possibly be represented on that computer. Just which of the (infinitely many) real numbers are representable? The answer is a particular set of *floating-point numbers*.

The particular set of representable numbers depends upon the *floating-point format* implemented on a computer system. A floating-point format together with rules for performing arithmetic operations on numbers in that format constitute a *floating-point system*. Most large computers have at least one floating-point system implemented in the machine's hardware. Most microcomputers do not have floating-point hardware, but make provision for floating-point computation in software. The system described here is based on a standard for binary floating-point arithmetic adopted in 1985 by the American National Standards Institute (ANSI) and formulated by a working group of the Institute of Electrical and Electronics Engineers (IEEE), often referred to as *IEEE arithmetic* (IEEE, 1985), and on a 1987

34   *BASIC NUMERICAL METHODS*

ANSI standard for general-radix floating-point arithmetic (Cody, et al, 1984). Floating-point standards are discussed in more detail in section 2.4.

The representable numbers that we shall discuss are those which can be represented in the form:

$$(-1)^s(d_0.d_1d_2\ldots d_{t-1}) \times \beta^e,$$

where $s = 0$ or 1 gives the sign of the number, each d_i is a base-β digit satisfying $0 \leq d_i \leq \beta - 1$, and $E_{min} \leq e \leq E_{max}$. Thus, four parameters—the *base* (or *radix*) β, the number of digits t (called the *precision*), and the range of allowable exponents $[E_{min}, E_{max}]$—specify the representable values in a floating point system. In addition, special quantities may be represented, including $+\infty$, $-\infty$, and symbolic entities called "NaNs" which represent the results of invalid operations such as dividing zero by zero. ("NaN" is an acronym for "Not a Number.")

A particular number represented in the form above has three components: an *arithmetic sign* (s), an *exponent*(e), and a *fraction*, or *mantissa*, $(d_0.d_1d_2\ldots d_{t-1})$.

Some numbers are enumerated redundantly in the definition above; for instance, the number 2 can be represented as 2.0×10^0 or as 0.2×10^1. The *normalized* representation of any redundantly enumerated quantity is the one for which $d_0 > 0$. Some representable numbers, namely those whose magnitude is less than $\beta^{E_{min}}$, cannot be normalized. Such numbers are said to be *subnormal*. Some floating-point hardware requires floating-point numbers to be normalized; subnormal numbers are then invalid on those systems. If the result of a computation is non-zero, but less than the smallest representable floating-point number, then *underflow* is said to have occurred. On systems which permit only normalized numbers, a subnormal result will cause underflow.

The normalized floating-point numbers are those that can be represented in the modified version of scientific notation described above. A useful notation for a floating point number is to write an ordered pair

$$(\pm d_0d_1 \cdots d_{t-1}, e),$$

where the radix point is assumed to lie to the *left* of the first digit. When the sign is positive, we shall omit it in this notation.

Example 2.1. Take $t = 3$ and $\beta = 10$. Then the fraction $1/4$ would be represented as (250,0).

COMMENT. How are the bits of an actual 32-bit machine divided up among sign, exponent, and fraction for floating-point purposes? The answer varies from machine to machine, and it is instructive to examine what floating-point numbers look like on two real machines of rather different architecture.

IBM mainframes such as the 4300 and 3080 series use *hexadecimal* floating-point numbers. The radix is taken to be $\beta = 16$, and for single-precision numbers, the number of digits in the fraction is $t = 6$. This means that the fraction can represent something over 7 decimal digits. Of the 32 bits in a single word, the first 7 bits are devoted to the exponent, which is an integer represented in *excess-64* notation; if the binary value of this integer is k, then the exponent it represents is $k - 64$. The next bit is the sign bit, and the remaining 24 bits contain six groups of four bits each—the six hexadecimal digits represented in binary. Valid floating-point numbers on these machines need not be normalized, so that the first hexadecimal digit of the fraction may or may not be nonzero. The radix point is assumed to precede the left-most digit. Representing the sign by S, the exponent by E, and the fraction by f, the layout of a single-precision floating-point number on these IBM machines is given by

S	E		f

0 1   7 8   31

where the digits underneath the box represent bit positions in the 32-bit word.

Other machines divide the available space in different ways. The VAX-11 computers made by Digital Equipment Corporation also use 32-bit single-precision floating-point numbers, but their internal representation is entirely different. VAXen also require that all numbers be normalized, but the first bit—the one that is always on— is not explicitly represented, so that a t-bit fraction actually represents $t + 1$ significant bits. The radix is $\beta = 2$, and the fraction carries $t = 23$ bits, plus a sign bit. The exponent occupies eight bits, represented in *excess-128* notation. A special value of the exponent is reserved as a special case; if the eight bits of the exponent are all zero, the exponent is not considered to be -128. Rather, the interpretation of the number depends on the value of the sign bit. If the sign bit is zero, then the number is taken to be floating-point zero regardless of the value of remaining bits in the fraction. If the sign bit is one, then the number is taken to be an "invalid number" which, when encountered, triggers a reserved-operand fault. This feature was a crude precursor of the NaNs defined by current floating-point standards. The arrangement of bits in the 32-bit word is peculiar on the VAX. Denote the sign bit by S, the exponent by E, and the fractional part by $.1f_1f_2$, where f_1 contains seven significant bits, and f_2 contains the last 16 bits. Then in memory, the floating-point number is stored as

f_2	S	E	f_1

31   16 15 14   7 6   0

where the integers underneath represent bit positions in the 32-bit word. The apparent peculiarity of the VAX's representation is a consequence of the VAX's historical antecedents. The VAX architecture is an extension of that of the PDP-11, which was a 16-bit machine. The floating-point format was chosen to maintain compatibility with the format used on the PDP-11. (The format doesn't seem to be quite so strange when it is thought of as a pair of 16-bit words pasted together.) Fortunately, the particular format used for floating-point numbers is rarely of direct concern to programmers.

As we have mentioned, not every real number can be expressed as a floating-point number, which means that when computations must be done with numbers that cannot be exactly represented, some approximation must be done. Very large numbers (those with exponents exceeding E_{max}) simply cannot be approximated at all. [When such numbers arise as the result of intermediate calculations, the computer is said to encounter (floating-point) *overflow*, which is usually a condition from which recovery is impossible.] Other numbers are represented by the "closest" floating-point number. In fact, one view of a floating-point number is that it represents the entire interval of real numbers to which it is closest. When we so represent a real number, a portion of its decimal (or binary, or octal, or hexadecimal) representation is lost, and we refer to this loss as *truncation error*.

> *Example 2.2.* For $t = 3$ and $\beta = 10$, $1/3 = (333,0)$. How much truncation error is there? The error is $1/3000$, which can itself be represented approximately as $(333,-3)$. The relative error is $1/1000$, or $(100,-2)$.

Truncation error is one source of error producing inexact computational results. Another has to do with the way arithmetic with floating-point numbers is carried out. Of course, the computed result of adding two floating-point numbers must itself be a floating-point number. That is, the floating-point numbers must be closed under floating addition. However, even if we had two numbers represented exactly in floating-point form, it is entirely possible that their sum is one of those unrepresentable real numbers, so that another approximation of a real number by a nearby floating-point number must be made. This latter approximation is a result of the addition operation, and is called *round-off error*.

Example 2.3. To add 100 and 1.01, the floating representations are exact: (100,3) and (101,1). The sum is 101.01, which is not representable with only three digits. The floating-point sum is (101,3).

We have illustrated two sources of error which arise because computers work with floating-point numbers instead of real numbers. It is useful to be able to say what effect these errors can have on the numerical results produced by an algorithm. To do so, however, requires a little more analysis which depends quite heavily on the precise details of how a machine computes its floating point arithmetic operations. We shall discuss arithmetic error and the way errors propagate in arithmetic operations in the context of a specific model for these computations called the *double-precision register model*. Not every computer performs floating arithmetic in exactly the way described here, but the analyses here are approximately correct for a wide variety of computer architectures. We shall illustrate with a decimal computer ($\beta = 10$) carrying $t = 3$ digits. The principles apply regardless of the base or precision. For more detailed discussion, including the effects of alternative architectures, see Householder (1953) or Henrici (1964).

The double-precision register model assumes that the result of any operation will be computed using twice as many digits as the operands themselves have. Moreover, it is assumed that the first $2t$ digits of the results can be computed correctly. Note that we are not asserting that the result will be good to $2t$ digits; this would require correct computation of at least $2t + 1$ digits, so that rounding could be done. Some computers do in fact compute with just such an extra digit, called a *guard digit*. Some computers simply truncate the result after the $2t^{\text{th}}$ digit. The new standards for floating-point arithmetic (Cody, et al, 1984) do require that double-precision carry at least one full *decimal* digit of accuracy beyond twice single-precision. Thus binary double precision which conforms to the standard must carry $2t + 4$ bits. In our examples with $t = 3$, the double-precision part of the computation will be assumed to be correctly rounded, so that intermediate results will be computed to 6 digits of accuracy. These intermediate results will then be rounded to produce the $t = 3$ digit floating point result. Some computers truncate rather than round at this stage as well. This changes the error analysis somewhat; once again our exposition follow the proposed standards for floating-point arithmetic, which are becoming increasingly available in calculators and on microcomputers.

Addition is accomplished by first copying both addends into a double-precision register, then "unnormalizing" the addend which has the smaller exponent (so that the two exponents are the same), then adding the double-precision results, and then normalizing (if necessary) and rounding to a single-precision result. Example 2.5 shows how a sum of the form $(x+y)+z$

is computed.

Example 2.4. The fraction 2/3 computed in floating arithmetic would be (200,1) divided by (300,1). The exact decimal result is .666666666.... The double precision result is either (666666,0) or (666667,0), depending on whether the machine we are using rounds or truncates these intermediate results. In either case, storing the single-precision result produces (667,0).

Example 2.5. Adding 105, 3.14 and 2.47 can be written as:
$$(105, 3) \Rightarrow (105000, 3)$$
$$(314, 1) \Rightarrow (003140, 3)$$
$$\overline{\qquad}$$
$$(108140, 3) \Rightarrow (108, 3);$$
$$(108, 3) \Rightarrow (108000, 3)$$
$$(247, 1) \Rightarrow (002470, 3)$$
$$\overline{\qquad}$$
$$(110470, 3) \Rightarrow (110, 3).$$

Multiplication is accomplished by first multiplying the fractions, adding the exponents, renormalizing, and then rounding to single precision.

Example 2.6. Multiplying 3 and 1/3 can be written as:
$$(300, 1) \Rightarrow (300000, 1)$$
$$(333, 0) \Rightarrow (333000, 0)$$
$$\overline{\qquad}$$
$$(099900, 1) \Rightarrow (99900x, 0)$$
$$\Rightarrow (999, 0).$$

The last digit in the renormalization is usually made zero. This depends, however, on the particular computer, which may carry an extra digit through all computations, but "off-camera" so to speak (the "guard digit").

Division is carried out just like multiplication, except that the fractions are divided, and the exponents subtracted.

Example 2.7. Dividing 1 by 3.14 can be written as:
$$(100, 1) \Rightarrow (100000, 1)$$
$$(314, 1) \Rightarrow (314000, 1)$$
$$\overline{\qquad}$$
$$(318471, 0) \Rightarrow (318, 0).$$

Under the double-precision register model, it can be shown that the result of a floating point operation done on exactly represented numbers x and y has errors of the following form.

$$(x \oplus y) = (x + y)(1 + \epsilon)$$
$$(x \otimes y) = (x \cdot y)(1 + \epsilon)$$
$$(x \oslash y) = (x/y)(1 + \epsilon),$$

where the operators $\oplus$, $\otimes$, and $\oslash$ denote, respectively, floating-point addition, multiplication, and division, and where in each case $|\epsilon| \leq \beta^{-t}$. These bounds are used in the error analyses of the next section.

EXERCISES

2.1. [07] On an IBM 3081, the first few bits of the fraction of a particular floating-point number is given by $00110101_2 \ldots$ Is this number normalized or unnormalized?

2.2. [00] What is the plural of "VAX"?

2.3. [10] Modify Example 2.5 to show that floating-point addition is not associative.

2.2 Rounding error and error analysis

The basic problem we are concerned with can be illustrated in the following framework. We wish to compute the value of a function $f(\cdot)$ at a particular point x. Here, x represents the input data, and may in fact be a vector of points $x = (x_1, \ldots, x_n)$. It is often the case that $f(\cdot)$ is given only implicitly, but may be represented as the limit of an iterative process, f_1, $f_2, \ldots$. Errors arise from two basic sources. First, the input data x generally can only be represented approximately in the floating-point notation of the computer, so we are doing computations on some new quantity x^* instead of x itself. We might call this error *representation error*. Second, f itself cannot generally be represented exactly, since the floating-point numbers are not closed even under simple operations such as addition and multiplication. Since the result of the algorithm must be a floating-point number, the algorithm must be closed under its arithmetic operations. Thus, instead of computing $f(x)$, we are left computing $f^*(x^*)$, and even if we had been able to express x exactly, we would introduce error simply due to the fact that $f^*(\cdot) \neq f(\cdot)$. This latter source of error is directly generated by the floating-point algorithm for $f(\cdot)$. We can write

$$f(x) - f^*(x^*) = [f(x) - f(x^*)] + [f(x^*) - f^*(x^*)],$$

so that the overall error is the sum of two components: the error *propagated* from representation, and the error *generated* by f^*. (Householder, 1953).

This generated error usually arises from one of two sources. The algorithm f^* is typically composed of basic floating-point operations, sometimes called *flops*. The term is used to refer both to the individual floating-point operations themselves and to the amount of work required to obtain the result. (When considered as a measure of computing effort, a flop represents the amount of work required to process one step of an inner-product calculation—about one floating multiply, one add, and a few array address calculations.) These floating-point computations, being approximations, introduce error at each step of the computation, typically through rounding or truncating intermediate results. Their cumulative effect is the accumulated *round-off error* due to f^*. When f is specified iteratively, a second source of error arises due to the fact that (infinite) limits can only be approximated by a finite number of iterations of the algorithm. Thus, the algorithm $f^*(x)$ may have the form $f^*(x) \equiv f^{**}_{s(x)}(x)$, where $s(x)$ is a stopping rule ("convergence criterion") which determines the last iteration in the sequence $f^{**}_1, f^{**}_2, \ldots$. We can then write the generated error as

$$f(x^*) - f^*(x^*) = [f(x^*) - f^{**}_\infty(x^*)] + [f^{**}_\infty(x^*) - f^{**}_{s(x^*)}(x^*)].$$

The first of these terms is associated with the inherent algorithmic error in using even the infinitely iterated algorithm f^{**}. The second term represents the error in approximating the limit by a finite iteration (*approximation error*).

For all algorithms, we should like to know whether we can, for given inputs, make some statement about the size of the error from computing f on a given computer with a given algorithm. In particular, we would like to be able to give *error bounds* for the computation. Also, we might well ask about how small changes (perturbations) in the input can affect the output. Can small changes in the input data, for instance, produce very large percentage changes in the output?

When we deal with iterative algorithms (as we shall in Chapter 4 and elsewhere), we shall also want to know the answers to additional questions: Under what conditions will the algorithm converge, and can convergence (to the right answer!) be guaranteed? How fast does convergence take place, that is, how much improvement do we get from each additional iteration? Are there bounds on the approximation error, that is, the error from early stopping?

Leaving aside iterative algorithms for the moment, we shall concentrate in the rest of this chapter on error analysis for propagation of floating-point errors. There are two basic approaches to error analysis of this kind. One approach is to start with the input x, and to follow through the computation of x one step at a time, keeping track of the error introduced as a result of each floating-point arithmetic operation. This type of direct, or *forward error analysis,* is extremely difficult to carry out except in the simplest of cases. We shall carry out such an analysis in the case of computing the raw ingredients of a sample mean. A second basic approach, which we shall resort to in the next chapter, is a *backwards error analysis*, or inverse analysis. This approach considers f^* to be the exact solution to a slightly modified problem, that is $f^*(x) = f(\tilde{x})$, for some $\tilde{x}$ "near" x. As long as $\tilde{x}$ is "near enough" to x, we will be satisfied with the result.

COMMENT. The ideas of forward and backwards error analyses became standard in numerical analysis work after publication of Wilkinson (1963), in which they were clearly explained and systematically exploited. Wilkinson credits the idea behind backwards analysis to papers of von Neumann and Goldstine and of Turing in 1947 and 1948. The first explicit description of a backwards analysis appeared in an unpublished technical report by Givens in 1954.

This last discussion leads us to a rough definition of stability of an algorithm, borrowed from Stewart (1973). Although we are interested in computing $f(x)$, we can at best hope to compute $f(x^*)$. Of course, x and x^* are very close; the latter is the best floating-point approximation to the former. Some computations, however, are such that $f(x)$ and $f(x^*)$ are not at all near to one another. In this case, small perturbations in the problem produce large perturbations *in the exact answer*. Such a computation is said to be *ill-conditioned*. When we go to a computer approximation to f, namely, f^*, we cannot hope to do better than f itself. But we do hope that f^* does not also introduce additional wild fluctuations. This leads to the following definition. If, for any input x, there is another input $\tilde{x}$ that is near x, for which $f^*(x) = f(\tilde{x})$, then the algorithm f^* is said to be *stable*. This definition states that a stable algorithm is one for which the computed answer is never very far from the exact answer for a slightly different problem.

While the definition of well-conditioned and ill-conditioned data rarely causes difficulty, the definition of stability may seem at first glance to be peculiar, perhaps even off the mark. The idea, however, is straightforward, and can be explained in the following terms. Suppose that the input data are measurements of people's heights, recorded to the nearest inch. People who are 183 cm (72.05 in) tall would be recorded as being 72 inches tall; so would someone 182 cm (71.65 in) in height. The algorithm "convert centimeters to inches and round" is stable in the sense we have discussed. Although $f(182) \neq 72$, we do have that $f^*(182) = f(182.88) = 72$, and the input 182 is "close" to 182.88. The algorithm does not produce the exact height in inches, but it does produce the exact height in inches for someone whose height in centimeters is only slightly different from that of the person whose height is of interest to us.

COMMENT. It is useful to extend the notion of stability somewhat along the following lines. Suppose that the function f of interest takes values in the *range* Ω. If an algorithm for f produces values outside Ω, then we often wish to say that the algorithm is unstable. Examples of situations in which this extended notion of stability is useful include computation of functions which are always positive, or which always produce integer results. An algorithm for computing a variance, for instance, which can produce a negative output value might well be classified as "unstable."

It may be helpful to illustrate these ideas. Consider real division, that is, $f(x) = 1/x$. What values of x are well-conditioned for this function? We compute $f(x^*)$ and show that it is close to $f(x)$. Closeness here means precisely that the relative error is small. Since $x^* = x(1 + \epsilon)$, where ϵ is itself small, we compute

$$
\begin{aligned}
f(x^*) &= f(x(1 + \epsilon)) \\
&= 1/[x(1 + \epsilon)] \\
&= (1/x)(1/(1 + \epsilon)) \\
&= (1/x)(1 - \epsilon + \epsilon^2 - \epsilon^3 + ...) \\
&\approx (1/x)(1 - \epsilon),
\end{aligned}
$$

where we have neglected terms of size ϵ^2 and smaller. This shows that, for all $x \neq 0$, x is well-conditioned with respect to this function.

Consider now the floating-point computation of $1/x$ using the double-precision register model; call this algorithm f^*. Is f^* stable? To find out, we compute $f^*(x)$ and see whether we can represent it as approximately the exact value f for a slightly different input $\tilde{x}$. Again we compute

$$
\begin{aligned}
f^*(x) &= (1/x)(1 + \epsilon) \\
&\approx (1/x)(1 + \epsilon + \epsilon^2 + \epsilon^3 + ...) \\
&= (1/x)(1/(1 - \epsilon)) \\
&= f(x(1 - \epsilon)) \\
&= f(\tilde{x}).
\end{aligned}
$$

The approximation in the second line above has relative error on the order of ϵ^2, which is negligible. Hence f^*, the floating-point reciprocal function, is stable for all non-zero inputs x.

As a third illustration, we give an example of an ill-conditioned problem. Let $f(x) = (1.01 + x)/(1.01 - x)$. The problem is well-conditioned for $x = 2$, since $f(2) = -3.0404$, and $f(2.01) = -3.02$, so changing x by 0.5% changes $f(x)$ by 0.67%. However, the problem is ill-conditioned for x near one; $f(1) = 201$, and $f(1.005) = 403$, so that changing x here by 0.5% changes f by more than 100%.

EXERCISES

2.4. [24] In section 2.2 we showed that the floating-point reciprocal function $f^*(x)$ defined by dividing 1 by x using floating-point division, is a *stable algorithm* for computing reciprocals. We also showed that the real reciprocal function $f(x) = 1/x$ is a *well-conditioned function* for all non-zero floating-point inputs x^*. Show that the *floating-point reciprocal function* defined by $f(x) \equiv 1 \oslash x$, where $\oslash$ denotes floating-point division, is also a well-conditioned function for all non-zero floating-point inputs.

2.5. [28] Show that, although the function $f(x) = (1.01 + x)/(1.01 - x)$ is ill-conditioned for x near 1.01 (as shown in section 2.2), the algorithm which computes this function by floating-point arithmetic is stable for all floating-point $x \neq 1.01$ in some neighborhood of $x = 1.01$.

2.6. [20] Let $f(x) \neq 0$ be twice-differentiable in a region about x, and let $\tilde{x} = x(1 + \epsilon)$ be a point in this region. Show that, for ϵ sufficiently small, the relative error in $f(\tilde{x})$ is approximately $\epsilon x f'(x)/f(x)$.

2.3 Algorithms for moment computations

We now apply the foregoing methods of analysis to some of the most basic of statistical computations: computing moments. The (sample) moments of a data set are defined to be:

$$M_j = \frac{1}{n} \sum_{i=1}^{n} (X_i)^j.$$

This definition has, as a special case, $\overline{X} = M_1$. For $j > 1$, of much greater general interest are the sample central moments, or the moments about the mean:

$$m_j = \frac{1}{n} \sum_{i=1}^{n} (X_i - M_1)^j = \frac{1}{n} \sum_{i=1}^{n} (X_i - \overline{X})^j.$$

Again, we recognize m_2 as the variance, another special case. Since the division by n is not of primary importance in the numerical analysis, and since in statistical applications we often substitute an alternative expression (such as $n - 1$) for this denominator, we shall concentrate on computing the numerator expressions for moments and central moments.

COMMENT. Related to the sample moments are the *k-statistics* introduced by R. A. Fisher (1929). These are symmetric functions of the data which can be expressed in terms of the moments m_j. The k-statistics are the symmetric sample functions whose expectations agree with the corresponding cumulants. By way of contrast, the mean of m_j is not in general equal to μ_j, the corresponding population moment, nor is it even expressible as a linear function of μ_j and population moments of lower order. From the theoretical perspective, then, k-statistics are attractive. In practice, however, they are computed from the m_j, so that their numerical properties are inherited from those of the sample moments.

2.3.1 Error analysis for means

Consider first the problem of computing the sample mean of n observations. What numerical error can arise? Are some ways of computing the mean better than others? Let's concentrate on computing the sum of the observations for the moment, using a computer program such as the following fragment of Pascal:

$$s := 0;$$
$$\textbf{for } j := 1 \textbf{ to } n \textbf{ do } s := s + x[j].$$

We accumulate the sum into a Pascal variable s, after appropriate initialization.

COMMENT. It is interesting to note that in the variable declaration portion of this program we would have had the Pascal statement "**var** s : **real**;". Of course, s is a floating-point number not a real number, and in most implementations of Pascal, a single-precision one at that.

The direct error analysis of the special case $f(x, y, z) = x + y + z$ is instructive. Let "$\oplus$" represent floating-point addition in the development below. The floating-point approximation f^* can be obtained in three steps:

$$s_0 \leftarrow x \oplus 0$$
$$s_1 \leftarrow s_0 \oplus y$$
$$s_2 \leftarrow s_1 \oplus z.$$

Using the double-precision register model and the error bounds known for that case, we can conduct an error analysis for this three-term sum.

$$
\begin{aligned}
s_0 &= x \oplus 0 = (x + 0)(1 + \epsilon_0) \\
s_1 &= s_0 \oplus y = (s_0 + y)(1 + \epsilon_1) \\
&= \big((x + 0)(1 + \epsilon_0) + y\big)(1 + \epsilon_1) \\
s_2 &= s_1 \oplus z = \big(((x + 0)(1 + \epsilon_0) + y)(1 + \epsilon_1) + z\big)(1 + \epsilon_2) \\
&= (x + y + z) + x(\epsilon_0 + \epsilon_1 + \epsilon_2) \\
&\quad + y(\epsilon_1 + \epsilon_2) \\
&\quad + z\epsilon_2 + other\ smaller\ terms.
\end{aligned}
$$

Throughout this development, the errors ϵ_i satisfy $|\epsilon_i| \leq \beta^{-t}$. Here, as in Section 3.1, β is the radix and t the floating-point precision. It seems clear that the first component of the sum x gets most distorted in the result, although the errors could, in practice, cancel one another. But in the worst case, when all of the ϵ's have the same sign, it is clear that the overall error receives its largest contribution from the first term, x. We can formalize this worst-case to produce a bound on the error:

$$
\begin{aligned}
|f - f^*| &\approx |x(\epsilon_0 + \epsilon_1 + \epsilon_2) + y(\epsilon_1 + \epsilon_2) + z\epsilon_2| \\
&\leq |x|\big(|\epsilon_0| + |\epsilon_1| + |\epsilon_2|\big) \\
&\quad + |y|\big(|\epsilon_1| + |\epsilon_2|\big) \\
&\quad + |z||\epsilon_2| \\
&\leq \big(3|x| + 2|y| + |z|\big) \cdot \beta^{-t}.
\end{aligned}
$$

From this analysis it is clearer that the first term does indeed have the greatest potential to contribute to the error, and this bound can be minimized by adding the terms of the sum in increasing order of magnitude.

In general, this is a good practice: round-off error accumulates more slowly when small terms are added first. There is another reason for doing so; if the sum is quite large and the variation in size of the terms is also large, then adding the small terms first gives them a chance to accumulate. Adding them one at a time to an already large partial sum would leave the partial sum essentially unchanged. For example, suppose we had $n = 101$ terms to add, the first 100 of which equaled 1 and the last term of which was 1000. With $t = 3$, and adding from small to large, the correct answer $(110, 4)$ is obtained. Adding large to small produces $(100, 4) + (100, 1) \Rightarrow (100, 4)$ at step 1, and at every step thereafter, producing a final result in error by 10%.

The general expression for the error bound in computing a floating-point sum Σ^* of n elements is given by

$$
|\sum{}^* x_i - \sum x_i| \leq \left(\sum_{i=0}^{n-1}(n - i)|x_{i+1}|\right)\beta^{-t}.
$$

Thus, for $f = \overline{x}$,

$$
|f^* - f| \leq \left(\sum_{i=0}^{n-1}(1 - \frac{i}{n})|x_{i+1}| + |\overline{x}|\right)\beta^{-t} \tag{2.3.1}
$$

$$
\leq \left(\frac{1}{2}(n + 3)|x_{\max}|\right)\beta^{-t}. \tag{2.3.2}
$$

We see from these expressions not only that it may be advisable to add items in increasing order, but also that in the worst case, the error depends on the absolute size of the addends and their number. In the worst case, the error in the mean can grow linearly in the number of addends.

This $O(n)$ behavior is a consequence of the propagation of rounding errors in the left-hand addends; by the time we add in the final term in the sum, the partial sum to which it is added has already been corrupted by $n-2$ previous additions. There is a marked asymmetry at this point, adding an unsullied single term (x_n) to a term with possibly substantial error. This asymmetry can be removed, to considerable numerical advantage using a variant of the divide-and-conquer paradigm. If n is even, the data set can be subdivided into two equal sets, whose sums can be computed separately and then added. The numerical error in each of these two terms contribute equally to that in the final result. The two terms are each sums of sets of numbers, which can in turn be computed by subdividing *those* sets. Clearly, the number of subdivisions required is $\lceil \log_2 n \rceil$. Detailed analysis also shows that the relative numerical error using this process is bounded by a constant times $(\log n)\beta^{-t}$. This algorithm is called the *pairwise method* for computing sums (Chan, Golub, and LeVeque, 1983). A one-pass version of the algorithm can be implemented using at most a constant plus $\lceil \log_2 n \rceil$ storage positions.

Neely (1966) undertook a study of alternative algorithms for computing means, variances, and correlation coefficients. One algorithm for means that is highly recommended when accuracy is of primary importance is the following two-pass algorithm:

Algorithm 2.3.1:

$sum := 0$
for $j := 1$ **to** n **do** $sum := sum + x[j]$
$xbar := sum/n$
$sum := 0$
for $j := 1$ **to** n **do** $sum := sum + (x[j] - xbar)$
$xbar := xbar + sum/n.$

This algorithm basically computes the overall size of $\overline{x}$ first, and then computes an adjustment which allows for deviations much smaller than $\overline{x}$ to accumulate. Although this adjustment term is algebraically equal to zero, it need not be numerically, and tends to be of the same size as the first-order term in the error analysis.

2.3.2 Computing the variance

Variances and standard deviations are based on sums of squared deviations about the mean, and there are many ways to compute these sums of squares. Two algebraically equivalent expressions for this sum of squares are given by

$$\sum_{i=1}^{n}(X_i - \overline{X})^2 = \sum_{i=1}^{n} X_i^2 - \frac{1}{n}\Big(\sum_{j=1}^{n} X_j\Big)^2. \qquad (2.3.3)$$

If every element in the set of X's is shifted by the same constant G, the variance, and hence (2.3.3), is unchanged. Thus, for any G we can also write

$$\sum_{i=1}^{n}(X_i - \overline{X})^2 = \sum_{i=1}^{n}(X_i - G)^2 - \frac{1}{n}\left(\sum_{j=1}^{n}(X_j - G)\right)^2. \qquad (2.3.4)$$

These expressions suggest two algorithms:

Algorithm 2.3.2:

$sum := 0$
for $j := 1$ **to** n **do** $sum := sum + x[j]$
$xbar := sum/n$
$s := 0$
for $j := 1$ **to** n **do** $s := s + (x[j] - xbar)^2.$

and

Algorithm 2.3.3:

$sum := 0$
$square := 0$
for $j := 1$ **to** n **do**
    $sum := sum + x[j]$
    $square := square + (x[j])^2$
end
$s := square - (sum)^2/n.$

The second of these algorithms ends with a subtraction of one positive quantity from another, and if the values in the data set are away from zero, the two terms can be quite large in size. A problem arises, however, if the variation in the data set is small, in that *cancellation of significant digits* can occur. This problem only arises in subtraction, that is to say, addition of two floating-point numbers with opposite signs. Cancellation can obliterate all of the significant digits in a result.

Example 2.8. $(474, 1) - (473, 1) = (001, 1) \Rightarrow (100, -1)$, which has only one significant digit, even though the two terms being subtracted each have three significant digits.

Let us examine these two alternative algorithms on a small data set using our hypothetical machine: $X = (19, 20, 21)$. Clearly, the mean is 20 and the sum of squared deviations is 2. Following Algorithm 2.3.2 on this data set produces

$$(-100, 1)^2 = (100, 1)$$
$$(000, 1)^2 = (000, 1)$$
$$(100, 1)^2 = (100, 1)$$

$$\overline{}$$
$$(200, 1).$$

which is the correct answer. Algorithm 2.3.3, on the other hand, produces the following results as the partial sums of the variable *square* are computed:

$$(190, 2)^2 = (361, 3) \ [+(000, 1) = (361, 3)]$$
$$(200, 2)^2 = (400, 3) \ [+(361, 3) = (761, 3)]$$
$$(210, 2)^2 = (441, 3) \ [+(761, 3) = (1202, 4)]$$

$$\Rightarrow (120, 4);$$

$$-(sum)^2/n = (120, 4) \ (\text{exactly})$$

$$= (000, 1).$$

The problem here is that Algorithm 2.3.3 suffers from cancellation which 2.3.2 avoids, through the device of *centering*. Whenever moments are calculated, it is advisable to center the calculation in order to achieve maximum accuracy. The pairwise method of summation outlined in the previous section can also be applied to good effect when accumulating sums of squares; the details are left as an exercise.

2.3.3 Diagnostics and conditioning

For the variance computation, it is easy to see when cancellation can be serious. Let S denote the sum of squared deviations, so that

$$S = \sum_{i=1}^{n} X_i^2 - n\overline{X}^2$$

or

$$\sum_{i=1}^{n} X_i^2 = S + n\overline{X}^2.$$

The problem occurs when the left-hand side and the second term on the right-hand side are approximately equal as floating point numbers, or equivalently, when the relative size of $n\overline{X}^2$ is too large compared to the relative size of S. Dividing both sides by S to express things in relative terms, the right-hand side becomes

$$1 + \frac{n\overline{X}^2}{S} = 1 + (CV)^{-2},$$

where CV is the coefficient of variation, the standard deviation divided by the mean. When $CV^2 < \beta^{-t}$, then all significant digits will be lost due to cancellation. Thus, the CV is a good *diagnostic* for the variance computation; when it is small, it indicates that there is likely to be numerical difficulty associated with the computation.

The coefficient of variation is closely related to a sensitivity measure associated with each data set, called the *condition number* of the data with respect to the variance computation. (In later chapters we shall discuss condition numbers associated with other fundamental computations. It is important to keep in mind that a condition number describes a data set relative to a particular function; it is *not* associated with any particular algorithm for computing that function.) The condition number for computing the variance, denoted by κ, bounds the amount by which relative errors introduced into the data set at hand are magnified in the variance computation. Thus, if relative errors of size u are introduced into the x_i's, the relative change in S cannot exceed κu. Data sets with large values of κ are said to be *ill-conditioned*. Under most circumstances, u is at least as large as β^{-t}, simply due to machine representation; in statistical problems u is often substantially larger, due to the imprecision with which the data are measured or recorded. Chan and Lewis (1978) showed that the condition number for this problem is given by

$$\kappa = \sqrt{1 + n\overline{X}^2/S} = \sqrt{1 + (CV)^{-2}},$$

the square-root of the previous expression. Unless CV is quite large, we have $\kappa \approx CV^{-1}$ as a useful approximation.

Although the condition number describes the stability of the data set with respect to the function S, it also figures prominently in deriving error bounds for particular algorithms for computing S (for details, see Chan, Golub, and LeVeque, 1983). If relative errors in the data are on the order of $u = \beta^{-t}$, then relative errors in S computed by Algorithm 2.3.2 are bounded by $nu + n^2\kappa^2 u^2$, those computed by Algorithm 2.3.3 are bounded by $n\kappa^2 u$. If Algorithm 2.3.3 is applied to $x[i] - xbar$ instead of $x[i]$, the error bound is given by $nu + \kappa^2(nu)^3$. In each case, if pairwise summation is employed, n is replaced by $\log_2 n$ in the expressions for the bounds. Provided $n < \beta^t$ (and κ isn't extremely small), the centered Algorithm 2.3.2 is superior numerically to the uncentered Algorithm 2.3.3. When pairwise summation is employed, centering wins provided that $n < 2^{2^t}$ (!).

Chan, Golub, and LeVeque (1983) introduce an extremely stable algorithm for variance computations based on *pairwise updating*; their method appears to have relative errors in S bounded by $\kappa u \log_2 n$.

2.3.4 Inner products and general moment computations

We have already examined several special cases of the computation of inner products. An *inner product* between two vectors $a = (a_1, \ldots, a_n)$ and $b = (b_1, \ldots, b_n)$, denoted by $\langle a, b \rangle$, is given by

$$\langle a, b \rangle \equiv a'b = \sum_{i=1}^{n} a_i b_i. \qquad (2.3.5)$$

The mean can be viewed as a special case with $a = \frac{1}{n}(1, \ldots, 1)$. The sum of squared deviations is also a special case with $a_i = b_i = (X_i - \overline{X})$. We have seen in our earlier discussion that, while rounding errors due to arithmetic errors often accumulate slowly, they need not do so. What can be said in their behalf, however, is that they occur in predictable and quantifiable ways. Unfortunately, the same cannot be said of cancellation. Equally unfortunate is the confluence of three facts: inner products are ubiquitous in statistical computation, the process of taking products often produces numbers which are large relative to the size of the ultimate result, and the signs of these large terms are not generally predictable *a priori* (since they depend upon the particular data set on which the computations are being done). This means that inner products are a breeding ground for cancellation error, and it is essential that everything possible be done in critical computations to minimize the possibility of catastrophic cancellation.

There are two primary weapons in the war on cancellation. The first is relatively simple, and often easy and cheap to implement: accumulate all sums such as those in inner products in double precision. This means to capture the double-precision result when multiplying the cross-product terms, and then to add these to a double-precision variable, retaining all of the digits in the result. Many programming languages have a "double precision" or "long real" data type, which in turn is implemented in hardware on many computers. The second weapon is to pay careful attention to computing formulæin order to avoid the situation in algorithm 2.3.3, in which cancellation occurs at the very last step. The expedient of first subtracting the mean before accumulating sums of squares (as in algorithm 2.3.2) produced a much more numerically stable result. This process can be carried out more generally by the device of *provisional centering*.

As we noted above, in the computation of $S = \sum(X_i - \overline{X})^2$ we can write

$$\begin{aligned} S &= \sum_{i=1}^{n}(X_i - G + G - \overline{X})^2 \\ &= \sum_{i=1}^{n}(X_i - G)^2 - n(G - \overline{X})^2. \end{aligned} \qquad (2.3.6)$$

Thus, Algorithm 2.3.2 results from choosing $G = \overline{X}$, while Algorithm 2.3.3 results from choosing $G = 0$. The further G is from $\overline{X}$, the more ill-conditioned the data are for the resulting algorithm. In fact, on average, the best possible choice of G is precisely $\overline{X}$. Expression (2.3.6) suggests an alternative algorithm for computing S when a one-pass algorithm is desirable (or essential), namely, to center the data about a provisional mean, G, selected so as to be close to the actual mean. After one complete pass through the data the value of $\overline{X}$ is known, so that the second term on the right-hand side of (2.3.6) can be computed.

One choice that often works well is to select G to be equal to the first observation in the data set. While this choice *can* actually be worse than using $G = 0$, it does very well on the average. If the data have already been sorted, however, this choice may well be unfortunate, since the provisional mean will be on the outskirts of the data set rather than in the middle, where it will do the most good. Even in this case, however, the centered data have condition number bounded by $\sqrt{1 + n}$, so that unless n is enormous the resulting error bound will be acceptable.

A more complicated alternative to selecting G as the first observation is to modify the current G as the computation progresses. If at any stage k the sum $\sum_{i=1}^{k}(X_i - G)$ is too far from zero, indicating that G is too far from $\overline{X}$, then (2.3.6) is applied (with $n = k$) to re-center the computation about the current value of $\overline{X}$. The term $\sum_{i=1}^{k}(X_i - G)^2$ is replaced by

$$\sum_{i=1}^{k}(X_i - G)^2 - k(G - \overline{X})^2, \qquad (2.3.7)$$

and then G is replaced by the current value of $\overline{X}$.

This last idea can be extended so as to update the mean and the sum of squares at *each* stage. While this algorithm may appear to be highly desirable by using the best possible estimate of the centering point (so far) as each item in the data set is processed, the rounding error introduced by all of the extra multiplications, divisions, and additions can quickly swamp the small savings from frequent updating unless great care is taken in the implementation. Carefully implemented updating formulæ have relative error bounds of $n\kappa u$, and consequently are superior to the uncentered algorithm under all circumstances, and to the mean-centered algorithm if n is larger than β^t/κ (which in practice rarely occurs). Applying pairwise summation to stable updating formulæ appears to reduce the bound to $\kappa u \log_2 n$. This method was proposed by Chan, Golub, and LeVeque (1983); other updating algorithms are outlined in their paper as well. Updating methods are quite useful in regression computations; they make it a simple matter, for instance, to add or delete cases from a regression model. Updating methods are discussed in some detail in Section 3.7.

For more general inner products than the variance computation, it is easy to show that

$$\sum_{i=1}^{n}(X_i - \overline{X})(Y_i - \overline{Y}) = \sum_{i=1}^{n}(X_i - G)(Y_i - H) - n(G - \overline{X})(H - \overline{Y}), \quad (2.3.8)$$

where the X's are centered about G and the Y's are centered about H. Once again, this accumulation of terms is most likely to be numerically stable when G and H are close to their respective means.

In general, the greatest numerical stability is achieved when accumulating products of terms if each of the terms in the product is first centered, and then, after accumulation, adjusted for the correct mean. When this final adjustment is small, then by definition cancellation cannot badly affect the accuracy of the result. This principle applies to computation of higher moments as well, although the adjustments required are less straightforward.

As a practical matter, it is usually most satisfactory to compute moments generally by first selecting a provisional mean that is likely to do well and then sticking with that choice throughout the computation, rather than applying updating. For most problems encountered in data analysis, provided that n is moderate (say, less than $10,000$), the gains in accuracy do not seem to be worth the additional cost of more complex and less efficient programs.

COMMENT. All of the inner product computations discussed here have *weighted* counterparts. Weights arise naturally in statistical computations; least-squares regression coefficients, for example, can be viewed as weighted averages of the response variable. Weighted means are often computed from summary data, and when data are obtained from sources with different precision, weighted analyses are appropriate. Weighted inner products play an important role in robust regression as well; possible outliers are assigned weights close to zero while those observations that are most trusted are assigned weights close to unity.

Let w_i denote the weight associated with the ith observation. Generally the weights are taken to be nonnegative. Without loss of generality, suppose that $\sum w_i = n$. The *weighted mean* of $X = (X_1, \ldots, X_n)$ is given by $\overline{X} = \sum w_i X_i / n$. The weighted sum of squares is given by $S = \sum w_i(X_i - \overline{X})^2$; note that the mean in this expression is the *weighted* mean. Corresponding to equation (2.3.6), we have the centering formula

$$S = \sum w_i(X_i - G)^2 - n(G - \overline{X})^2,$$

similarly for the cross-product formula (2.3.8). The proof is left as an exercise.

COMMENT. The inner products that we have described in this section are really special cases of more general quadratic forms. If Q is a symmetric positive semidefinite matrix, then we can define an inner product with respect to Q as

$$\langle a, b\rangle_Q \equiv a'Qb = \sum_{i=1}^{n}\sum_{j=1}^{n} q_{ij}a_ib_j.$$

Thus, equation (2.3.5) is a special case with $Q = I$, and weighted sums of squares and cross products are also special cases with Q being the diagonal matrix of weights. The discussion above applies equally to inner products generalized in this way. Centering formulæ such as (2.3.8) are developed in the exercises.

COMMENT. Sample means (first moments) can be viewed as *linear forms*, that is, singly-subscripted weighted sums of the observations. Sample variances and covariances (based on second moments) are computed from sums of squares and cross products; these are expressible in terms of *quadratic forms*, as we have noted above. Third moments are computed as *cubic forms* (higher moments as higher-order forms) which can be thought of computationally as sums with triply-subscripted weights. There is nothing special numerically about linear or quadratic forms that does not apply equally to higher-order calculations—they all should be centered first. Although centering formulæ analogous to (2.3.8) can be derived, they are not as straightforward to apply. The cubic case is considered in the exercises.

EXERCISES

2.7. [27] Let $f(x) = (\sum_{i=1}^{n} X_i)/n$. Using the standard algorithm (call it f^*), show that the absolute error satisfies

$$|f^* - f| \le \Big(\sum_{i=0}^{n-1}(1 - \frac{i}{n})|X_{i+1}| + |\overline{X}|\Big) \cdot \beta^{-t}.$$

2.8. [16] (Continuation) Prove that (2.3.2) follows from (2.3.1).

2.9. [31] Write a detailed algorithm implementing the one-pass pairwise summation method, utilizing at most a constant plus $2 \cdot \lceil \log_2 n \rceil$ storage positions. Can one get by with only a constant plus $\lceil \log_2 n \rceil$ storage positions?

2.10. [17] Prove that Algorithm 2.3.1 computes a result that is algebraically equivalent to the sample mean.

2.11. [24] Using $\beta = 10$, $t = 3$, and the double-precision register model, what does Algorithm 2.3.1 produce for the data set with $n = 101$, $X_1 = 1000$, and $X_2 = X_3 = \ldots = X_{101} = 1$? Compare this result to that obtained by the standard algorithm, which stops after the first three lines of algorithm 2.3.1.

2.12. [21] (Continuation) What does Algorithm 2.3.1 produce for the data set with $n = 101$, $X_1 = X_2 = \ldots = X_{100} = 1$, and $X_{101} = 1000$, under the conditions of the previous exercise? How does the standard algorithm measure up under these circumstances?

2.13. [30] Let $x = (X_1, \ldots, X_n)$, and let $f(x) = \sum_{i=1}^{n}(X_i - \overline{X})^2$. Let $f^*(x)$ be an implementation of Algorithm 2.3.2 discussed in the text. Neglecting terms of order smaller than β^{-t}, show that

$$f^*(x) \approx f(x)(1 + \epsilon),$$

where

$$|\epsilon| \leq \left(3 + \left(n(n+1)\right)/2\right)\beta^{-t}.$$

In particular, conclude that Algorithm 2.3.2 is stable provided that the number n of inputs is not too large.

2.14. [32] Show that, using Algorithm 2.3.3, computing the sum of squared deviations as in the exercise above is not stable when $\left(\sum_{i=1}^{n} X_i\right)^2$ is large.

2.15. [24] Modify Algorithm 2.3.2 to incorporate pairwise summation.

2.16. [19] Using the double-precision register model with $t = 3$, but assuming that partial sums are truncated rather than rounded, compute the sum of squares of the data $(302, 300, 298)$ using algorithm 3.2.3. This is equivalent to a single-precision register computation.

2.17. [47] Establish whether the pairwise-updating variance algorithm given in Chan, Golub, and LeVeque (1983) has relative error bounded by $\kappa u \log_2 n$.

2.18. [18] Prove (2.3.6).

2.19. [20] Prove that data provisionally centered about $G = X_1$, the first data point, has a smaller condition number than the uncentered data if and only if $\overline{X}$ is closer to X_1 than it is to zero.

2.20. [24] Prove that data provisionally centered by choosing G to be one of the data points has condition number no larger than $\sqrt{1 + n}$.

2.21. [23] (Chan, Golub, and LeVeque, 1983). Show that if G is set equal to one of the data points X_i chosen at random, then $E(\kappa^2) = 2$ for the centered data.

2.22. [20] Prove that expression (2.3.7) is correct.

2.23. [18] Prove (2.3.8).

2.24. [46] Write a program that will automatically conduct a forward error analysis for non-iterative algorithms for scalar functions, such as Algorithms 2.3.1 and 2.3.2.

2.25. [40] Write an interactive computer program that will assist in examining examples of rounding and truncation error and error propagation such as those used in the text. This program should begin by reading in the number of digits to be used in floating-point representation (t), and perhaps β as well. It is desirable also to be able to specify the rounding and truncation rules, and whether a guard digit is carried. The program should then read expressions of the form

$$(nnn, e) \oplus (mmm, f),$$

where $\oplus$ is one of the elementary operations $(+, \times, /)$, and it should produce *a)* the t-digit floating-point representation of the result, *b)* the intermediate calculations on the pseudo-machine being emulated (for instance, the $2t$-digit floating-point register result), and *c)* the absolute error and the relative error of the result.

You may assume that t is relatively small, and you may assume that the (double-precision) floating point result on your computer is the "true" result for use in computing the absolute and relative errors.

2.26. [38] (Continuation) Modify the program in the previous exercise to conduct a step-by-step error analysis for a statistical algorithm for computing the mean.

2.27. [41] (Continuation) To examine alternative statistical algorithms using a program such as the one in the previous exercise, it is easiest to emulate a more complete floating-point machine. Assume that this machine has 100 floating-point locations, addressed as 00, 01, ..., 99. Assume that there are four basic operations, $+$, $-$, $*$ (multiplication), and $/$ (division), that floating-point constants can be assigned to a memory location, and that values can be read in (in floating-point form) and printed out. A suitable "assembly language" might have instructions of the form:

Operation Destination Argument ;comment.

Write a program which will accept a program written in this assembly language, and which will then execute this program, providing the intermediate results of the previous two exercises.

A sample input to this program which adds three numbers and prints their mean might be:

```
T 3                  ;set a 3-digit machine
E 10                 ;set exponent range to -10<=e<=10
= 00 (000,1)         ;store zero in location 00
R 01                 ;read first number into location 01
+ 00 01              ;add first number to 00
R 01                 ;read second number
+ 00 01              ;add it to the sum
R 01                 ;get last number
+ 00 01              ;add it into the sum
= 02 (300,1)         ;store number of observations
/ 00 02              ;divide by number of observations
W 00                 ;print result
  .                  ;end of program
(361,3)              ;
(400,3)              ;input data
(441,3)              ;
```

A refinement of this program might add a construct for looping to facilitate examination of more general algorithms.

2.28. [30] Develop the analog of equation (2.3.8) for the sum of products of three terms

$$\sum_{i=1}^{n}(X_i - \overline{X})(Y_i - \overline{Y})(Z_i - \overline{Z}).$$

2.29. [22] For nonnegative weights w_i, define the weighted mean by the formula $\overline{X}_w = \sum_1^n w_i X_i / (\sum w_j)$. Prove the centering formula

$$\sum w_i(X_i - \overline{X}_w)(Y_i - \overline{Y}_w) = \sum w_i(X_i - G)(Y_i - H)$$
$$- (\sum w_i)(G - \overline{X}_w)(H - \overline{Y}_w). \qquad (2.3.9)$$

2.30. [26] Let Q be a symmetric positive-definite $n \times n$ matrix, and let x and y be $n \times 1$ vectors. Denote the $n \times 1$ vector of ones by $e = (1, 1, \ldots, 1)$, and let G and H denote arbitrary fixed scalar constants. Develop a centering formula analogous to (2.3.9) for the quadratic form $(x - \overline{x}e)'Q(y - \overline{y}e)$ in terms of suitably defined "means" $\overline{x}$ and $\overline{y}$ for x and y, and the centered vectors $(x - Ge)$ and $(y - He)$.

2.4 Floating-point standards

With the advent of the powerful and inexpensive microprocessor, an increasing amount of numerical computation is being performed on microcomputers. Most of the microprocessors on which personal computers and scientific workstations are based have only limited fixed-point arithmetic operations as part of the basic instruction set. As a consequence, floating-point computations must be carried out either in software, or in special-purpose numerical data processors. Recognizing the need for a uniform environment within which quality numerical software could be developed for use on microcomputers, a working group of the Institute of Electrical and Electronics Engineers (IEEE) began in the late 1970s to develop a standard for binary floating-point arithmetic which could be implemented with relatively little expense in the microcoprocessor setting and which, at the same time, could be the foundation on which portable numerical software of high quality could be based.

The result of the efforts of this group was ANSI/IEEE standard 754, approved in 1985, which spells out a standard programming environment for floating-point computation. It is a notable document in several respects. First, it specifies a *format* for binary floating-point numbers. Second, it describes the *results of arithmetic operations* on numbers represented in this format. Third, it describes in complete detail how exceptional results, such as division by zero, overflow, underflow, and illegal arguments or operations, must be handled. A system that conforms to this standard may be implemented entirely in software, entirely in hardware, or in any combination; it is the actual environment as seen by the programmer that conforms to the standard. A detailed description of the standard in its nearly final form can be found in Stevenson (1981); the final version can be obtained from the IEEE (IEEE,1985).

The IEEE standard 754 has many features that are noteworthy from the standpoint of statistical computing. It provides not only for the four basic arithmetic operations of addition, subtraction, multiplication, and division, but also for remainder, square root, and conversion between binary and decimal representations. All of these operations are required to produce a result that is accurate to within half a unit in the last place of the destination. The square-root operation is particularly important in matrix computations, as we shall see in Chapter 3. The requirement that format conversion be implemented and done accurately virtually ensures that floating-point computations done on different microcomputers will produce the same output from the same input, provided both conform to the standard, as nothing will be lost or distorted due to differences in converting numbers in decimal representation to internal form on input, or in converting floating-point numbers to decimal form for output.

The floating-point standard differs in several important respects from the hardware floating-point systems contained in most mainframe computers. One example is the standard's requirement that *three* extra bits beyond the floating-point fraction be carried through the arithmetic operations; these are sufficient to guarantee that the computations produce results good to at least one-half bit in the last place. The three bits are called, respectively, the *guard bit,* the *round bit,* and the *sticky bit.* The guard bit functions as we described in section 2.1; it is simply the first bit past the fraction bits in the destination. The round bit is required to assure accurate rounding after post-normalization following certain subtraction operations. Finally, the sticky bit is the logical OR of all of the bits following the round bit. It can be shown that these three extra bits are sufficient to make it possible to round correctly in all instances as if the operations were first carried out to infinite precision.

A second innovation is the requirement for "gradual underflow." Consider a floating-point number with the smallest possible exponent $E_{\min}$, and smallest possible fraction. For concreteness, suppose that the number of bits in the fraction is $t = 3$ and that the radix (of course) is $\beta = 2$. As in section 2.1, we require the first bit of the fraction to be non-zero, so that this smallest possible (normalized) floating-point number will have fraction equal to .100. When we subtract this number from $(101, E_{\min})$, the intermediate result is exactly $(001, E_{\min})$. However, the normalized form of this result is $(100, E_{\min} - 2)$, which has an exponent that is smaller than the format allows. When this occurs, the computation is said to *underflow.* Most systems set the result to zero and raise a flag indicating that underflow has occurred. The IEEE standard takes a different route; it allows the denormalized expression $(001, E_{\min})$ to be a valid result in this case. Note that this result is still good to within half a bit in the last place, but that the operation as a whole produces a result with only one significant bit, as the leading zeros really simply augment the exponent. This approach is what is termed *gradual underflow.* It has the pleasant side effect that the expression $(X - Y) + Y$ always produces X exactly when $X - Y$ underflows; the common alternative of flushing underflows to zero would produce Y exactly as the result. Cody (1981) notes that this amounts to preserving the associative law to within rounding error. The use of gradual underflow by permitting denormalized numbers is not without controversy, although the benefits seem to outweigh the disadvantages in practice (see Coonen, 1981). The effects of using different underflow schemes on the reliability of numerical software has been studied by Demmel (1984). He examined a number of calculations, including Gaussian elimination and eigenvalue extraction; he concluded that gradual underflow made it substantially easier to write good numerical programs than do other approaches (most notably flushing to zero).

A third feature is that, in addition to a basic 32-bit format (with 8-bit exponent and 23-bit fraction) called *single precision,* a system conforming to the standard must provide an extended format *(extended single precision)* with greater exponent range and number of digits in the fraction. Extended-single formats require a minimum of 36 bits to implement. The *double precision* format has 64 bits—11 for the exponent and 52 for the fraction. This allocation of exponent and fraction is sufficient to provide accurate rounding of single-precision products using double-precision arithmetic. The "double precision" found on most mainframe computers from the 1960's through the mid 1980's typically concatenates two adjacent computer words; the second word either contains a duplicate of the exponent found in the first word, or the entire second word is allocated to the fraction.

Some additional features incorporated in the standard include explicit representations for, and definitions for computations with, positive and negative infinities; representations for non-numerical results (the result of $0/0$ is a NaN—"not a number"—which can be trapped and dealt with by the programmer); and a choice of rounding modes.These rounding modes include both ordinary rounding and what is commonly called truncation—rounding toward zero. In addition, rounding modes which support interval arithmetic (rounding toward plus or minus infinity) are also included.

A number of numeric data processor chips (NDPs) conform largely or completely with the standard. Those which do not conform completely predate the adoption of the final standard in 1985. The NDPs which statisticians are most likely to encounter are those designed to be mathematical co-processors for the CPUs found in major personal computers. The two most common CPUs used in personal computers right now are the Intel 8086 series (on which the IBM PC family is based), and the Motorola M68000 series (on which the Apple Macintosh is based). The first NDP which approached the standard, commonly called "IEEE floating point," was the Intel 8087, designed to work in tandem with the 8086. Adding the 8087 coprocessor to an IBM PC greatly increased both speed and accuracy of floating-point computations. The 32-bit member of the 8086 family, the 80286, also has a mathematical co-processor, the 80287. The latter provides accurate floating-point computation, but little speed advantage. A co-processor for the Motorola 68000 family also exists, the 68881. It is relatively less common, as it has only recently become available.

In August 1983, a second IEEE working group (p854) proposed a standard for floating-point arithmetic that is independent of word-length and radix. (Actually, the standard provides only for two radices, $\beta = 2$ and $\beta = 10$, having found no defensible reason for implementing floating-point arithmetic using any other radix.) The draft proposal was published in August, 1984, (Cody, *et al*, 1984) and was adopted on March 12, 1987. This proposal is a generalization of the binary floating-point standard discussed above; it, too, provides for single- and double-precision basic formats, as well as extended precision formats. Unlike so-called double precision or extended precision on existing mainframes, the generalized standard requires that these formats have greater precision *and* wider exponent range than the basic single-precision format. As a consequence, such operations as exponentiation and conversion between floating-point representations and decimal strings can be carried out both more efficiently and more accurately. The requirements for wider exponent range in double precision entail that products of up to eight single-precision factors can be formed with neither overflow or underflow. Double precision must, in addition, provide a fraction that is at least a full decimal digit wider than twice single precision; this provides full protection for double-precision formation of inner products of single-precision data.

Although the requirements for a floating-point system to conform to the standards appear to be complex and expensive, it turns out in practice that such systems do not slow down floating-point computation. Quite the reverse seems to be the case. The standards have been carefully thought out so that all of the basic numerical building blocks are provided in a coherent and well-defined manner within any system conforming to the standard. As a consequence, much less special programming at the level of, say, the FORTRAN programmer needs to be done to deal with peculiarities of a floating-point system or to guard against numerical problems.

3 NUMERICAL LINEAR ALGEBRA

This chapter examines the fundamental computational ideas associated with multiple regression analysis and related computations for the generalized linear model, including the multivariate case. The emphasis in the section is not to provide a complete set of algorithms, but rather to provide a unified background against which alternative algorithms can be weighed. We focus on particular algorithms only to the extent that they are useful for understanding subsequent computational methods or are helpful in illuminating underlying statistical ideas. To this end we shall describe one particularly valuable approach to the regression computations, namely, the Householder algorithm, and shall use that algorithm as a starting point from which to compare other algorithms. This will make it possible to compare both the computational details of various algorithms and the relative advantages of each over the others.

We assume that the reader is already familiar with the linear model, particularly multiple regression analysis, from the statistical and algebraic standpoint. Standard textbooks such as Draper and Smith (1981), Weisberg (1980), and Neter and Wasserman (1974) can provide this background. The computational details of the algorithms discussed here are covered at greater length in Golub and Van Loan (1983), Kennedy and Gentle (1980), Maindonald (1984), and Stewart (1973). Most of the standard distributional results from linear models theory can be deduced immediately from the matrix decompositions discussed in this chapter. Ansley (1985) gives a particularly clear account.

3.1 Multiple regression analysis

The problem we first address is that of a multiple linear regression fit to a set of data. The idea is to approximate a vector Y of responses as well as possible by a vector of *fitted values* $\hat{Y}$ that is a linear function of a set of p predictor vectors. If we set these predictors into an $n \times p$ matrix X, we can write

$$\hat{Y} = Xb,$$

and the linear regression problem is to find a "good" choice for the vector of coefficients b. (The matrix X will ordinarily contain a column of ones; if so we shall assume that this column is the left-most column of the matrix. Later in this section it will be helpful to distinguish this column from the others, but for the moment it will make no difference whether the X matrix contains a column of ones or not.) One approach is to choose that value of b which minimizes the sum of squared deviations of fitted values from the observed counterparts, $\sum(Y_i - \hat{Y}_i)^2$; that minimizing value, which we shall denote by $\hat{\beta}$, is called the least-squares solution.

A closely related problem is that of estimating coefficients in a linear model,

$$Y = X\beta + \epsilon,$$

where Y is an $n \times 1$ (or later, an $n \times q$) vector of responses, X an $n \times p$ matrix of predictors, and ϵ an $n \times 1$ matrix of unobservable errors. These errors are assumed to have mean zero and covariance matrix $\sigma^2 I$. The notation $\epsilon \sim (0, \sigma^2 I)$ is used to indicate that ϵ has mean zero and variance matrix $\sigma^2 I$; when we assume that ϵ has, in addition, a normal distribution, we shall use the notation $\epsilon \sim \mathcal{N}(0, \sigma^2 I)$. We shall assume throughout that $n \geq p$. The vector β is called the vector of regression coefficients, and it is β about which we would like to make statistical inferences.

The least-squares estimate for b in the linear regression problem, the best linear unbiased estimate for β in the linear model, and the maximum-likelihood estimate for β in the linear model with normal errors all coincide. The estimator is given by the solution $\hat{\beta}$ to the *normal equations:*

$$(X'X)\hat{\beta} = X'Y, \tag{3.1.1}$$

which gives the standard formula

$$\hat{\beta} = (X'X)^{-1}X'Y. \tag{3.1.2}$$

This suggests an obvious approach to carrying out the computations, namely, compute $X'X$ and $X'Y$, invert $X'X$, and then form the matrix product of $(X'X)^{-1}$ and $X'Y$. For problems in which n and p are both small, and for which a "one-shot" computer program is required, this approach may be suitable. The approach is straightforward to implement—with the possible exception of the matrix inversion step—and is easy to understand. Some programming and computational effort can be saved by noting that $X'X$ and $X'Y$ can be computed in the same loop if Y is stored as the $(p+1)$st column of X.

The obvious approach has several major deficiencies, however, which should rule it out for general regression programs. Note, for instance, that the matrices $X'X$ and $X'Y$ are both matrices of inner products of columns from X and Y. We noted in Chapter 2 that inner products are notorious for being the source of numerical errors, so we should prefer an approach to the regression problem that minimizes the computation of inner products, particularly without some form of centering. Another source of numerical error and possible computational inefficiency arises in the matrix inversion step. Thus, we might expect the straightforward approach of forming and then solving the normal equations to be subject to numerical instability, as indeed it is.

We should also note that simply obtaining $\hat{\beta}$ is rarely enough—we are also interested in the goodness of fit of the regression model, indications of lack of fit, and measures of variability of $\hat{\beta}$. In particular, for purposes of statistical inference, we must compute such quantities as the residual variance s^2, the analysis of variance ("ANOVA") table, the coefficient of determination R^2, and the variance matrix $V(\hat{\beta}) = s^2(X'X)^{-1}$. We now examine some computational approaches to obtaining these quantities of interest.

3.1.1 The least-squares problem and orthogonal transformations

Consider an $n \times 1$ vector x, and define its *squared (Euclidean) length* by $|x|^2 = \sum_{i=1}^{n} x_i^2$. Then $\hat{\beta}$ solves the *least-squares problem:*

$$\min_{\beta} |Y - X\beta|^2 = |Y - X\hat{\beta}|^2. \tag{3.1.3}$$

In addition, we have that $s^2 = |Y - X\hat{\beta}|^2/(n-p) = RSS/(n-p)$. RSS stands for *residual sum of squares*, the squared length of $Y - X\hat{\beta}$, the vector of *residuals*. To obtain these and other related regression quantities, it is possible to transform the original least-squares problem using linear transformations into an equivalent problem whose solution is identical to the original, but which affords both numerical and computational advantages.

We now turn, then, to the effect of general linear transformations on the components of a regression problem. Let Y denote an $n \times 1$ random vector with $E(Y) = \mu$ and $V(Y) = \Sigma$. If A is an arbitrary $k \times n$ matrix, then $E(AY) = A\mu$, and $V(AY) = A\Sigma A'$. An $n \times n$ matrix Q is said to be *orthogonal* if Q is composed of real elements, and if

$$|Qx| = |x| \qquad \forall x \in \mathcal{R}^n. \tag{3.1.4}$$

This defining relation can be expressed as saying that orthogonal transformations preserve length. It is easily deduced from (3.1.4) that $QQ' = Q'Q = I$.

The effect of an orthogonal transformation Q on the regression problem is easily seen. If

$$Y = X\beta + \epsilon, \qquad \epsilon \sim (0, \sigma^2 I) \tag{3.1.5}$$

then

$$QY = QX\beta + Q\epsilon$$

or

$$Y^* = X^*\beta + \epsilon^*, \qquad \epsilon^* \sim (0, \sigma^2 I). \qquad (3.1.6)$$

Equations (3.1.5) and (3.1.6) are identical in form, and it is clear that a solution for β in one of them must also be a solution for β in the other. It is possible to use this fact to select an orthogonal transformation Q which makes the corresponding least-squares problem easier to solve and at the same time will provide a solution which is numerically more stable than simply solving the normal equations directly.

How might we choose such a transformation Q? In general, the $n \times p$ matrix of regressors X has no special structure. Suppose that it were possible to choose Q so that the upper $p \times p$ block of $QX = X^*$ were upper triangular, and the lower $(n - p) \times p$ block were identically zero, thereby introducing a great deal of special structure into the transformed matrix of regressors X^*. Under these circumstances, we can write

$$Y^* = X^*\beta + \epsilon^*$$

in the form

$$\begin{pmatrix} Y_1^* \\ Y_2^* \end{pmatrix} = \begin{pmatrix} X_1^* \\ X_2^* \end{pmatrix}\beta + \begin{pmatrix} \epsilon_1^* \\ \epsilon_2^* \end{pmatrix} \qquad (3.1.7)$$

or

$$\begin{pmatrix} Y_1^* \\ Y_2^* \end{pmatrix} = \begin{pmatrix} X_1^* \\ 0 \end{pmatrix}\beta + \begin{pmatrix} \epsilon_1^* \\ \epsilon_2^* \end{pmatrix} \qquad (3.1.8)$$

where X_1^* is an upper triangular $p \times p$ matrix.

Since Q is orthogonal, it preserves lengths, so that we can decompose the squared length of Y into two components, writing

$$|Y|^2 = |QY|^2 = |Y^*|^2 = |Y_1^*|^2 + |Y_2^*|^2. \qquad (3.1.9)$$

Moreover, the quantity to be minimized in (3.1.3) can now be decomposed in a similar fashion by writing

$$\begin{aligned}
|Y - X\beta|^2 &= |Q(Y - X\beta)|^2 \\
&= |QY - QX\beta|^2 \\
&= \left| \begin{pmatrix} Y_1^* \\ Y_2^* \end{pmatrix} - \begin{pmatrix} X_1^*\beta \\ X_2^*\beta \end{pmatrix} \right|^2 \\
&= |Y_1^* - X_1^*\beta|^2 + |Y_2^* - X_2^*\beta|^2 \\
&= |Y_1^* - X_1^*\beta|^2 + |Y_2^*|^2. \qquad (3.1.10)
\end{aligned}$$

The choice of β which minimizes (3.1.10) clearly also minimizes (3.1.3), but note that the second term of (3.1.10) does not involve β. Thus, we need only minimize $|Y_1^* - X_1^*\beta|^2$. This is a very easy problem. Under mild conditions (namely, that rank$(X) = p$), the formal solution to this minimization problem is given by

$$\hat{\beta} = (X_1^*)^{-1} Y_1^*. \tag{3.1.11}$$

However, it is not necessary to invert X_1^* to obtain a numerical solution to (3.1.11), since X_1^* has upper triangular form and can be solved directly by back-substitution. That is, to obtain $\hat{\beta}$, solve first for $\hat{\beta}_p$, then use this solution to solve for $\hat{\beta}_{p-1}$, and so forth; at each step one must solve only a single equation in a single unknown. It follows immediately from (3.1.11) that the residual sum of squares from the regression problem is given by $RSS = |Y_2^*|^2$.

It is a remarkable fact that such an orthogonal matrix Q can be found which transforms the least-squares problem in the fashion described above, *and* this transformation can be accomplished in a computationally efficient and numerically stable fashion using *Householder rotation* or *Householder reflection* matrices.

COMMENT. In general, an $n \times p$ matrix X, $n \geq p$, can be decomposed into the product of an $n \times n$ orthogonal matrix (Q' in our example), and an $n \times p$ block upper-triangular matrix (X^* in our notation). Alternatively, since the last $n - p$ rows of X^* are zero, we can write $X = Q'X^* = Q_1' X_1^*$, where Q_1' is $n \times p$ and X^* is $p \times p$ upper triangular. This latter decomposition is usually written $X = QR$ in the numerical analysis literature, and is called the QR *decomposition*.

3.1.2 Householder transformations

The matrix Q required in the regression problem will actually be built by composing a collection of several more elementary transformations, whose construction we examine in this section. Since we desire a transformation X^* of X whose columns contain mostly zeros, it is natural to ask first whether we can construct a transformation matrix, say H_t, which acts on an $n \times 1$ vector x by rotating it to make all but the first t components of the result zero. Clearly, H_t must itself be a function of x, and we shall write $H_t^{(x)}$ when it is necessary to emphasize this dependence. More specifically, we require that the components h_i of $H_t x$ satisfy the following three requirements:

$$(H_t x)_i = (H_t^{(x)} x)_i = h_i = \begin{cases} x_i, & i < t; \\ h_t, & i = t; \\ 0, & i > t. \end{cases} \tag{3.1.12}$$

The choice of h_t is not arbitrary, of course, but must be chosen so that H_t preserves lengths. Since we must have $|H_t x|^2 = |x|^2$, it follows from

(3.1.12) that

$$\sum_{i=1}^{n} h_i^2 = \sum_{i=1}^{t-1} x_i^2 + h_t^2 = \sum_{i=1}^{n} x_i^2, \qquad (3.1.13)$$

from which we can in turn deduce that $h_t^2 = \sum_{i=t}^{n} x_i^2$. These requirements will be sufficient to allow us to construct the matrix Q as a product of the form $H_p H_{p-1} H_{p-2} \cdots H_2 H_1$

How can such a transformation be constructed? For each choice of a vector u, consider the transformation matrix $H = H(u)$ defined by

$$H = \left(I - 2\frac{uu'}{|u|^2} \right). \qquad (3.1.14)$$

H acts on x to produce

$$Hx = x - 2u\left(\frac{u'x}{|u|^2} \right)$$
$$\equiv x - 2cu. \qquad (3.1.15)$$

Now, $H = H'$ and $HH' = H'H = I$. Moreover, $|Hx|^2 = |x|^2$ for all x, so that for each u, $H^{(u)}$ is an orthogonal transformation, or rotation. Finally, it is possible to select such a u—and hence H—so that the transformation meets the requirements of (3.1.12). To do this, set

$$u_i = \begin{cases} 0, & i < t; \\ x_t \pm s, & i = t; \\ x_i, & i > t, \end{cases} \qquad (3.1.16)$$

where $s^2 = \sum_{j=t}^{n} x_j^2$. If we divide x into subvectors x_U, x_t, and x_L, representing the first $t - 1$ components, the tth component, and the last $n - t$ components, respectively, then we can write

$$x = \begin{pmatrix} x_U \\ x_t \\ x_L \end{pmatrix}, \quad \text{and} \quad u = \begin{pmatrix} 0 \\ x_t \pm s \\ x_L \end{pmatrix}.$$

Note that $|u|^2 = x_t^2 \pm 2x_t s + s^2 + |x_L|^2 = 2(\pm x_t s + s^2)$, so that $|u|^2$ is a multiple of s. Since u must be a non-zero vector, s must be non-zero in order to proceed further. The quantity s is called the *pivot value*. Provided that $s \neq 0$, so that $|u|^2 \neq 0$, the constant c in (3.1.15) is given by

$$c = u'x/|u|^2$$
$$= (0'x_U + (x_t \pm s) \cdot x_t + |x_L|^2)/|u|^2$$
$$= (\pm x_t s + s^2)/|u|^2$$
$$= \frac{1}{2}.$$

Thus,

$$Hx = x - 2cu = x - u = \begin{pmatrix} x_U \\ \mp s \\ 0 \end{pmatrix}, \qquad (3.1.17)$$

which is precisely what we required. We shall defer discussion of the case $s = 0$ to section 3.5.

Choosing the sign for s. There remains one small item of business which must be resolved before we have completely specified H, and that is the matter of selecting the sign of s in equation (3.1.16). *Algebraically,* either choice will produce a rotation matrix with the desired properties, however *numerically* the choice is crucial. If s is chosen so that its sign is opposite to that of x_t, then disastrous cancellation can occur. Consequently, the sign of s in (3.1.16) should be chosen so that it agrees with the sign of x_t.

3.1.3 Computing with Householder transformations

The Householder transformations we have constructed are $n \times n$ matrices, and the transformation Q we require is the product of p such matrices. Computing such a large matrix product would appear to be quite expensive in terms of computation time (on the order of pn^3 multiplications and additions), computation space (on the order of n^2 storage locations just to save the H_t matrices alone), and numerical precision (since most of the matrix multiplications involved are really just series of inner products). Fortunately, the Householder transformations have so much special structure that it is never necessary to form the H_t matrices explicitly, and the result of most of the implicit inner products can be deduced exactly, without having to compute them—enabling us to avoid the associated numerical instability of the inner products.

In order to convert X to the upper triangular block form which we have called X^*, we apply the matrix Q which in turn is a product of Householder transformations $H_p H_{p-1} H_{p-2} \cdots H_2 H_1$. As we proceed from right to left, each H_t is applied to a collection of column vectors resulting from earlier computation. For an arbitrary vector v, we have that

$$\begin{aligned} H_t v &= \left(v - 2\left(\frac{uu'}{|u|^2}\right)v \right) \\ &= \left(v - 2\left(\frac{u'v}{|u|^2}\right)u \right) \\ &\equiv (v - fu), \end{aligned} \qquad (3.1.18)$$

where $f = 2u'v/|u|^2$. There are several important aspects to note. First, components $1, \ldots, t-1$ of the result need not be computed, they can simply be *copied*, since they are unchanged. Second, for $j \geq t$, the result has the form $v_j - fu_j$, so that f can be computed just once, and then each remaining component involves just a single multiplication and a single subtraction. Third, the constant f is simply the product of $2/|u|^2$ —which can be computed once when H_t is first constructed and then saved—and a reduced inner product $u'v = \sum_{j=t}^{n} u_j v_j$. Finally, if we wish we can renormalize the vector u so that, say $|u|^2 = 2$, so that $2/|u|^2 = 1$, and $f = u'v$. Thus, to apply H_t to a column, we need only have t and u to do the job. Since $t - 1$ of the components of u are zero, we don't ever have to store more than $n - t + 1$ numbers to represent H_t. Nor do we have to compute more than a single inner product to apply H_t to a vector. Using this data structure, the p transformation matrices require no more space than X itself to represent, and no more than p^2 inner products to apply.

3.1.4 Decomposing X

We now examine how to construct the particular matrices H_t which will accomplish the upper-triangular block decomposition of the regression matrix X. Begin by writing the X matrix in terms of its columns, which we shall label X_1 through X_p. Because in the decomposition process we shall modify these columns iteratively, when it is desirable to emphasize their role as the initial values for this iterative procedure we shall also write $X_j \equiv X_j^{(0)}$. It is often useful, both computationally and conceptually to consider the response vector Y to be appended as the final column of X, and so we may denote it by X_{p+1}. When there are multiple responses, that is, when Y is an $n \times k$ matrix, we can simply append all q columns of Y to the end of X. This involves no change whatever in notation, so that the multivariate regression problem can be handled simultaneously with the more familiar univariate multiple regression problem. Writing

$$X^{(0)} \equiv X = \begin{bmatrix} X_1 & X_2 & \cdots & X_p \end{bmatrix}$$

$$\equiv \begin{bmatrix} X_1^{(0)} & X_2^{(0)} & \cdots & X_p^{(0)} \end{bmatrix},$$

we first choose a Householder transformation H_1 to zero all but the first position of column X_1, and we then apply this transformation to all of the columns of X (and Y). H_1 clearly must be a function of the contents of X_1. Applying H_1 to X we obtain a new matrix $X^{(1)}$, with columns $X_j^{(1)}$:

$$X^{(1)} = H_1 X^{(0)} = \begin{bmatrix} X_1^{(1)} & X_2^{(1)} & \cdots & X_p^{(1)} \end{bmatrix},$$

and $X_1^{(1)}$ has the form

$$X_1^{(1)} = \begin{pmatrix} x_1^1 \\ 0 \\ \vdots \\ 0 \end{pmatrix}.$$

The next step is to construct a Householder matrix H_2 which zeros the last $n-2$ positions of $X_2^{(1)}$, and which leaves the first position unchanged. We apply this matrix to each column of $X^{(1)}$ to obtain

$$X^{(2)} \equiv H_2 X^{(1)} \equiv \left[X_1^{(2)} \ \ X_2^{(2)} \ \ \cdots \ \ X_p^{(2)} \right].$$

Note that, since H_2 leaves unchanged the first position of any vector to which it is applied, $H_2 X_1^{(1)} = X_1^{(1)}$. Thus, the first two columns of $X^{(2)}$ have the form

$$[X_1^{(2)} \ X_2^{(2)}] = \begin{pmatrix} x_1^1 & x_2^1 \\ 0 & x_2^2 \\ 0 & 0 \\ \vdots & \vdots \\ 0 & 0 \end{pmatrix}.$$

Continuing in this fashion, at stage j we form the Householder matrix H_j which zeros the last $n-j$ positions of $X_j^{(j-1)}$. Applying this to $X^{(j-1)}$, we obtain $X^{(j)}$, which has the property that its upper left $j \times j$ block is upper triangular and its lower left $(n-j) \times j$ block is zero. Note particularly that H_j *does not change any of the first $j-1$ columns to which it is applied.* At the end of this process, we have $X^* \equiv X^{(p)} = H_p H_{p-1} \cdots H_2 H_1 X$, and

$$X^* = \begin{pmatrix} x_1^1 & x_2^1 & x_3^1 & \cdots & x_p^1 \\ 0 & x_2^2 & x_3^2 & \cdots & x_p^2 \\ 0 & 0 & x_3^3 & \cdots & x_p^3 \\ 0 & 0 & 0 & \cdots & x_p^4 \\ \vdots & \vdots & \vdots & \ddots & \vdots \\ 0 & 0 & 0 & \cdots & x_p^p \\ 0 & 0 & 0 & \cdots & 0 \\ \vdots & \vdots & \vdots & \ddots & \vdots \\ 0 & 0 & 0 & \cdots & 0 \end{pmatrix} = \begin{pmatrix} X_1^* \\ 0 \end{pmatrix}. \tag{3.1.19}$$

Note that, in equation (3.1.19), the subscripts indicate the column and the superscripts the row, but also, the superscripts indicate the step at which that value was computed. Thus, the first row of the matrix X^* was computed in step 1, and all further computations only involved rows 2 through n.

The transformation matrix Q is just the product of the H matrices, but note that we need never actually calculate Q explicitly, indeed, using the data structure and algorithm for Householder transformations described above, we outlined above, we need not even compute the H matrices and perform the explicit matrix multiplications. There are even more computational short-cuts which can be taken in this process: since H_j doesn't affect the first $j - 1$ columns, they need not be recomputed, and applying H_j to $X_j^{(j-1)}$ merely involves replacing the jth value in that column by $\mp s$, and the last $n - j$ positions by zero. Finally, note that the order in which columns are selected to construct the H_j matrices need not correspond to the order in which the columns of X appear in the computer memory (or the FORTRAN program) so that, with minor additional bookkeeping, variables can be introduced into the regression equation in any desired order.

3.1.5 Interpreting the results from Householder transformations

Many of the quantities needed in a multiple regression analysis can be obtained directly from the computations outlined above. In every case, the computation proceeds by making a Householder transformation which "pivots" at one lower position each time. If the regression model contains a constant term, then generally the first column of X will be a column of ones, $X_1 = (1, 1, \ldots, 1)'$. In this case it is easy to see the effect of H_1. Applied to the column of ones, it produces a new vector which consists entirely of zeros except in the first position and, since H_1 preserves length, that first position must contain $\sqrt{n}$. The effect on Y can be obtained by straightforward computation. Denote $H_1 Y$ by $Y^{(1)}$, and its components by $(y_1^1, y_1^2, \ldots, y_1^n)'$. (Here, the subscript indicates the stage number, and the superscript the original row number.) The first position of $Y^{(1)}$ is just $y_1^1 = (\sum y_j)/\sqrt{n}$. Since H_1 must preserve the length of Y, too, it follows that the squared length of all but the first component of $Y^{(1)}$ must be $|Y|^2 - (y_1^1)^2 \equiv |Y|^2 - (\sum y_j)^2/n \equiv TSS$, the total sum of squares of Y adjusted for the mean. And $(y_1^1)^2$ is precisely the sum of squares associated with the mean of Y. Note that exactly the same argument applies to each column of X (except the first, of course), so that after applying H_1 the first row of $X^{(1)}$ contains the square roots of the constant-term sums of squares for each of the remaining predictors and for the response vectors. What is more, *this row will be unaltered by subsequent steps of the computation.* This means that, for the constant term at any rate, we need not identify in advance which column is to be the response vector.

Now let's see the effect of H_1 on the rest of the regression problem. To do this, we shall partition $Y^{(1)}$, $X^{(1)}$, β, and $\epsilon^{(1)}$ into two pieces corresponding to the first component of $Y^{(1)}$ and all the rest of $Y^{(1)}$. The notation below is slightly cumbersome, but unavoidable. We shall use the superscript or subscript "*" to denote a vector whose components vary over the remaining values of the corresponding super- or subscript. For example, the first row of $X^{(1)}$ will be written (x_1^1, x_*^1), where $x_* \equiv (x_2^1, x_3^1, \ldots, x_p^1)$. We can now write

$$Y = X\beta + \epsilon,$$

and

$$Y^{(1)} = H_1 Y = H_1 X\beta + H_1 \epsilon$$
$$= X^{(1)}\beta + \epsilon^{(1)},$$

which expands to become

$$\begin{pmatrix} y_1^1 \\ Y_*^{(1)} \end{pmatrix} = \begin{pmatrix} x_1^1 & x_*^1 \\ 0 & X_*^{(1)} \end{pmatrix} \begin{pmatrix} \beta_1 \\ \beta_* \end{pmatrix} + \begin{pmatrix} \epsilon_1^1 \\ \epsilon_*^1 \end{pmatrix}. \qquad (3.1.20)$$

Since ϵ and $\epsilon^{(1)}$ have the same distribution, the least-squares problem can then be rewritten as

$$\min_{\beta_1, \beta_*} \left| y_1^1 - \left(x_1^1 \beta_1 + x_*^1 \beta_* \right) \right|^2 + \left| Y_*^{(1)} - X_*^{(1)} \beta_* \right|^2.$$

Note that β_1 appears only in the first term, so that we immediately obtain $\hat{\beta}_1 = (y_1^1 - X_1^* \hat{\beta}_*)/(x_1^1)$. What is more, it follows that $\hat{\beta}_*$ must be the least-squares solution to the reduced problem consisting of all but the first row of (3.1.20). We can either solve this reduced problem directly, or we can carry along the first row, since by construction the subsequent Householder transformations do not affect the first row. Clearly, this reduced problem corresponds to the original least-squares problem after removing the mean.

Proceeding to the next step, we obtain $\hat{\beta}_2$, which only depends upon the first column of $X_*^{(1)}$, (or equivalently, $X_2^{(1)}$). Since both $Y_*^{(1)}$ and $X_*^{(1)}$ have already been adjusted for the first column of the original X matrix, since subsequent transformations will leave y_2^2 unchanged, and since $|Y|^2 = |Y^{(j)}|^2$ for each step J, it follows that $(y_2^2)^2$ is the reduction in sum of squares due to X_2 after adjustment for X_1. Similarly, we can write

$$(y_1^1)^2 = (y_p^1)^2 = \text{SS}(\beta_1)$$
$$(y_2^2)^2 = (y_p^2)^2 = \text{SS}(\beta_2|\beta_1)$$
$$\vdots \qquad\qquad\qquad (3.1.21)$$
$$(y_p^p)^2 = \text{SS}(\beta_p|\beta_1, \beta_2, \ldots, \beta_{p-1})$$
$$|Y_*^{(p)}|^2 = \text{RSS}(\beta_1, \beta_2, \ldots, \beta_p).$$

This is simply the sequential analysis of variance table for the multiple regression problem; the model sum of squares (or regression sum of squares) after the first j columns of X have been included in the regression model is simply the sum of the squares of the first j components of Y^* up to that point. If the first column of X is a constant column, it is traditional to separate its sum of squares from the remainder. Thus, the traditional analysis of variance table for the full regression would be

Source	Degrees of freedom	Sum of squares
Regression (adjusted for mean)	$p-1$	$\sum_2^p (y_p^j)^2$
Residual	$n-p$	$\sum_{p+1}^n (y_p^j)^2$
Total (adjusted)	$n-1$	$\sum_2^n (y_p^j)^2$
Constant	1	$(y_p^1)^2$
Total (unadjusted)	n	$\sum_1^n (y_p^j)^2 = |Y|^2$

It follows directly from (3.1.8) and (3.1.21) that $Y_*^{(p)} = (y_p^{p+1}, y_p^{p+2}, \ldots, y_p^n)'$ $\sim \mathcal{N}(0, \sigma^2 I_{n-p})$, that RSS$\sim \sigma^2 \chi_{n-p}^2$, and that RSS and $\hat{\beta}$ are independent.

3.1.6 Other orthogonalization methods

The Householder-transformation algorithm for the linear least-squares problem is an example of a more general set of algorithms called *orthogonalization methods*. The effect of all orthogonalization methods is the same, namely to project Y into the subspace $C(X)$ of $\mathcal{R}^n$ spanned by the columns of X, and to compute $\hat{\beta}$ from the coefficients of Y expressed in terms of some basis for $C(X)$. Each method expresses the columns of X in terms of a new orthogonal basis, contained in the columns of a new matrix Q. We write the QR *decomposition* of X as $X = QR$, where R is generally chosen to have special structure such as upper-triangularity. The methods can differ with respect to the particular basis which is constructed for the latter subspace, and the order in which the basis vectors are computed. The primary differences among them, however, are in the nature of the intermediate results which are computed.

3.1.6.1 Gram-Schmidt methods

The Householder method does not form the new basis Q explicitly. Rather, the basis exists implicitly in the factored series of transformation matrices H_t. By contrast, *Gram-Schmidt orthogonalization* operates by forming a new basis vector orthogonal to the rest at each step. After the jth step of Gram-Schmidt, the first j columns of X have been replaced by j mutually orthogonal vectors which span the same space as the first j columns of the original matrix. The remaining $p - j$ columns have been replaced by new columns which are *residuals* from the projections of the last $p - j$ variables onto the space spanned by the first j, and they are therefore each orthogonal to the first j columns, but not necessarily orthogonal to each other. The

Gram-Schmidt procedure is easy to understand and to program, and it has the advantage that an explicit basis for X is obtained. The procedure we have described is actually what is called *Modified Gram-Schmidt (MGS)*, which is numerically much more stable than the classical method which does not update the last $p - j$ columns at each step. The following algorithm computes the QR decomposition of X by Modified Gram-Schmidt, overwriting the original X matrix with the new basis matrix Q.

Algorithm MGS:

> **for** $j := 1$ **to** p **do**
> **begin**
> $r_{jj} := \sqrt{\sum_{i=1}^{n} x_{ij}^2}$
> **for** $i := 1$ **to** n **do** $x_{ij} := x_{ij}/r_{jj}$
> **for** $k := j + 1$ **to** p **do**
> > **begin**
> > $r_{jk} := \sum_{i=1}^{n} x_{ij} x_{ik}$
> > **for** $i := 1$ **to** n **do** $x_{ik} := x_{ik} - x_{ij} r_{jk}$
> > **end**
> **end.**

Although Modified Gram-Schmidt is stable numerically for obtaining the regression coefficients $\hat{\beta}$, numerically the basis vectors themselves can be badly nonorthogonal unless particular care is taken. If X is particularly ill-conditioned, it may be necessary to re-orthogonalize the resulting columns of Q, and Gram-Schmidt may even become totally inadequate. Obtaining an orthonormal basis for $C(X)$ is about twice as expensive using Householder transformations, since the basis vectors need to be constructed from the factored form of Q. However, the Householder basis so obtained essentially will be orthogonal to within machine precision, whereas the MGS basis will in general deviate from orthogonality by a matrix whose size grows with the condition number of X. The numerical properties of Gram-Schmidt methods were analyzed by Björck (1967) and discussed in Golub and Van Loan (1983). Matrix conditioning is dealt with in section 3.5.

3.1.6.2 Givens rotations

Another orthogonalization method also obtains the Q matrix in factored form. Using *Givens rotations*, the matrix X is reduced to upper triangular form $\begin{pmatrix} R \\ 0 \end{pmatrix}$ by making exactly one subdiagonal element equal to zero at each step. In effect, these elementary transformations rotate each column of X in the same 2-dimensional subspace through the same angle θ. The

Givens rotations have the form of an identity matrix except in two rows, say i and j. The rows are zero except in the ith and jth position, where $g_{ii} = g_{jj} = \cos(\theta)$, and $g_{ij} = -g_{ji} = \sin(\theta)$. Applying this matrix, which we could denote by $G[i, j, \theta]$, rotates a vector through an angle θ in the i, j coordinate plane. Givens transformations are equally stable numerically as Householder transformations, but require roughly twice as much work to compute the QR decomposition of a general matrix X. They are most valuable in situations where X is large but sparse, that is, most of the elements in X are already zero. Since Givens rotations can zero out these elements one at a time, they can quickly reduce X to upper triangular form without additional computation. This property also makes Givens rotations a valuable tool in computing the intermediate steps of matrix computations which reduce an input matrix to a special form (such as bi- or tri-diagonal form). Givens rotations play a featured role in algorithms for the singular value decomposition and extraction of eigenvalues.

EXERCISES

3.1. [13] Suppose that the linear model $Y = X\beta + \epsilon$ holds, that $\epsilon \sim \mathcal{N}(0, \sigma^2 I)$, and the β is estimated using (3.1.2). What is the distribution of $\hat{\beta}$?

3.2. [22] Prove that $QQ' = Q'Q = I$ whenever (3.1.4) holds.

3.3. [25] Show that an orthogonal matrix has the effect of rotating vectors through a constant angle.

3.4. [08] Show that $RSS = |Y_2^*|^2$.

3.5. [01] Justify the last step in (3.1.10).

3.6. [18] Show that no single rotation matrix can have property (3.1.12) for all x.

3.7. [18] Describe geometrically the class of transformations which satisfy (3.1.14) for given u.

3.8. [19] Show that $H = H'$ and $HH' = H'H = I$.

3.9. [11] Show that $|Hx|^2 = |x|^2$ for all x.

3.10. [15] Show that $H^{(x)}$ leaves the first $t - 1$ components of any vector v unchanged, that is, that $[H^{(x)}v]_i = v_i$, $i = 1, \ldots, t - 1$.

3.11. [25] Describe in detail the "additional bookkeeping" required to allow for entry and deletion in arbitrary (memory) order. (Note that if variables are deleted, they must be deleted in reverse order from the order they were entered.)

3.12. [12] How does one reverse a step in the Householder transformation described above?

3.13. [20] Prove that $(y_p^{p+1}, y_p^{p+2}, \ldots, y_p^n)' \sim \mathcal{N}(0, \sigma^2 I_{n-p})$, that the residual sum of squares RSS $\sim \sigma^2 \chi^2_{n-p}$, and that RSS and $\hat{\beta}$ are independent.

3.14. [21] Prove that the t-statistics for testing the hypothesis that $\beta_p = 0$ is simply y_p^p/s, where $s^2 = RSS/(n-p)$.

3.15. [14] Let $A = \begin{pmatrix} a_{11} & a_{12} \\ a_{21} & a_{22} \end{pmatrix}$. What are the entries in the Givens rotation matrix $G[1, 2, \theta]$ which converts a_{21} to zero?

3.2 Solving linear systems

A linear system of equations can be written in the form

$$A\beta = c, \qquad (3.2.1)$$

where β is a $p \times 1$ vector of unknowns, A is a $p \times p$ matrix of known coefficients, and c is a $p \times 1$ vector (known quaintly as "the right-hand side"). Quite general methods for solving such systems efficiently and in a numerically stable fashion are known, and the problem has been extensively analyzed. Linear systems arise constantly in statistical computations, so it is worth the effort to learn some of the basics in computing their solution. However, statistical problems usually exhibit special structure which makes it possible to take advantage of more specialized algorithms. We shall concentrate on the techniques most useful in statistical computation. As a first step, suppose that A in (3.2.1) can be written as the product of a lower-triangular matrix L and an upper-triangular matrix R. Then the solution β of (3.2.1) can be obtained in two steps, first by writing

$$A\beta = (LR)\beta = L(R\beta) \equiv L\gamma = c, \qquad (3.2.2)$$

and solving for γ, and then writing

$$R\beta = \gamma \qquad (3.2.3)$$

and solving for β. Since each of these steps involves solving a *triangular* system, each can be solved using the efficient and numerically stable procedure of *back-substitution*. In the upper-triangular case of (3.2.3), this amounts to computing, for $j = p, p-1, \ldots, 1$,

$$\beta_j = \frac{\gamma_j - \sum_{k=j+1}^p R_{jk}\beta_k}{R_{jj}}. \qquad (3.2.4)$$

The computation for the lower-triangular case is similar. This process hinges on being able to obtain the factorization $A = LR$ efficiently and in a stable fashion.

A collection of portable FORTRAN subroutines for obtaining efficient and accurate solutions to such problems has been prepared by a team of outstanding numerical analysts. This package, called *LINPACK* (1979), contains separate routines to deal with different cases of (3.2.1) which exhibit various sorts of special structure. *LINPACK* is in the public domain—as it was developed under contract to the U. S. government—and can be easily obtained for little more than the cost of duplication. Indeed, *LINPACK* can even be obtained in its entirety for such computers as the IBM PC. The details of numerical linear algebra are fascinating in and of themselves, and the interested reader should consult Householder (1964) for an early description of some of the mathematical structure in approaches to (3.2.1), Stewart (1973) for an exceedingly clear account of matrix computations and associated numerical considerations, and Golub and Van Loan (1983) for an advanced up-to-date source on currently available methods. Forsythe and Moler (1967) is a jewel of a book, suitable for self-study, which describes most of the important issues which arise in solving general linear systems, including floating-point error analysis. Lawson and Hanson (1974) treat the numerical and computational issues involved in solving general least-squares problems. Since the general problem has been treated extensively, and since high-quality computer software is available, we shall not attempt to duplicate that effort. Rather, we shall concentrate on some of the special structure and special requirements often found in statistical problems which should be taken into account when taking advantage of the resources available for numerical solution of linear systems.

In the previous section we showed that the least-squares problem can be reduced to solving a set of p simultaneous equations in p unknowns, $\beta_1, \beta_2, \ldots, \beta_p$, and we also saw that these equations could be represented in more than one way, either through the normal equations of equation (3.1.1)

$$(X'X)\hat{\beta} = (X'Y), \tag{3.2.5}$$

or through the reduced form of (3.1.11)

$$X_1^*\hat{\beta} = Y_1^*, \tag{3.2.6}$$

where X_1^* is upper triangular in form. From the construction of X_1^*, it follows that $X'X = (X_1^*)'(X_1^*)$, so that X_1^* is an upper-triangular *matrix square root* of $X'X$. Given that such a matrix exists—and we have shown by construction that one will—we can view (3.2.6) as being derived algebraically from (3.2.5) by multiplying both sides of (3.2.5) on the left by $(X_1^{*\prime})^{-1}$, although we computed X_1^* directly without having to form $X'X$

first. We can also view the procedure which produces (3.2.6) as one which automatically obtains for us the triangular factorization of $X'X$ which enables us to go directly to the second back-substitution step (3.2.2) in solving the normal equations (3.2.5).

Alternative methods for solving (3.2.1) include *Gaussian elimination*, the *Crout-Doolittle* method, and others. The latter method is a square-root-free modification of the Cholesky factorization for general (not necessarily symmetric) square matrices; see Forsythe and Moler (1967, Section 12) or Golub and Van Loan, 1983). For statistical purposes the Cholesky and SWEEP methods, discussed respectively in Sections 3.3 and 3.4, are generally superior.

Since the transformation matrix Q of the last section is orthogonal, we can write $Q'Q = I$, and thus since $X^* = QX$, we can also write $X = Q'X^*$. Because the lower $n - p \times p$ half of X^* is zero, we can then write

$$X = Q^{(1)'}X_1^*, \qquad (3.2.7)$$

so that each column of X is represented as a linear combination of the (orthogonal) columns of $Q^{(1)'}$. Moreover, because of the triangular structure of X_1^*, the jth column of X is written as a linear combination of the first j columns of $Q^{(1)'}$. We can interpret the first j coefficients in the kth column of X_1^* to be the regression coefficients of X_k (the kth original X variable) on the first j columns of $Q^{(1)}$. Since those columns can all be expressed as linear combinations of $X_1, X_2, \ldots, X_j$, we need only carry out that re-expression to obtain the regression coefficients of X_k on $X_1, X_2, \ldots, X_j$. Once again, this re-expression is easy to carry out because the triangular structure makes it possible to use back-substitution, which in turn means that each step involves only a single new substitution and the solution of a single equation in a single unknown.

Most of what we have said up to this point has emphasized two aspects of the regression problem, namely solving the least squares problem by minimizing the squared length of the residuals, and obtaining the regression coefficient estimates. The Householder computations produce as a by-product such auxiliary quantities as RSS, the analysis of variance decomposition, and (implicitly) and orthogonal basis for $\mathcal{C}(X)$. There are two other quantities which are important to have available in a regression, namely, an estimate for the variance matrix of the regression coefficients, $\mathrm{Var}(\hat{\beta}) = \sigma^2 (X'X)^{-1} \equiv \sigma^2 C$, and the vector of residuals $r = Y - X\hat{\beta}$. While neither of these quantities is a direct by-product of the Householder decomposition, neither is particularly difficult to obtain. The primary reason for computing the C matrix is to obtain standard errors for the estimated regression coefficients. The standard error for $\hat{\beta}_j$ is simply $\hat{\sigma}$ times $\sqrt{c_{jj}}$, which is the square root of the jth diagonal element of $(X'X)^{-1}$. The latter can be written as $(X_1^{*-1})'(X_1^{*-1})$. Because X_1^* is upper triangular, so is its inverse.

If this inverse matrix is obtained, then c_{ij} is just the inner product of the ith and jth columns of X_1^{*-1}, so that the standard errors of the regression coefficients can be obtained from the lengths of the columns of (X_1^{*-1}).

The inverse of an upper triangular matrix can be obtained by a variant of back substitution; if T is upper triangular, then its inverse U is obtained by computing,

> **for** $j = p$ **downto** 1 **do**
> **begin**
> $u_{jj} := 1/t_{jj}$
> **for** $k = j - 1$ **downto** 1 **do** $u_{kj} := -\left(\sum_{i=k+1}^{j} t_{ki} u_{ij}\right)/t_{kk}$
> **end.**   (3.2.8)

This operation requires $p^3/6$ floating-point operations to accomplish the back substitution against the identity matrix. An additional $p^3/6$ operations are required to obtain $(X'X)^{-1}$, should that be necessary, by multiplying $(X_1^*)'$ by its transpose.

The vector of residuals from Y at the jth step of the Householder decomposition can be obtained from components $j + 1$ through n of $Y^{(j)}$; indeed, these components give the coordinates of the residual vector with respect to the basis consisting of the last $n-j$ columns of $Q^{(j)}$. The residuals can be converted back into the original space by applying the Householder transformations to the vector $(0, \ldots, 0, y_j^{j+1}, y_j^{j+2}, \ldots, y_j^n)'$ in the order H_j, $H_{j-1}, \ldots, H_1$. Thus, the effort needed to compute the residuals at step j is essentially proportional to j. We defer further details to section 3.6.

Other quantities, called *regression diagnostics*, are statistics which can help to identify unusual variables (columns), unusual observations (rows), possible outliers, or linear dependencies in the predictor variables. Diagnostics are the subject of Section 3.6. Some of the most useful diagnostics can be computed from the quantities readily available either as the end product or as intermediate results in the regression computation. Measures of *leverage*, and measures of *sensitivity* such as Cook's distance, are in this category. For a full treatment of such measures consult Cook and Weisberg (1982). Other regression diagnostics require computation of the singular values of X, which require computations beyond what we have developed up to this point; we defer further discussion of such these diagnostics to the section on collinearity.

The n observations can be thought of as n points (y_i, x_i) in $\mathcal{R}^{p+1}$; here x_i denotes the ith row of X. From this perspective, the fitted values $\hat{Y} = X\hat{\beta}$ in the linear regression model lie on a p-dimensional plane in this $(p + 1)$-dimensional space. A plane that fits well will pass through the "center" of the (y_i, x_i)'s; the least-squares fit passes through the point of mean values $(\bar{y}, \bar{x})$. A small change in the coefficients of this plane will cause it to rotate or "tilt" about such a point in the middle of the distribution of the predictor variables. As a consequence, the further a predictor value $(x_{i1}, x_{i2}, \ldots, x_{ip})$ is from the center, the more its corresponding fitted value $\hat{y}_i$ changes as a result of introducing a small "tilt" in the plane. Let's examine what happens to the fitted values and the regression coefficients when a single new observation is added to the data set at an x-point that is far from the center of the x_is. The least-squares criterion penalizes large deviations much more heavily than small deviations, so that if the new observation y lies very far from the current plane of fitted values, it would add a considerable amount to the residual sum of squares. But since the x-value is far from the center of the other x_i's, we could tilt the plane by pulling it toward the new y, without affecting the size of residuals corresponding to x_i's in the middle. Thus, the least-squares criterion implies that single points whose predictor values are far from the center of the predictors may have considerably more *influence* on the resulting estimated model than points near to the center. A measure of this potential can be based on the distance (in some appropriate sense) an observation is from the predictor values of the other observations. One such measure is the *leverage* of an observation, so named because of the "tilting" description above.

Instead of thinking of the data as n instances of $(p + 1)$-dimensional points, we can think of the $p + 1$ variables (columns) as being points in $\mathcal{R}^n$. If we write

$$\hat{Y} = X\hat{\beta} = X(X'X)^{-1}X'Y \equiv PY, \qquad (3.2.9)$$

we can immediately recognize P as the projection of Y onto the space (in $\mathcal{R}^n$) spanned by the columns of X. This matrix P is usually called the *hat matrix*, since is is the operator which "puts the hat on Y" (Hoaglin and Welsch, 1978). In terms of the Householder decomposition, we can write $X = Q'X^* = Q^{(1)'}X_1^* + Q^{(2)'} \cdot 0 = Q^{(1)'}X_1^*$, where $Q^{(1)}$ consists of the first p rows of Q. Since X_1^* is triangular and, in practice, nonsingular, it follows that

$$\hat{Y} = Q^{(1)'}Q^{(1)}Y$$

or

$$P = Q^{(1)'}Q^{(1)} \qquad (3.2.10)$$

There is rarely a reason to compute this $n \times n$ matrix explicitly. The useful quantities obtained from P are its diagonal elements; p_{ii} is called the *leverage* of the i-th observation. It follows immediately from (3.2.9) that $\text{Var}(\hat{y}_i) = \sigma^2 p_{ii}$, and that the variance of the ith residual is given by $\sigma^2(1 - p_{ii})$. These two variance expressions imply that $0 \le p_{ii} \le 1$, so that the leverage value can be interpreted as the proportion of the variance in $\hat{y}_i$ determined solely by the ith observation itself. If Q were available—or even if just $Q^{(1)}$ were—p_{ii} could be computed as the sum of squares of the first p elements in row i of Q (which is equivalent to the squared length of the ith column of $Q^{(1)}$). As in computing residuals, however, the computation of leverage values is complicated by the fact that the Householder approach never forms the Q matrix explicitly. It exists in coded form in the transformation matrices H_t, and can be implicitly reconstructed.

We recommend that, for later use in the auxiliary computations associated with regression, at each stage of the Householder procedure the current basis for X, that is, the first j rows of Q, be computed explicitly and saved. Golub and Van Loan (1983) have a brief discussion. Routines for accomplishing this task efficiently can be found in the LINPACK (1979) and ROSEPACK (1980) subroutine libraries.

COMMENT. Linear systems such as (3.2.1) can be solved using several general methods. The class of methods emphasized up to this point are *orthogonalization methods;* they operate by transforming the problem into an equivalent, simpler problem by applying orthogonal transformations. Another class of methods in wide use are the *elimination methods.* They operate by transforming the problem into an equivalent one by applying *elementary operations,* or *Gauss transformations.* These have the advantage that at each step exactly one row or column is modified, and the updates required at each step are nearly trivial to accomplish. The disadvantage is that elimination methods are not nearly so stable numerically as the orthogonalization methods. Examples of elimination methods include *Gaussian elimination,* the *Crout-Doolittle* algorithm, and the *Gauss-Jordan* algorithm. Many of the variants among elimination methods are schemes for conserving the space required to store intermediate results, or to eliminate mathematical operations (particularly square roots) that were onerous in the days of mechanical calculators.

When the left-hand side matrix A of (3.2.1) is dense and has no special structure, then elimination methods are quite useful, and Gaussian elimination is highly recommended. When A does have special structure such as symmetry and nonnegative definiteness, as is often the case in statistical computations, the *Cholesky* algorithm and the *Gauss-Jordan (SWEEP)* algorithm are preferable among elimination methods due to the statistical interpretation that can be given to the intermediate results. The next two sections treat these two important methods in some detail.

EXERCISES

3.16. [20] Write out the back-substitution algorithm for solving (3.2.2) for γ in terms of L and c. This is the lower-triangular counterpart to (3.2.4).

3.17. [22] Show that the inverse of a nonsingular upper-triangular matrix is upper triangular.

3.18. [07] Show that the elements c_{ij} of the C matrix can be obtained as the inner product of the corresponding columns of X_1^{*-1}.

3.19. [16] Prove that expression (3.2.8) gives the inverse of an upper-triangular matrix.

3.20. [13] Prove that the matrix P of equation (3.2.9) has the properties $P = P'$ and $P^2 = P$.

3.21. [15] Prove that $\mathrm{Var}(\hat{y}_i) = \sigma^2 p_{ii}$, and that the variance of the ith residual is given by $\sigma^2(1 - p_{ii})$, where p_{ii} is the ith diagonal element of P in equation (3.2.9).

3.22. [35] Write an algorithm which computes the Householder decomposition of X and which, at each stage maintains the current matrix Q explicitly.

3.3 The Cholesky factorization

In the previous section we saw that the upper triangular matrix X_1^* from the Householder reduction plays a central role in regression computations. This is not accidental. In the normal equations formulation of the solution to the least-squares problem, $X'X$ plays the role of the $p \times p$ matrix of coefficients in a linear system. Since $X'X = (X_1^*)'(X_1^*)$, we say that X_1^* is a triangular square-root of the matrix $X'X$. Note that this gives us an $L - R$ decomposition of the coefficient matrix which can be used to solve for $\hat{\beta}$ as in equations (3.2.2) through (3.2.4). This decomposition is called the *Cholesky factorization* or *Cholesky decomposition* of $X'X$, and X_1^* is called the *Cholesky triangle* The matrix $X'X$ of coefficients in the normal equations system has considerable structure beyond that of a general left-hand side of a linear system. This structure can and should be exploited.

COMMENT. As elsewhere, it is convenient computationally and helpful conceptually to append the q columns of Y to those of X, and to speak of Y_j when we wish to emphasize the role of a particular column as the jth response variable, and as X_{p+j} when we wish to emphasize that the computations being done on the column are identical to those being done on the (other) X's.

Let S denote the $(p + q) \times (p + q)$ matrix of sums of squares and cross-products of columns of X and Y, so that

$$S = \begin{pmatrix} S_{XX} & S_{XY} \\ S_{YX} & S_{YY} \end{pmatrix}. \qquad (3.3.1)$$

More precisely, $s_{ij} = x_i' x_j$, where x_i denotes the ith column of X. S has several special properties. First, it is it symmetric, that is, $s_{ij} = s_{ji}$. Second, it is *positive semidefinite*.

Definition. *A symmetric $p \times p$ matrix A is said to be positive semidefinite if, for all vectors $v \in \mathcal{R}^p$, $v'Av \geq 0$. In this case we write $A \geq 0$. If, in addition, $v'Av = 0$ only when $v = 0$, A is said to be positive definite, and we write $A > 0$.*

The existence of a real-valued matrix square root is guaranteed by positive semidefiniteness, and we have essentially established that fact by our construction of X_1^* by Householder reduction of S. Note that a matrix square root is not unique; if Q is any $p \times p$ orthogonal matrix then QX_1^* is also a square root of S.

COMMENT. The definition of positive definite and positive semidefinite that we have used here presupposes that the matrix involved is symmetric. This definition is at odds with that prevalent in the numerical analysis literature, and readers who may consult that literature need to be aware of this fact. Asymmetric positive definite matrices arise only very rarely in statistics, so that there is little to be gained by insisting that the term "symmetric positive definite" be used to describe the matrices we deal with. It is the definiteness property which will generally distinguish them from other matrices of interest and not the symmetry property.

These matrix properties have a statistical interpretation. Every covariance matrix Σ of a random vector Y is positive semidefinite; and for every positive semidefinite matrix, a random vector can be constructed with the given matrix as its covariance. For this reason, it is often helpful to think of $X'X$ as a matrix of variances and covariances, even though this is literally true for a data set at hand only if the columns have first been centered. By slightly modifying the argument of section 3.1, we can interpret the diagonal elements of X_1^* as conditional standard deviations, given the linear regression on the preceding column variables of X. From this interpretation it is easy to deduce that the diagonal elements of the Cholesky triangle must be non-negative, and that if any is zero, it means that the corresponding variable is a linear combination of its predecessors (since its residual variance about the regression on its predecessors is zero).

Given the matrix S, the Cholesky decomposition can be obtained by direct computation on s. Indeed, it is most convenient to overwrite S (or its upper triangle) by the Cholesky factorization. Let U represent the

Cholesky triangle, so that $U'U = S$. This implies that, for $i \geq j$,

$$s_{ij} = \sum_{k=1}^{j} u_{ki} u_{kj}. \tag{3.3.2}$$

By rearranging these terms, we obtain

$$u_{ij} = \frac{1}{u_{ii}} \left(s_{ij} - \sum_{k=1}^{i-1} u_{ki} u_{kj} \right) \qquad i < j$$

and

$$u_{ii} = \left(s_{ii} - \sum_{k=1}^{i-1} u_{ki}^2 \right)^{\frac{1}{2}}. \tag{3.3.3}$$

Doing this a row at a time, starting with the first row of S, we can simply replace s_{ij} by u_{ij}, and the needed values of U will have already been computed when required on the right-hand sides of equations (3.3.3). This gives us the in-place Cholesky algorithm as

for $i := 1$ **to** p **do**
  begin
  $s_{ii} := \left(s_{ii} - \sum_{k=1}^{i-1} s_{ki}^2 \right)^{\frac{1}{2}}$
  for $j := i + 1$ **to** p **do** $s_{ij} := \left(s_{ij} - \sum_{k=1}^{i-1} s_{ki} s_{kj} \right) / s_{ii}$
  end. $\tag{3.3.4}$

The algorithm runs to completion provided that no s_{ii} ever becomes zero or negative. As in section 3.1, the divisor s_{ii} is called the ith *pivot*. The case in which zero pivots occur is the topic of section 3.5, and for the moment we shall assume that each $s_{ii} > 0$. Negative values for the argument of the square root in the third line of (3.3.4) cannot occur (except due to rounding error).

While the algorithm above is suitable for programming, to some extent it obscures the statistical interpretation. After p steps of the row operations outlined above, the S matrix (ignoring the symmetric lower left corner) will have been transformed in the following way:

$$\begin{pmatrix} S_{XX} & S_{XY} \\ & S_{YY} \end{pmatrix} \Rightarrow \begin{pmatrix} S_{XX}^{1/2} & S_{XX}^{-1/2} S_{XY} \\ & \sqrt{S_{YY} - S_{YX} S_{XX}^{-1} S_{XY}} \end{pmatrix}. \tag{3.3.5}$$

The statistical interpretation is a little clearer at this point. The upper right corner of the resulting matrix (which corresponds to Y_1^* in the Householder formulation) is simply $S_{XX}^{1/2}$ times the matrix of regression coefficients of X on Y, and $S_{XX}^{1/2}$ is itself in the upper left corner. The lower right element is the conditional standard deviation of Y about its regression on X. In the case where Y is q-dimensional, the lower right piece is actually the upper triangular square root of the residual variance matrix.

The regression computations can be developed completely in terms of the Cholesky factorization; Maindonald's (1984) development of the linear model is almost completely built around the Cholesky factorization of S. This approach to the linear model has the great advantage of numerical stability (once the S matrix has been computed, of course!); Golub and Van Loan (1983) provide details. The main advantage, however, is conceptual and not computational. If the Cholesky triangle is actually desired, the Householder transformations of X provide it in a more stable fashion numerically. If the goal is to obtain the regression quantities efficiently from S, the *SWEEP* operator discussed in the next section is superior.

EXERCISES

3.23. [13] Prove that S is positive semidefinite.

3.24. [11] Prove that all of the principal $j \times j$ blocks of S are positive semidefinite.

3.25. [08] Show that, if A is any orthogonal matrix, then AX_1^* is a square root of S whenever X_1^* is.

3.26. [20] Prove that if $\mathrm{Cov}(Y)=\Sigma$, then $\Sigma \geq 0$, and that if $\Sigma \geq 0$, then there is a random variable Y whose covariance is Σ.

3.27. [23] Show that, in the absence of rounding errors, the argument of the square root, that is, $(s_{ii} - \sum_{k=1}^{i-1} s_{ki}^2)$, is nonnegative at each step of (3.3.4).

3.4 The SWEEP operator

Although it is preferable not to form the normal equations and then to solve them, both for reasons of numerical accuracy and efficiency, it is often necessary to perform the regression computations in that form, either because the data have already been reduced to the sums of squares and cross-products, or because the available computer memory is not sufficient to store all of the data and intermediate results simultaneously. (This latter concern is of decreasing importance. Random-access memory is becoming less and less expensive, so that personal workstations with a megabyte or more of main memory are commonplace. In addition, computer architectures are increasingly designed to support virtual memory, so that the amount of memory accessible from a program can exceed by a substantial factor the actual amount of physical random-access memory without loss of much efficiency.)

When the regression computations are done using the sums of squares and cross-products (SSCP) matrix, it is possible to arrange the computations themselves and the data on which they operate in an elegant fashion. This arrangement is particularly well suited to stepwise regression, since at each step the current regression coefficients, their standard errors, and the residual sums of squares are immediately available, and each step is easily reversed. The operation which adds (or removes) one variable to the current model corresponds to an operator called the *SWEEP operator*, which operates on matrices with a particular structure. This data structure, and the algorithm which implements the SWEEP operator on it, are the topics of this section.

An excellent tutorial on the SWEEP operator, its relationship to Gaussian elimination and the Cholesky factorization, and its role in computing generalized inverses is Goodnight (1979). The term *SWEEP operator* was introduced by Beaton (1964). Dempster (1969) gives a detailed examination of the SWEEP operator and other related orthogonalization methods. He also shows how linear regression and other multivariate computations can be carried out using SWEEP.

COMMENT. At the end of the sequence of SWEEP operations, the SSCP matrix will have been replaced by the negative of its inverse matrix. When viewed solely as an algorithm for matrix inversion, the SWEEP algorithm is a space-conserving variant of Gaussian elimination known as the *Gauss-Jordan algorithm*. In keeping with Stigler's law of eponymy, Jordan seems to have had nothing to do with devising this algorithm. The first association of the algorithm with Jordan appears in a revised edition of his *Handbuch der Vermessungskunde* which was prepared after his death. Householder (1964) attributes the algorithm to Clasen in 1888.

There are in fact several SWEEP operators which are closely related to one another. The one developed here differs from that of Goodnight's (1979) development in that the matrix tableau we use is symmetric at every step, so that little more than half of the implicit matrix need be stored, and the inverse operator is virtually identical to the forward operator. The price we pay is that the C matrix we obtain will be multiplied by -1, which introduces only a minor wrinkle in subsequent computations.

The notation for this section will be slightly different from that used earlier in this chapter, so as to emphasize certain features of the SWEEP algorithm. In this context, it is particularly easy to compute simultaneously the regressions of a set of p predictors X on each of a set of q predictors Y. Indeed, it is actually preferable to think of the problem in these terms, because at each step those predictor variables which have not yet been entered into the model are treated just as if they were themselves response variables. The regression computation can be thought of as a series of steps at the beginning of which are a set of variables which we shall call *candidate variables,* and an empty set of *model variables.* The effect of each step in the regression process is to select a candidate variable and then move it into the set of model variables, adjusting the remaining candidates for the linear effects of the new model variable. At the end of step k in the process, there are exactly k variables in the model set, and the remaining $p + q - k$ variables, adjusted for the model, remain as candidates.

In stepwise regression computations, we actually treat the *response* variables no differently from the *predictor* variables; we have lumped them all into the "candidate" set. Of course, only predictor variables will be allowed to be moved from the candidate set to the model set, that is, to *enter the model,* and the order in which they do so will be determined by their effect on the response variables. At a given stage in the proceedings, it may also become apparent that one or more of the model variables are no longer needed, given the other variables that have subsequently been entered into the model. In such circumstances it is desirable to remove a variable from the model, that is, to move it from the model set back into the set of candidates.

3.4.1 The elementary SWEEP operator

Consider the two-variable regression problem, in which we regress a single variable y on a single variable x. The SSCP matrix in this simple case is given by the 2×2 matrix

$$\begin{pmatrix} x'x & x'y \\ y'x & y'y \end{pmatrix}. \tag{3.4.1}$$

The regression coefficient is given by $\hat{\beta} = x'y/x'x$, the C matrix (from which the standard error of $\hat{\beta}$ is obtained) is $1/x'x$, and the residual sum of square is given by $RSS = y'y - (x'y)^2/x'x$. Thus, we can represent the results of the regression in the 2×2 matrix

$$\begin{pmatrix} -(x'x)^{-1} & (x'y)/(x'x) \\ (y'x)/(x'x) & y'y - (x'y)^2/(x'x) \end{pmatrix} = \begin{pmatrix} -C & \hat{\beta} \\ \hat{\beta}' & RSS \end{pmatrix}. \tag{3.4.2}$$

It is clear that matrices (3.4.1) and (3.4.2) can easily be computed from one another, provided only that $x'x > 0$. There is a slight wrinkle in the representation in (3.4.2), namely, we have preceded the C matrix with a minus sign. This is both a bookkeeping and a computational device; using this representation we can use essentially the same computations to go from (3.4.2) to (3.4.1) as we did to obtain (3.4.2) in the first place.

The elementary SWEEP operator performs just the calculations required to obtain (3.4.2), only in the context of several additional candidates. This may be viewed either as adjusting other potential predictors for the linear effect of the x currently being added to the model, or as computing simultaneously the regressions of several responses on the given x predictor. For this large regression problem, begin with a sum of squares and cross-products matrix S, with elements s_{ij}. The effect of the *elementary SWEEP operator on the k-th column* is a new matrix $\tilde{S}$, which we write

$$\tilde{S} = SWEEP[k]S, \qquad (3.4.3)$$

and whose elements are given by

$$\tilde{s}_{kk} = -\frac{1}{s_{kk}};$$

$$\tilde{s}_{ik} = \frac{s_{ik}}{s_{kk}}, \qquad i \neq k;$$

$$\tilde{s}_{kj} = \frac{s_{kj}}{s_{kk}}, \qquad j \neq k; \qquad (3.4.4)$$

$$\tilde{s}_{ij} = s_{ij} - \frac{s_{ik}s_{kj}}{s_{kk}}, \qquad i \neq k, j \neq k.$$

The diagonal element s_{kk} is called the kth *pivot* for the SWEEP operator. The SWEEP operator is undefined if $s_{kk} = 0$. Note that $\tilde{s}_{ii}$ is the residual sum of squares for the ith variable after adjustment for variable k (from which it follows the $s_{kk} \geq 0$ in the absence of rounding error), and that $\tilde{s}_{ij}$ is the residual sum of cross-products for variables i and j, after adjustment for variable k. Note also, that $\tilde{S}$ is symmetric, so that the lower triangular part need not be either computed or stored.

It is interesting and useful to note that the inverse of this operation, which we might write as

$$S = RSWEEP[k]\tilde{S},$$

is given by

$$s_{kk} = -\frac{1}{\tilde{s}_{kk}};$$

$$s_{ik} = -\frac{\tilde{s}_{ik}}{\tilde{s}_{kk}}, \qquad i \neq k;$$

$$s_{kj} = -\frac{\tilde{s}_{kj}}{\tilde{s}_{kk}}, \qquad j \neq k; \qquad (3.4.5)$$

$$s_{ij} = \tilde{s}_{ij} - \frac{\tilde{s}_{ik}\tilde{s}_{kj}}{\tilde{s}_{kk}}, \qquad i \neq k, j \neq k.$$

If S is a matrix of sums of squares and cross-products, then $\tilde{s}_{jk}$ simply gives the regression coefficient when the jth variable is regressed on the kth. Suppose, however, that variables j and k had already been adjusted for a third variable, say variable i, so that the jth and kth rows and columns of S actually contained *residual* sums of squares and cross-products. It is well-known that the regression coefficient of a variable in multiple regression is simply the simple regression coefficient of the residuals of y (adjusted for the other x's) and the residuals of the x variable in question (adjusted for the other x's). Thus, in this case, $\tilde{s}_{jk}$ would give the coefficient for x_k in the regression of x_j on x_k and x_i. This suggests that it is possible to compute multiple regressions by repeated application of the SWEEP operator, and this is in fact the case.

3.4.2 The matrix SWEEP operator

Denote repeated application of the SWEEP operator by $SWEEP[i, j, \ldots, k]$, so that

$$SWEEP[i, j, \ldots, k] \equiv SWEEP[k] \cdots SWEEP[j]SWEEP[i].$$

Without loss of generality, let us examine the effect of sweeping on the first m elements of S. Assume that S is partitioned so that

$$S = \begin{pmatrix} S_{11} & S_{12} \\ S_{21} & S_{22} \end{pmatrix},$$

where S_{11} is $m \times m$. The effect of this m-fold sweep can be expressed as

$$\tilde{S} \equiv SWEEP[1, 2, \ldots, m]S = \begin{pmatrix} -S_{11}^{-1} & S_{11}^{-1}S_{12} \\ S_{21}S_{11}^{-1} & S_{22} - S_{21}S_{11}^{-1}S_{12} \end{pmatrix}. \quad (3.4.6)$$

This matrix version of the SWEEP operator is defined provided that S_{11} is nonsingular, that is, that S_{11}^{-1} exists. This will be the case if and only if none of the elementary SWEEP operations require division by zero. Section 3.5 deals with this eventuality in detail.

The inverse $SWEEP^{-1}$ operator is closely related to the SWEEP operator itself, that is,

$$SWEEP^{-1}[1, 2, \ldots, m]S = \begin{pmatrix} -S_{11}^{-1} & -S_{11}^{-1}S_{12} \\ -S_{21}S_{11}^{-1} & S_{22} - S_{21}S_{11}^{-1}S_{12} \end{pmatrix}. \quad (3.4.7)$$

The statistical interpretation of (3.4.6) is straightforward. The upper-right partition of the matrix is the matrix of regression coefficients; each column of $\tilde{S}_{12}$ gives the multiple regression coefficients of one of the $p+q-m$ candidates on the first m variables. The covariance matrix of each of these columns is proportional to $\tilde{S}_{11}$, the constant of proportionality being the residual variance for the candidate column. Since $\tilde{S}_{22}$ is the residual SSCP matrix, $\tilde{S}_{22}/(n-m)$ is the usual estimator for the residual variance matrix of the last $p + q - m$ variables conditional on the first m variables. Note that the *partial correlations* of the candidates given the model are obtained by dividing rows and columns of $\tilde{S}_{22}$ by the square roots of the diagonal elements.

From the computational standpoint, using SWEEP has many advantages. At each stage the most important regression quantities have already been computed. The amount of storage required is minimal; indeed, at each step the matrix is symmetric, so that only the upper triangular part need be computed or retained. The process of removing a variable is nearly identical to that of adding one; only a single sign need be changed in the computation. At any stage of the computation, the kth diagonal element will be positive if, and only if, the kth variable remains in the set of candidates. Thus, the diagonals of the matrix itself can be used to keep track of which variables are in the model and which are not. Multiple response variables, that is, multiple y's can be handled easily; response variables are merely those candidates which will not be allowed to become model variables.

3.4.3 Stepwise regression

Stepwise regression proceeds by selecting one variable at a time from the set of candidates to add to the model. Variants of stepwise regression also allow variables to be removed from the model at certain steps. The SWEEP algorithm is particularly well-suited to such methods. Stepwise methods usually select the candidate to enter at a particular step by choosing the variable which optimizes some measure of fit of the model to the response. Examples include selecting the variable which maximizes the increase in R^2, which produces minimum residual sum of squares, which produces the largest t-statistics upon entry, or which has the largest incremental sum of squares. All of these criteria produce the same choice.

The SWEEP algorithm provides us with everything we need to make this selection; the kth candidate enters the model provided that it maximizes s_{ky}^2/s_{kk}, which gives the amount by which the residual sum of squares is reduced by adding variable k to the model. Because the residual sum of squares of the response variable at each step is always available, we need only record the initial sum of squares (corrected for the mean if all of our models are constrained to include a constant) in order to have the current value of R^2 at each step.

Jennrich (1977) gives a lucid account of the computational details of stepwise regression based on the SWEEP operator, and discusses in depth construction of a general-purpose stepwise regression program.

3.4.4 The constant term

Since most regression models have at least a constant term (corresponding to a column of ones as a model variable), and since such measures as R^2 are expressed in terms of the residual sum of squares after adjusting for this "variable," it is worthwhile to examine what happens to the constant term in the SWEEP computations. Shifting notation slightly, let 0 represent the subscript corresponding to the column of ones, and let y denote the subscript of the response variable. We shall write column zero at the left, and column y at the right, so that our matrix of regressors can be written

$$Z = \begin{bmatrix} 1 & X & Y \end{bmatrix}.$$

The initial cross product matrix is $S = Z'Z$, which has elements

$$s_{00} = n,$$

$$s_{0j} = \sum_{1 \leq k \leq n} X_{kj},$$

$$s_{ij} = \sum_{1 \leq k \leq n} X_{ki} X_{kj}. \qquad (3.4.8)$$

We shall denote the matrix resulting from sweeping out the constant term as $S^{(0)}$, that is, $S^{(0)} \equiv SWEEP[0]S$. Computing the elements of $S^{(0)}$ we have

$$s_{00}^{(0)} = -\frac{1}{n},$$

$$s_{0j}^{(0)} = \frac{1}{n} \sum_{1 \leq k \leq n} X_{ik} = \overline{X}_i,$$

$$s_{ij}^{(0)} = s_{ij} - \frac{1}{n} \left(\sum_{1 \leq k \leq n} X_{ki} \right) \left(\sum_{1 \leq k \leq n} X_{kj} \right)$$

$$= \sum_{1 \leq k \leq n} X_{ki} X_{kj} - \frac{1}{n} \left(\sum_{1 \leq k \leq n} X_{ki} \right) \left(\sum_{1 \leq k \leq n} X_{kj} \right) \qquad (3.4.9)$$

$$= \sum_{1 \leq k \leq n} (X_{ki} - \overline{X}_i)(X_{kj} - \overline{X}_j). \qquad (3.4.10)$$

Writing $S^{(0)}$ in partitioned form (omitting the symmetric lower half), we have

$$S^{(0)} = \begin{pmatrix} -\frac{1}{n} & \overline{X} & \overline{Y} \\ & S_{xx} & S_{xy} \\ & & S_{yy} \end{pmatrix}, \qquad (3.4.11)$$

where S_{xx}, S_{yy}, and S_{xy} represent respectively the sums of squares of X and of Y, and the cross-products of X with Y, *corrected for the mean*. The matrix omitting the 0th row and column is called the *corrected sum of squares and cross-products* matrix, or *CSSCP*. It is also the matrix of the inner products of the centered columns of X and Y, and we have seen that numerically (3.4.10) is generally much more accurate than (3.4.9), despite the algebraic equivalence of the two expressions. Indeed, substantial numerical inaccuracy can be introduced by actually forming the raw SSCP matrix and then sweeping on column 0; it is preferable, then to perform this first SWEEP operation implicitly, by computing $S^{(0)}$ directly. Note that the 0th row of $S^{(0)}$ must be computed anyway in order to compute S_{xx}, S_{yy}, and S_{xy}. The remainder of the SWEEP operations can then be performed on $S^{(0)}$.

It is instructive to note that the 0th row of the S array can almost be dispensed with. After having swept out the 0th element, we are left with the array $S^{(0)}$ of (3.4.11). In subsequent SWEEP operations, the ij-th elements of the S array are adjusted, and the adjustment when variable k is being swept depend only on the current value of s_{ij} itself, and on the ith, jth, and kth elements of the kth column of S. Since i, j, and k are all greater than zero, we can conclude that the result of sweeping the CSSCP matrix is the same, whether or not the 0th row and column are appended (since no adjustment depends on what happens in row 0 of the S array).

EXERCISES

3.28. [15] Prove (3.4.5).

3.29. [27] Prove that (3.4.6) correctly gives the result of sweeping on the first m elements of S.

3.30. [25] Prove that a "multiple sweep" can be done one at a time, in any order.

3.31. [10] Show that $SWEEP^{-1}$ is the inverse of SWEEP.

3.32. [12] Show that $SWEEP^{-1} = SWEEP^3$.

3.33. [10] Show that positive diagonals correspond to candidates and negative diagonals to model variables.

3.34. [20] Show that the criteria of Section 3.4.3 for choosing the next candidate to enter are equivalent, in the sense that they always produce the same selection.

3.35. [40] Describe in detail a stepwise regression program based on Householder transformations of X rather than on the SSCP matrix. Your program should handle both forward and reverse stepping. What advantages does this approach have over SWEEPing? What disadvantages?

3.36. [23] Let S_{XX} denote the $(p+1) \times (p+1)$ matrix of sums and cross-products of the columns of $(\begin{array}{cc} 1 & X \end{array})$. Prove that S_{xx}^{-1} and the lower $p \times p$ block of S_{XX}^{-1} are equal.

3.37. [27] (Dempster, 1969). Let X be an $n \times p$ matrix of regressor variables (perhaps containing an initial column of ones), and let Y denote an $n \times 1$ vector of responses. Consider the $(n+p+1) \times (n+p+1)$ SSCP matrix Q formed for the augmented matrix $(\begin{array}{ccc} X & Y & I_n \end{array})$, where I_n is the $n \times n$ identity matrix. Show that, after applying the SWEEP operator to the first p rows of Q, the $p + 1$st column contains a) the regression coefficients of Y on X in the first p positions, b) the residual sum of squares for Y in the $p + 1$st position, and c) the ith residual in position $p + i + 1$, for $1 \leq i \leq n$.

3.38. [28] (Continuation) After sweeping on the first p rows in the exercise above, SWEEP[p+1+i] is applied, with $1 \leq i \leq n$. Show that this has the effect of deleting the ith observation from the regression.

3.5 Collinearity and conditioning

In our discussions of the Householder transformations, the Cholesky decomposition, and the SWEEP operator we noted that, in each case, it was possible in principle to encounter a zero divisor. These divisors are called pivots, and they play a central role in each of the algorithms for solving linear systems. As one might expect, the pivotal quantities in the three algorithms are related to one another. When the pivots are small or zero, difficulties both in numerical computation and statistical interpretation arise. This section examines the source of those difficulties, means for dealing with them computationally, and some useful tools for thinking about and analyzing collinear predictors.

3.5.1 Collinearity

The statistical interpretation of a zero pivot is clear, and we shall examine it first in the case of the Householder computations. If one of the columns (say X_k) of X can be expressed exactly as a linear combination of other columns, then X_k is said to be *collinear* with the other columns of X. Let X_k be the first column of X that is collinear with its predecessors. Then we can write

$$X_k = \sum_{j=1}^{k-1} \alpha_j X_j. \tag{3.5.1}$$

Applying the first $k - 1$ steps of the Householder procedure, we obtain

$$
\begin{aligned}
X_k^{(k-1)} &= (H_{k-1} H_{k-2} \cdots H_1) X_k \\
&= \sum_{j=1}^{k-1} \alpha_j (H_{k-1} H_{k-2} \cdots H_1) X_j \\
&= \begin{pmatrix} a_1 \\ \vdots \\ a_{k-1} \\ \hline \mathbf{0} \end{pmatrix},
\end{aligned}
\tag{3.5.2}
$$

for some constants $a_1, \ldots, a_{k-1}$. This last equality follows from the properties of the Householder transformations on the right-hand sides of (3.5.2), which zero out positions k through n of the first $k-1$ columns of X. Because of this, $X_k^{(k-1)}$ is zero below the $(k-1)$-st component, so that (3.1.16) implies that $u = 0$ when forming X_k! Thus, when X_k is collinear with its predecessors (equivalently, when the matrix X exhibits collinearity), a zero value for u, and hence for s, will appear at the kth step of the Householder transformations. It is easily seen that the converse is true as well; if $s = 0$ at step k, then X_k must be collinear with X's appearing earlier in the matrix. Because the Householder procedure produces the Cholesky decomposition as a by-product, these comments apply equally to the direct Cholesky factorization of $X'X$.

COMMENT. The term *multicollinearity* is often used to describe the situation in which one predictor is a linear combination of others. Although some authors distinguish between collinearity and multicollinearity (the latter involving several X variables), there seems to be no strong reason for doing so. We shall adopt the simpler term for our discussion here.

Precisely the same thing happens in the SWEEP operations, and in precisely the same circumstances. Whenever $s = 0$ in the Householder computations, or $s_{kk} = 0$ in the Cholesky decomposition, $s_{kk} = 0$ in the kth SWEEP step as well. Indeed, the pivot element in the Cholesky decomposition is simply the square root of the same pivot element in the SWEEP operation.

The statistical interpretation of quantities in the SWEEP operation provides another way of seeing that zero pivots are caused by collinearity. Provided that the regression model contains a constant term, s_{kk} before SWEEPing can be interpreted as a conditional variance of X_k about its linear regression on $X_1, X_2, \ldots X_{k-1}$. Thus, the SWEEP algorithm encounters a zero pivot if the conditional variance of X_k given its predecessors is zero, and this conditional variance can be zero if, and only if, X_k is a linear combination of its predecessors.

While the interpretation is clearest when a pivot is precisely zero, this almost never happens in practice, due to accumulated numerical errors, due to rounding or measurement errors in the variables themselves, or due to relationships among the variables that are not precisely linear.

3.5.2 Tolerance

The effects of collinearity are seen most easily in terms of the SWEEP representation for the regression computations. Throughout this section, let us assume that X_k exhibits substantial, but not perfect, collinearity with $X_1, X_2, \ldots, X_{k-1}$. Since s_{kk} is the conditional sum of squares of X_k

about its linear regression on the previous variables, and since X_k is nearly equal to a linear combination of those variables, s_{kk} must be close to zero. When we pivot on s_{kk} the kth diagonal element, which is the reciprocal of s_{kk}, becomes very large.

The other elements in the kth row and kth column are also divided by s_{kk}, which can make them very large also. If variable i is not already in the model, then S_{ik} is just the conditional cross-product of X_i and X_k, so that this quantity must already have been fairly small by virtue of the collinearity of X_k with the variables already in the model, so that dividing by s_{kk} amounts to little more than renormalization. However, for X_i in the model, s_{ik} represents the regression coefficient of X_i in the regression of X_k on the model variables, so we can expect s_{ik} to be moderately large for those variables X_i involved in the collinearity with X_k. Thus, the off-diagonal elements of the new $(X'X)^{-1}$ matrix after adding X_k, as well as the kth diagonal element, will typically be quite large.

Finally, we can note what happens to the diagonal elements of the new $(X'X)^{-1}$ matrix when X_k is introduced into the model. Let s_{kk}^c represent the sum of squares of X_k about its mean. Then prior to entering X_k into the model, $R^2 = 1 - s_{kk}/s_{kk}^c$ is the multiple correlation coefficient of X_k with the model variables, since s_{kk} is the residual sum of squares from the regression of X_k on the model variables. If X_k is entered, then the SWEEP computations show that the kth diagonal element of the new $(X'X)^{-1}$ matrix is simply $1/s_{kk}$. This leads to an important interpretation of the diagonal elements of $C = (X'X)^{-1}$. If variable k is in the model, then $1 - c_{kk}^{-1}/s_{kk}^c = R_k^2$, the squared multiple correlation of X_k with the other model variables.

The quantity c_{kk}^{-1}/s_{kk}^c (after entry), or s_{kk}/s_{kk}^c (before entry), is called the *tolerance* of the kth variable. If this tolerance is small, then X_k is nearly collinear with the other variables in the model. This interpretation holds equally well for the other diagonal elements of the C matrix; c_{ii} will be large whenever X_i is substantially collinear with any subset of the other variables in the model.

COMMENT. If we write the regression model as $Y = \alpha + X\beta + \epsilon$, (separating the constant term from the others) we can rewrite the model in the form

$$Y = \alpha + X\beta + \epsilon$$
$$= \bar{Y} + (X - \bar{X})\beta + \epsilon \qquad (3.5.3)$$
$$= \bar{Y} + (X - \bar{X})D^{-1}D\beta + \epsilon,$$

or

$$Y - \bar{Y} = (X - \bar{X})D^{-1}(D\beta) + \epsilon$$
$$\equiv \tilde{X}(D\beta) + \epsilon,$$

or

$$\frac{Y - \bar{Y}}{\sqrt{s_{yy}^c}} = \tilde{Y} = \tilde{X}(D\beta \cdot \frac{1}{\sqrt{s_{yy}^c}}) + \epsilon$$

$$\equiv \tilde{X}\tilde{\beta} + \epsilon,$$

where D is any invertible matrix. If we choose D to be diagonal with elements $d_{ii} = \sqrt{s_{ii}^c}$, then we have re-expressed the regression model in *correlation form*. The regression coefficients $\tilde{\beta}_i$ are the regression coefficients on the centered and scaled X's; they are sometimes called *standardized regression coefficients*, or *path coefficients*. The former usage is actually inappropriate, as it is the regressor variables and not the coefficients that are standardized. In this form, the sum of squares of the columns of $\tilde{X}$ are all equal to unity, so that the kth diagonal element of $(\tilde{X}'\tilde{X})^{-1}$ is precisely $1 - 1/R_k^2$. This is an important property of correlation matrices which plays a central role in factor analysis models.

Since the standard error for $\hat{\beta}_i$ is proportional to c_{ii}, introducing a new variable which exhibits substantial collinearity with other variables already in the model can have the undesirable effect of markedly reducing the precision with which the ith regression coefficient is measured. Thus, it is helpful to monitor tolerances of candidates for entry into the regression model, and to exclude at any step those candidates whose tolerance is sufficiently small. This is particularly valuable in stepwise regression programs, where the regression models are examined "automatically," with little or no opportunity to examine the alternative variables to enter at a given step.

There are numerical as well as statistical reasons for using caution when allowing variables with low tolerance to enter a regression model. When the tolerance is small, entering the variable necessarily involves division by a quantity close to zero, and this division affects every element of the S matrix. A small error in s_{kk} can represent quite a large *percentage* error, and this would affect all of the subsequent computations.

3.5.3 The singular-value decomposition

One of the fundamental results in numerical analysis, the *singular-value decomposition*, or *SVD*, is also a particularly useful tool in characterizing and understanding the nature of numerical and statistical properties of linear models. The importance of the singular-value decomposition in statistics is two-fold. First, the singular values are the solution to an important variational problem which gives them a direct statistical interpretation, which in turn leads to a clearer understanding of the linear model and computations associated with it. Second, the singular-value decomposition can be used to assess the numerical stability of various algorithms as a function of the input data by quantifying the extent to which perturbations in the input affect perturbations in the output. As a consequence, we can gain insight into both the numerical and the statistical stability of quantities computed in the context of the linear model. In this section we shall focus on the existence of the SVD and on its statistical interpretation. The role of the SVD in regression diagnostics is discussed in section 3.6, and the relationship of the SVD and principal components methods are discussed in section 3.8. Computational details of the singular-value decomposition are deferred until section 3.9. The development in this section follows that of Thisted (1976, 1978) and of Nelder (1985).

In section 3.1 we showed how the linear least-squares problem could be transformed into that of an upper-triangular system by applying an orthogonal rotation to the columns of X. Slightly modifying the notation of earlier sections (to maintain consistency with the bulk of the numerical literature on the SVD), we may transform the expression

$$Y = X\beta + \epsilon$$

into

$$U'Y = U'X\beta + U'\epsilon$$
$$= X^*\beta + \epsilon^*$$
$$= \begin{pmatrix} X_1^* \\ 0 \end{pmatrix} \beta + \epsilon^* \qquad (3.5.4)$$
$$= \begin{pmatrix} X_1^*\beta \\ 0 \end{pmatrix} + \epsilon^*,$$

where as before, X_1^* is upper triangular and U is an orthogonal matrix. The solution to the least-squares problem is obtained by equating the left-hand side to the expectation of the right-hand side (in effect, just dropping the ϵ^*), and then applying back-substitution to the resulting triangular system. A natural question to ask is whether further simplification is possible by transforming the vector of coefficients as well as the matrix of regressors, so that X_1^* becomes a diagonal matrix. The answer to this question is affirmative; there always exists an orthogonal $p \times p$ matrix V such that $X_1^*V = D$, where D is a diagonal matrix with non-negative elements, say $d_1 \geq d_2 \geq \ldots \geq d_p$. Thus, we can continue the transformations in expression (3.5.4) to obtain

$$Y^* = \begin{pmatrix} X_1^* \\ 0 \end{pmatrix} \beta + \epsilon^*$$
$$= \begin{pmatrix} X_1^* \\ 0 \end{pmatrix} VV'\beta + \epsilon^* \qquad (3.5.5)$$
$$= \begin{pmatrix} D \\ 0 \end{pmatrix} \theta + \epsilon^*,$$

where $\theta = V'\beta$.

The simultaneous existence of the two orthogonal matrices U and V is guaranteed by the following result:

The Singular-Value Decomposition. *Let X be an arbitrary $n \times p$ matrix with $n \geq p$. Then there exist orthogonal matrices $U : n \times n$ and $V : p \times p$ such that*

$$U'XV = \mathcal{D} = \begin{pmatrix} D \\ 0 \end{pmatrix},$$

where D is the $p \times p$ diagonal matrix

$$D = \begin{pmatrix} d_1 & & & \\ & d_2 & & \\ & & \ddots & \\ & & & d_p \end{pmatrix},$$

and where $d_1 \geq d_2 \geq \ldots \geq d_p \geq 0$.

A proof can be found in Golub and Van Loan (1983). The values d_i are called the *singular values* of X, and the columns u_i and v_i of U and V, are called the *left* and *right singular vectors*, respectively. When $n = p$ and X is (symmetric) positive semidefinite, the singular values of X coincide with the *eigenvalues* of X, $U = V$, and the columns of U are the *eigenvectors* of X. We shall discuss the computational aspects of obtaining V (since obtaining U has already been discussed) in section 3.9.

Note that the least-square estimates for θ in expression (3.5.5) have a particularly nice form obtained by equating the left-hand side to the expectation of the right; $\hat{\theta}_i = y_i^*/d_i$, and $SE(\hat{\theta}_i) = \sigma/d_i$. We can easily interpret these new parameters θ. Since $U'XV = \begin{pmatrix} D \\ 0 \end{pmatrix} \equiv \mathcal{D}$, we can write $X = U\mathcal{D}V'$. Thus, the expected value of the response variable Y can be written

$$E(Y) = X\beta = U\mathcal{D}V'\beta = U\mathcal{D}\theta \tag{3.5.6}$$

$$= XV\theta. \tag{3.5.7}$$

Both expressions (3.5.6) and (3.5.7) are useful. Since the columns of U are orthogonal, expression (3.5.6) shows that θ is simply the vector of regression coefficients on a new collection of p orthogonal regressors, and that these coefficients are in turn merely rotated versions of the original p regression coefficients. As (3.5.7) indicates, θ_i is the regression coefficient on a particular linear combination of the original X_i's, where the coefficients of the linear combination are given by the ith column v_i of V. When d_i is small, then $\hat{\theta}$ has a large standard error, and we can interpret this to mean that the particular data set at hand gives us little information about the regression coefficient on a particular linear combination of the X_i's, namely, Xv_i.

The geometrical interpretation of the singular-value decomposition is worth mentioning. We can always think of the X matrix as being either a collection of n points (vectors) in p dimensional space, or a collection of p vectors in n dimensional space. The former characterization is the collection of multivariate observations, or cases, and is the set of points from which we would construct a scatterplot. We might call this space "scatterplot space"; two variables exhibit correlation here if the plotted points fall more or less along a straight line in this space. Moreover, the stronger the correlation, the greater the slope of this line, when the variables are expressed in standardized units. The latter characterization has one vector ("point") for each variable, and it is in this space that the familiar Euclidean geometry of the analysis of variance operates. We can call this "variable space," and two variables exhibit association in this picture when the plotted points corresponding to them are close to one another, that is, the angle between them (viewed from the origin) is substantially less than 90°.

The matrix U in the SVD rotates the points in variable space. Since this operation preserves angles, it means that the new (rotated) points retain their correlation structure. In effect, the only change is a change of basis, and it is one which is so constructed that the kth variable is a linear combination only of the first k basis vectors, thus producing the now familiar block-triangular form. Since $X = UU'X = U(U'X)$, we have merely re-expressed X in terms of the basis formed by the first p columns of U, so that the vector of regression coefficients on these (unaltered) variables must be the same, namely, β.

The matrix V rotates points in scatterplot space, and it does so in such a way as to remove correlation. In this space, correlation is manifested by a cloud of points lying along a non-horizontal line. V merely rotates this point cloud so that the line along which the points fall is a flat one. After this rotation, the regression coefficients are merely scale factors associated with the now (uncorrelated) set of predictors.

Another geometric interpretation is often useful. Consider the points on the unit sphere in $\mathcal{R}^p$. The image of these points under X is a hyperellipsoid in $\mathcal{R}^n$, the semi-axes of which have lengths equal to the singular values of X.

When will d_i be small? Since $|Xv_i|^2 = v_i'X'Xv_i = v_i'VD'U'UDV'v_i = e_i'D^2e_i = d_i^2$, we have that $|d_i|$ is small if, and only if, the squared length of Xv_i is small, that is, if a particular linear combination of the columns of X is nearly zero. The latter statement is, of course, a restatement of the definition of collinearity. Thus, we could define collinearity in the matrix X in terms of small singular values of X.

We earlier mentioned a variational characterization of the singular-value decomposition, which we now present. The proof can be found in Golub and Van Loan (1983, p. 286).

Theorem 3.5–1. *If X is an $n \times p$ matrix with $n \geq p$, and singular-values $d_1 \geq d_2 \geq \ldots \geq d_p \geq 0$, then*

$$
\begin{aligned}
d_k &= \max_{\dim(S)=k} \min_{u \in S} \frac{|u'X|}{|u|} \\
&= \max_{\dim(S)=k} \min_{\substack{u \in S \\ |u|=1}} |u'X| \qquad (3.5.8) \\
&= \max_{\dim(T)=k} \min_{\substack{v \in T \\ |v|=1}} |Xv|.
\end{aligned}
$$

S and T here are meant to range over all subspaces of $\mathcal{R}^n$ and $\mathcal{R}^p$, respectively.

If the d_i's are distinct and nonzero, the vectors u and v in the last expression of (3.5.8) are precisely the kth columns of U and V, respectively. The last expression leads to a statistical interpretation for d_1, which is the maximum possible value for $|Xv|$. This will be maximized when $|Xv|^2$ is, so that v_1 is simply the vector of coefficients for which the corresponding linear combination of the columns of X has greatest sum of squares, and d_1 is the square root of this maximum sum of squares. If the columns of X have had their means removed, then $|Xv|^2$ is proportional to the largest possible variance obtainable by taking linear combinations of the X's, so that d_1^2 is simply proportional to the largest variance possible by taking linear combinations of the X's.

A useful statistical interpretation of the largest singular value is given by Nelder (1985). If we write $\mathcal{I}(\theta : y)$ for the Fisher information on θ in the data y, then up to the multiple σ^2, $\mathcal{I}(\beta : Y) = X'X$, $\mathcal{I}(v'\beta : Y) = v'X'Xv$, and $\mathcal{I}(\beta : u'Y) = X'uu'X$. These are, respectively, the information in Y about the regression coefficient vector, the information in Y about a particular linear combination of the regression coefficients, and the information about β in a particular weighted sum of the underlying observations. Note that, by choosing $u = e_j$ we can extract the information about β in the jth observation. Finally, we can write the information about a weighted sum of the regression coefficients in a particular linear combination of the observations as $\mathcal{I}(v'\beta : u'Y) = v'X'uu'Xv$. In these terms, the information about $v'\beta$ in $u'Y$, for unit vectors u and v, is maximized when u and v are respectively the first left- and right-singular vectors, and d_1^2 is the maximum obtainable information.

Consider now d_p. The maximization part of (3.5.8) is trivial, since there is only one p-dimensional subspace of $\mathcal{R}^p$, namely $\mathcal{R}^p$ itself. Thus,

$$d_p = \min_{\substack{v \in \mathcal{R}^p \\ |v|=1}} |Xv|.$$

If X has full rank, then no linear combination of the columns of X is zero, so we must have $d_p > 0$ if, and only if, X has no exact collinearity. Further, d_p^2 is the minimum possible sum of squares obtainable by taking linear combinations of the columns of X. If the columns of X have had their means removed, then d_p^2 is proportional to the smallest possible variance obtainable by taking linear combinations of the X's. Another way of stating exactly the same result is that D_p^2 is the *minimum information* in the data about a single contrast in the parameters, that is, $v_p' \beta \equiv \theta_p$ is the linear combination of the underlying parameters about which the data are least informative. Note particularly that if $d_p = 0$, then there is exact collinearity in the X's, and the vector v_p gives the coefficients of the linear combination that is zero. This relationship between collinearity and the singular values of X was first examined carefully by Silvey (1969).

Intermediate singular values have similar interpretations as constrained maximum (or minimum) root mean squares or standard deviations of the predictors in the regression.

3.5.4 Conditioning and numerical stability

The *condition number* is a measure of the relative size of the singular values of a matrix which has a direct interpretation in terms of numerical stability of the regression computations using that matrix as the predictor. Condition numbers are defined in terms of a matrix norm; for most purposes the 2-norm defined by

$$\|X\|_2 = \sup_{|v|=1} |Xv|,$$

where $|u|$ represents the ordinary Euclidean 2-norm of u, will be most appropriate for our purposes, since the computations we shall be examining are defined in terms of minimizing Euclidean 2-norms.

Definition. *If X is a square matrix, then its condition number κ with respect to the norm $\| \cdot \|$ is given by*

$$\kappa(X) = \|X\| \cdot \|X^{-1}\|.$$

COMMENT. The term "conditioning" as used in this section comes from the literature in numerical analysis. It has nothing to do with the statistical notion of conditional distribution. Fortunately, context will always be sufficient to distinguish between these two important, but distinct, uses of the term "conditioning."

From our discussion of the singular-value decomposition, it is clear that, with respect to the matrix 2-norm, $\kappa = d_1/d_p$, and is a measure of the relative elongation of the axes under X. In particular, if κ is large, then the image of the unit sphere under X is a very long and flattened cigar-shaped ellipsoid. [Try to imagine a four-dimensional cigar!]

When X is not a square matrix, but rather has dimensions $n \times p$ with $n \geq p$, we define the condition number analogously as $\kappa = d_1/d_p$; the image of the unit sphere is once again a p-dimensional ellipsoid, but it is embedded in a p-dimensional subspace of $\mathcal{R}^n$.

The smallest value that $\kappa(X)$ can take on is $\kappa(X) = 1$; in particular, this occurs whenever X is orthogonal, since orthogonal matrices have unit singular values. If X exhibits exact collinearity, $\kappa(X)$ becomes infinite, so the condition number measures the closeness of X to singularity (or rank-deficiency). This can be made precise: $\kappa(X)^{-1} = \|E\|/\|X\|$, where E is the smallest matrix (in the $\|\cdot\|$ sense) for which $X + E$ is singular.

The condition number plays a central role in obtaining bounds on the errors in least-squares computations which occur when small perturbations are introduced in X or Y. For instance, let X be square and nonsingular, and consider the solution to the problem $X\beta = Y$. If X and Y are slightly altered, by how much can the solution β change?

Theorem 3.5–2. *Let X be $p \times p$ nonsingular, let ΔX and ΔY be perturbations to X and Y, and let β and γ denote, respectively, the solutions to*

$$X\beta = Y$$

and

$$(X + \Delta X)\gamma = (Y + \Delta Y). \tag{3.5.9}$$

If there exist $\delta \geq 0$ and $f < 1$ such that

$$\|\Delta X\| \leq f \cdot d_p(X)$$
$$\|\Delta X\| \leq \delta\|X\|$$
$$\|\Delta Y\| \leq \delta\|Y\|,$$

then (3.5.9) is a nonsingular system and

$$\frac{|\beta - \gamma|}{|\beta|} \leq \frac{2\delta\kappa(X)}{1 - f},$$

provided that β is non-zero.

See Golub and Van Loan (1983) for an outline of the proof. This bound, which can be attained, shows that introducing relative changes in the inputs of size δ can cause relative changes in the solution by a multiple of $\kappa(X)\delta$. Thus, roughly speaking, if the input data to a linear system are "good to t decimal places," then the solution to the linear system may only be good to $t - \log_{10}(\kappa(X))$ decimal places. When $\kappa(X)$ is large, the linear system is said to be *ill-conditioned*.

Two results are of particular interest in the context of the linear model, the first is a general theorem concerning the numerical stability of the least-squares problem, and the second is an error analysis of the Householder regression algorithm. The following result is valid when $\text{rank}(X) = r < p$ provided that $\kappa(X)$ is defined to be d_1/d_r in that case.

Let $\hat{\beta}$ be a solution to the least-squares minimization problem $\min |Y - X\beta|^2$ and let $r = Y - X\hat{\beta}$ be the vector of residuals from the exact least-squares regression. Denote the regression sum of squares and residual sum of squares, respectively, by SSM and SSR, and the total sum of squares by $|Y|^2 = SSM + SSR$. Let $R^2 = SSM/SST = (1 - SSR/SST)$.

Theorem 3.5–3. *Suppose that, in the least-squares problem, X and Y are perturbed by small amounts ΔX and ΔY, respectively, such that*

$$\epsilon = \max\left(\frac{\|\Delta X\|}{\|X\|}, \quad \frac{\|\Delta Y\|}{\|Y\|} \right) < \frac{1}{\kappa(X)}.$$

Denote by $\hat{\gamma}$ and $\hat{r}$, respectively, the coefficient estimates and the residuals from the perturbed problem. Then provided $R^2 < 1$,

$$\frac{|\hat{\beta} - \hat{\gamma}|}{|\hat{\beta}|} \leq \epsilon\left(\frac{2\kappa(X)}{R} + \frac{\sqrt{1 - R^2}}{R}\kappa(X)^2 \right) + O(\epsilon^2) \qquad (3.5.10)$$

and

$$\frac{|\hat{r} - r|}{|Y|} \leq \epsilon(1 + 2\kappa(X))(n - p) + O(\epsilon^2). \qquad (3.5.11)$$

Proof. Golub and Van Loan (1983, Theorem 6.1-3). It is possible to construct examples in which these bounds are attained.

Note in particular, that when R^2 is less than unity, so that the residual vector is not exactly zero, the sensitivity of the least-squares solution, given by (3.5.10), is a function of the square of the condition number of X, that is, by d_1^2/d_p^2. At the same time, the estimation of the residual vector itself is perturbed by a linear term in the condition, as (3.5.11) shows.

COMMENT. The condition number of X can be affected by rescaling the columns of X. For a discussion of the effects of scaling on collinearity, see Stewart (1986, 1987).

The previous theorem gives a bound on the relative error in the regression coefficients due to small changes in the regressors or response. This result describes the sensitivity of the *problem;* the obvious next question concerns the stability of various *algorithms* for solving it.

Solving the least-squares problem by first forming the normal equations and then using a Cholesky factorization is, in effect, solving the least-squares problem $\min |(X'X)\beta - (X'Y)|^2$ with zero residual. As we might expect, it is the condition of $X'X$ that matters in this computation, and $\kappa(X'X) = \kappa^2(X)$. Assuming that $X'X$ and $X'Y$ can be computed without error, it can be shown that the relative error in the resulting Cholesky computation $(|\hat{\beta} - \hat{\gamma}|)/|\hat{\beta}| \approx \kappa^2(X) \cdot \eta$, where $\eta = base^{-t}$, and where $base$ is the number base of the floating point representation and t is the number of digits in the floating-point mantissa. The Cholesky computation begins to break down when $\kappa(X) \approx \eta^{-1/2}$. Moreover, when R^2 is close to one and $\kappa(X)$ is large, the Cholesky procedure applied to the normal equations becomes unstable in the sense of Chapter 2.

By contrast, the Householder algorithm applied to the original least-squares problem is stable, and an inverse error analysis is available (Lawson and Hanson, 1974). In particular, the computed solution (which we have called $\hat{\gamma}$) is the solution to the perturbed problem

$$\min |(X + \Delta X)\gamma - (Y + \Delta Y)|^2,$$

where

$$\|\Delta X\| \le (6np - 3p^2 + 41p)p^{\frac{1}{2}} \|X\|\eta + O(\eta^2)$$

and

$$|\Delta Y| \le (6np - 3p^2 + 40)|Y|\eta + O(\eta^2).$$

The Householder method begins to break down only when $\kappa(X) \approx \eta^{-1}$. Thus, when the columns of X are nearly collinear, the Householder procedure will be much more robust numerically than methods based on forming the normal equations.

It is worth noting that the deficiencies of the Cholesky procedure are really due to having to form the normal equations, and not with the triangular factorization *per se*. When one is given a matrix of sums of squares and cross-products to start with, the Cholesky procedure is generally quite good. When it is necessary or desirable to work with the normal-equations form of the problem, it is nearly imperative that the *corrected* SSCP matrix be used, and that the CSSCP itself be accumulated separately by taking products about (possibly provisional) means. In particular, Theorem 3.5–3 can then be applied to the reduced problem (in which the regression on the column of ones is first removed).

3.5.5 Generalized inverses.

In our discussions of the linear regression problem, we have assumed throughout that the matrix X is of full rank, or equivalently, that $d_p > 0$, or that the Householder, Cholesky, and SWEEP operations never encounter a zero pivot. As we have noted, however, numerical problems arise when pivots or tolerances are close to zero. When X is not of full rank, $X'X$ is singular, and the vector of coefficients β is not estimable, although some linear combinations of the β's are estimable. The estimable contrasts are precisely those for which $I(c'\beta : Y)$ is non-zero. In order to proceed with computations involving the estimable functions, we can modify each of the computations we have discussed.

The general situation is most easily seen in terms of the singular-value decomposition. If the rank of X is $k < p$, then the singular values $d_{k+1}, \ldots d_p$ will all be zero, and $X_1^* = DV'$ will be "upper trapezoidal", in the sense that the last $p - k$ rows of X_1^* will be zero. This suggests that whenever a zero pivot is found in either the Householder or the Cholesky procedures, that the current variable be removed from the current computation and be renumbered as the last candidate. In the Householder case this can be achieved by exchanging columns; in the Cholesky case, by exchanging row and column. In each case, the exchange can be accomplished by a simple bookkeeping entry rather than by actual rearrangement of computer memory. When all of the columns involving non-zero pivots have been entered, the resulting X_1^* matrix will be upper trapezoidal.

At this point, any $p - k$ of the β's can be solved for in terms of the remaining k. The first k elements of θ in equation (3.5.5) are estimable, so that the estimable functions of β are simply linear combinations of $v_i'\beta$, for $i = 1, 2, \ldots, k$.

Mathematically, when X is of full rank the solution to the least-squares problem is given by $\hat{\beta} = X^+Y$, where $X^+ \equiv V D^+ U'$, and $D^+ = \begin{pmatrix} D^{-1} \\ 0 \end{pmatrix}$. This characterization suggests, when X is rank-deficient, that the quantity X^+ computed with D^{-1} replaced by

$$D^+ \equiv \begin{pmatrix} d_1^{-1} & & & & & & \\ & d_2^{-1} & & & & & \\ & & \ddots & & & & \\ & & & d_k^{-1} & & & \\ & & & & 0 & & \\ & & & & & \ddots & \\ & & & & & & 0 \end{pmatrix}$$

may be a useful quantity to examine. Indeed in the rank-deficient case, $\hat{\beta} = X^+Y$ is the unique solution to the least-squares problem which has minimum squared length, which in turn is the same as setting θ_{k+1} through θ_p equal to zero. The quantity X^+ is called the *Moore-Penrose pseudoinverse* of X, and the resulting estimate of β, the Moore-Penrose estimate.

Mathematically, the least-squares estimates so obtained can also be obtained from a generalized inverse solution to the normal equations: $\hat{\beta} = (X'X)^+ X'Y$. That these estimates are equivalent is left as an exercise. Because of the rank deficiency, the least-squares problem has infinitely many solutions; the Moore-Penrose solution is unique because of the added minimum-norm constraint. In general, let $\tilde{X}$ be an $n \times p$ matrix with $n \le p$ and rank $k \le p$. From the SVD, $\tilde{X}$ can be rotated by premultiplying with an orthogonal matrix U so that it has the form

$$U\tilde{X} \equiv W = \begin{pmatrix} A & B \\ 0 & 0 \end{pmatrix}, \tag{3.5.12}$$

with A nonsingular. If we define

$$W^- = \begin{pmatrix} A^{-1} & 0 \\ 0 & 0 \end{pmatrix},$$

then W^- is said to be a generalized inverse of W. This matrix will not, in general, be the Moore-Penrose inverse of W. It satisfies the conditions that $WW^-W = W$ and $(WW^-)' = WW^-$. We define $\tilde{X}^- \equiv W^-U$. Furthermore, in the regression context, $\hat{\beta} = X^-Y$ is a solution to the least-squares problem. Obtaining a generalized inverse of $X'X$ is relatively easy in the context of the SWEEP operator; it involves setting the corresponding row and column to zero when a SWEEP is to be done on a zero diagonal element. This version of SWEEP is nonreversible; for a full discussion see Goodnight (1979). It does not produce the Moore-Penrose estimates, but rather the estimates obtained by setting the last $p-k$ regression coefficients to zero and solving for the remaining k coefficients.

EXERCISES

3.39. [24] Show that if $s = 0$ at step k of the Householder procedure, then X_k can be written as $X_k = \sum_{j=1}^{k-1} \alpha_j X_j$.

3.40. [15] Show that equation (3.5.3) follows from the first step of the Householder (or Cholesky) computation.

3.41. [09] Deduce that, in the singular-value decomposition, we can write $X = UDV'$.

3.42. [17] If $U'XV = D$ gives the SVD of X, show that $Xv_i = d_i u_i$ and that $u_i'X = d_i v_i'$.

3.43. [12] Show by counter-example that if X is symmetric $(p \times p)$ but is not positive semidefinite, then U need not equal V.

3.44. [32] Show that d_2 is the maximum value of $|Xv|$, subject to the constraints that $|v| = 1$, and $v'v_1 = 0$, where v_1 is the first right singular vector.

3.45. [28] Prove that

$$d_k = \max_{\substack{\dim(S)=k \\ \dim(T)=k \\ u\in S,\, |u|=1}} \min_{\substack{v\in T \\ |v|=1}} u'Xv,$$

where S and T are subspaces of $\mathcal{R}^n$, and $\mathcal{R}^p$, respectively.

3.46. [08] Show that the singular values of a matrix with orthonormal columns are all equal to one.

3.47. [19] Prove that the rank of X is equal to the number of non-zero singular values of X.

3.48. [23] Show that when X is of full rank the solution to the least-squares problem is given by $\hat{\beta} = X^+Y$, where $X^+ \equiv VD^+U'$, and $D^+ = \begin{pmatrix} D^{-1} \\ 0 \end{pmatrix}$.

3.49. [20] Show that the Moore-Penrose solution $\hat{\beta} = X^+Y$ is the minimum norm least-squares estimator.

3.50. [20] Show that $WW^-W = W$ and $(WW^-)' = WW^-$.

3.51. [21] Show that $\tilde{X}^-$ satisfies $XX^-X = X$ and $(XX^-)' = XX^-$.

3.52. [10] Show that every $n \times p$ matrix X has a generalized inverse.

3.53. [23] Show that $\tilde{\beta} = X^-Y$ is a solution to the least-squares problem.

3.54. [17] Show that $X^+Y = (X'X)^+X'Y$.

3.55. [15] Show that, if $\hat{\beta}$ is the Moore-Penrose regression estimate, then the variance matrix of $\hat{\beta}$ is given by $(X'X)^+$.

3.6 Regression diagnostics

Until relatively recently, the computations associated with multiple linear regression were very expensive to carry out because computing resources were so limited. Those associated with other (possibly nonlinear) models were almost prohibitively expensive. Regression diagnostics are means of checking the adequacy and sensitivity of the least-squares solution in linear models. The computation involved may be ten to a hundred times as much as that required just to obtain the least-squares regression quantities, and so it is not surprising that such methods have gained currency only in the last decade, in which computing costs have plummeted as fast as computing speeds have increased. This section cannot be a comprehensive introduction to diagnostic methods for regression models; fortunately there are already excellent monographs which serve this purpose. Notable among recent works are Belsley, Kuh, and Welsch (1980), and Cook and Weisberg (1982). We shall focus on the computational aspects of obtaining regression diagnostics either as a matter of course in the regression analysis proper, or by carrying out auxiliary computations based on quantities already obtained.

From the computational standpoint, regression diagnostics fall into three categories: those that are based on the *residuals* from a fitted model, those that are based on the structure of the columns of X (that is, on the geometrical structure of the *predictor variables*), and those based on the structure of the rows (that is, on the geometric structure of the *observations*).

Notation. In this section, we shall denote the least-squares estimates of the regression coefficients by $\hat{\beta}$, the fitted values by $\hat{Y} = X\hat{\beta}$, and the residuals by $e = Y - \hat{Y}$. The ith row of the X matrix will be denoted by x_i. When it is necessary to represent the result of a calculation performed after omitting the ith point in the data set, the omitted variable will be placed in parentheses after the usual value. Thus, $\hat{\beta}(i)$ denotes the vector of regression coefficients computed using all save the ith observation in the data set, and $\hat{Y}(i)$ denotes the fitted values computed from $\hat{\beta}(i)$. Note, in particular, that $\hat{y}_i(i)$ is well-defined; it is the predicted value of y_i based on all except the ith observation and is equal to $x_i\hat{\beta}(i)$.

3.6.1 Residuals and fitted values

The numerical error analysis in the previous section indicates that the stability of computing residuals from a least-squares fit is much greater than that of computing the estimated coefficients. This is easily understood in terms of collinearity. Denote the vector of residuals by $e(\beta) = Y - X\beta$. If one of the predictors is exactly collinear with other predictors so that $Xv_k = 0$, say, then $X\gamma \equiv X(\beta + c \cdot v_k) = X\beta$, so that the vector of regression coefficients is indeterminate. However, $e(\gamma) = Y - X\gamma = Y - X\beta = e(\beta)$, so the vector of residuals is completely determined. Similar comments apply when the collinearity is not exact, in which case, there is a near indeterminacy in β, but a very slight sensitivity for $e(\beta)$. For these reasons the primary focus on algorithms for computing residuals is on minimizing computing time and storage requirements, rather than on numerical stability, although occasionally the latter plays a role.

When one is doing the basic regression computations using the SSCP or CSSCP and the SWEEP algorithm (or Cholesky), there is little choice in computing the residuals. In this case it is necessary to take the estimated coefficients $\hat{\beta}$ and compute $e \equiv e(\hat{\beta})$ on a row-by-row basis, first computing the *fitted values* $\hat{y}_i = x_i\hat{\beta}$, where here x_i denotes the ith row of X, and then computing $r_i = y_i - \hat{y}_i$. Such an approach may be necessary when the X matrix is large, so that it cannot fit all at once into primary storage, necessitating work with the cross-product matrices, or when the residuals and fitted values are to be computed only after several steps of a stepwise regression analysis.

A second approach to computing the residuals is available when the basic regression computations are carried out using orthogonalization methods. In this case, the response vector Y is typically appended to the X's and as each new candidate variable is added to the model, that column can be replaced by the residuals from its regression on the newly entered X. This is done automatically when orthogonalization is carried out using the Modified Gram-Schmidt (MGS) algorithm.

A third approach must be adopted when Householder (or Givens) methods are used to obtain the regression solution, since these methods do not produce explicitly either the orthogonal basis for the columns of X or its complement. Instead, the current basis is stored in factored form in terms of the elementary Householder reflections. After the kth step, the first k components of Y contain the coordinates of the projection of Y into the span of the columns X_1, X_2, ..., X_k, in terms of the orthogonal basis. The fitted values $\hat{Y}$ can be obtained by applying the Householder reflections to $(y_p^1, y_p^2, \ldots, y_p^k, 0, \ldots, 0)'$ in the reverse order. This quantity can then be subtracted from Y, which must be separately retained.

3.6.2 Leverage

We have briefly introduced the notion of leverage in section 3.2, where we also introduced the "hat matrix" $P = X(X'X)^{-1}X'$, so called because when applied to Y it "puts the hat on Y"—since $\hat{Y} = PY$. The diagonal elements of P, which we denote by p_{ii}, play an essential role in diagnostic measures concerned with relative positions of observations considered as points in $\mathcal{R}^p$, the "scatterplot space" of the predictors X. Computation of the leverage values p_{ii} is quite easy. From equation (3.2.10), we can write

$$p_{ii} = \sum_{j=1}^{p} q_{ij}^2, \qquad (3.6.1)$$

where q_{ij} is the ijth element of $Q = H_p H_{p-1} \ldots H_1$. Since at each step of the Householder computations, one new column of Q is implicitly constructed, the values p_{ii} can be accumulated by forming the current column of Q, squaring its elements, then adding them to the running total of the leverage values. Thus, the entire Q matrix need not be retained, as noted by Peters (1980).

3.6.3 Other case diagnostics

One of the most important diagnostic measures is a measure of residual size which takes account of the fact that some fitted values are more precisely determined than others. The fitted values with greatest precision are just those computed from X values near the center of the X's; this follows immediately from the observation in section 3.2 that $\text{Var}(\hat{y}_i) = \sigma^2 p_{ii}$. The variance of the ith residual e_i is just $\text{Var}(e_i) = \sigma^2(1 - p_{ii})$. This suggests that when assessing constancy of variance, for example, the raw residuals e_i may be misleading. The *Studentized residual* is computed by dividing e_i by an estimate of its standard deviation:

$$r_i = \frac{e_i}{s\sqrt{1 - p_{ii}}}. \tag{3.6.2}$$

If the model is correct, r_i should have approximately a t distribution on $n - p$ degrees of freedom, so that if this quantity is large (say, bigger than about two in magnitude) the corresponding observation may bear further investigation.

COMMENT. The terms *standardized residual* and *Studentized residual* do not have universally accepted meanings. Some authors use the term "standardized" to mean divided by any measure of spread, others use it to denote the residual divided by an estimate of residual standard deviation, others use it to refer to r_i. Similarly with the term "Studentized"; which has been applied both to r_i (Cook and Weisberg, 1982), and to the quantity

$$t_i = \frac{e_i}{s(i)\sqrt{1 - p_{ii}}} \tag{3.6.3}$$

(Belsley, Kuh, and Welsch, 1980); here, $s(i)$ is the residual standard deviation estimated without the ith case. Cook and Weisberg (1982) refer to the quantity t_i as the "externally Studentized" residual and to r_i as the "internally Studentized" residual. It is important for the student to understand that "Studentization" is not applied in a standardized way.

Other case diagnostics measure the effect of including or excluding the ith case on such quantities as the estimated regression coefficients or the ith fitted value. An example of the former is $DFBETA_i$ (Belsley, Kuh, and Welsch, 1980), which measures the change in $\hat{\beta}$ caused by omitting the ith observation.

$$DFBETA_i = \hat{\beta} - \hat{\beta}(i) = \left(\frac{e_i}{1 - p_{ii}}\right)(X'X)^{-1}x_i', \tag{3.6.4}$$

or, in matrix form, $DFBETA = (X'X)^{-1}X'W$, where W is a diagonal matrix whose ith diagonal element is equal to $e_i/(1 - p_{ii})$. An example of the latter kind of diagnostic is the change in fitted values obtained by omitting the ith observation, denoted by $DFFIT_i$ in Belsley, Kuh, and

Welsch (1980), given by

$$DFFIT_i = \hat{y}_i - \hat{y}_i(i) = \frac{p_{ii}e_i}{1 - p_{ii}}. \qquad (3.6.5)$$

It is easy to construct a modified version of $DFFIT$ which gives the effect of omitting the ith observation on the jth fitted value.

Closely related to $DFFIT$ is a much earlier and perhaps more easily interpreted diagnostic, the *predicted residual*, $e_i(i) = y_i - \hat{y}_i(i) = y_i - x_i\hat{\beta}(i)$. This quantity can be thought of as an estimate for prediction error, since it is precisely the error obtained by using all of the data except the ith observation to predict the ith. The predicted-residual sum of squares, $PRESS = \sum e^2(i)$ has been proposed as a measure of model adequacy in assessing alternative models. (Allen, 1974). It is easy to show that $e_i(i) = e_i/(1 - p_{ii})$. It is interesting to note that $e_i(i)$ is the estimated regression coefficient for a dummy variable consisting of a single non-zero value of unity in the ith position when added to the existing model; this fact is occasionally computationally useful.

Other quantities based upon the residuals e_i and the leverage values p_{ii} can be constructed which assess the influence of the ith case. These include the sample influence curve and Cook's distance measure $D_i = r_i^2 p_{ii}/(p - 1)(1 - p_{ii})$.

Belsley, Kuh, and Welsch (1980) and Cook and Weisberg (1982) discuss generalizations of such measures as $DFFIT$ and $DFBETA$, including standardized versions of these diagnostics and extensions to the case of multiple-row deletions. The computational details involved in obtaining these diagnostics are also discussed in these references. The multiple-row diagnostics are particularly problematic computationally, since the best known algorithms require computation time which grows exponentially in the sample size if all subsets of rows are to be examined. Even if only subsets of rows up to size k are to be considered, the computational effort grows proportionally to n^k, which can be prohibitively expensive even if k is small. Branch-and-bound techniques have often been suggested as a means for trimming the cost of computing diagnostics for all subsets of the observations (see Chapter 9), but this avenue has not been systematically explored. If the experience in using branch-and-bound methods for variable selection is any guide, such methods will not prove fruitful for general problems, since in the variable selection case even these methods break down when n is much larger than 25. Most regression problems have more than 25 observations.

3.6.4 Variable diagnostics

The previous sections have dealt with diagnostics that identify unusual or influential cases, or observations. These correspond to the rows of the

X matrix in regression; in this section we turn briefly to the problem of identifying unusual or influential variables (or columns of the X matrix). At first glance, this problem may appear to be the transpose, so to speak, of the previous one, but the concerns are quite different. The rows of X are usually taken to represent independent observations, particularly when X and Y are considered to be sampled jointly. No such structure is generally imposed on the columns of X, the underlying variables. The sorts of influence that columns of X can have on the regression result primarily from their dependence on other columns of X.

Roughly speaking, a column of X can be influential in the sense of introducing a large change in the fitted values $\hat{Y}$, or in the estimated regression coefficients $\hat{\beta}$, or in both. Since adding a variable to a regression model *must* reduce the residual sum of squares, any measure of effect on the fitted values must be measured with respect to the effect of other variables already in the model. Measures such as incremental R^2, incremental one-degree-of-freedom sums of squares, and changes in estimates of residual variance all accomplish this and are obtained in a straightforward manner from the standard regression computations.

It is possible, however, for a variable to have little effect on the fitted values, but a large effect on the estimated regression coefficients. This will occur when the columns of X are nearly collinear, as discussed in section 3.5. A primary measure of this dependence is the *tolerance* of the ith variable, given the other variables in the model. The tolerance is, in effect, $1 - R_i^2$, where R_i^2 is the squared multiple correlation of the ith column of X with the other variables in the model. Because the tolerance is closely related to the inverse of the diagonal elements of $(X'X)^{-1}$, these quantities can easily be computed at each step of the SWEEP operation. The reciprocal of $1 - R_i^2$ is sometimes called the ith *variance inflation factor*, and is often denoted by VIF_i.

Although the variance inflation factor (or tolerance) is a readily interpretable measure of the dependence of one variable on the others already in the model, a large VIF does not indicate which of the other variables is involved in the dependence, nor does it indicate what effect the variable in question has on such things as the standard errors of other regression coefficients in the model. These issues are directly addressed by quantities derived from the singular-value decomposition, which are the topic of the next section.

3.6.5 Conditioning diagnostics

When candidate variables in a regression problem are collinear or nearly so, it is useful to have measures which indicate the extent to which the variables are collinear, the extent to which they affect statistical properties of the resulting model, and the nature of the collinearity, that is to say, which variables are involved in the collinearity. Precisely such concerns have led to a plethora of papers on measures of collinearity, and even on tests for collinearity. Of course, collinearity is a matter of degree, and tests for whether it is "present" or "absent" are likely to be futile. We have seen in section 3.5 that collinear variables are those for which $\sum_{i=1}^{k} \alpha_i X_i \equiv 0$, and that this defining condition can be translated into a statement that one or more of the singular values of X is small.

When we say that a singular value is "small", we typically mean that it is small relative to the others; such an interpretation leads naturally to quantities such as the condition number, which for a general matrix is defined to be the ratio of the largest to the smallest nonzero singular value. We shall consider any measure of *relative* size of the singular values to be a *conditioning diagnostic for X*. It should be noted that, when the X variables are commensurate, there may be a natural definition of a "small" linear combination that arises from the nature of the subject matter; in such cases, simply looking at the smallest singular value by itself may suffice.

COMMENT. Innumerable collinearity measures can be (and have been) constructed from the singular values of X. Any such quantity has some legitimate claim to be a measure of collinearity. However, most such quantities either fail to capture some important aspect of the collinearity problem (the determinant of $X'X$, for instance, is dependent on the scale of the X's), or have no useful direct interpretation. We shall focus on a few measures which seem genuinely useful.

Some authors argue that $\sum \alpha_i X_i \equiv 0$ is not enough to define collinearity. Rather, *at least two* of the α_i's must be substantially nonzero. From the statistical standpoint, this is a distinction without a difference. The interpretation of the difficulty is much more straightforward when a single α_i dominates the expression, but the problem is the same — the data have little information about the effect of one linear combination of the variables on the mean response. If we can identify a single variable about whose effect we have little information, so much the better. Indeed, after rotation by the V matrix from the singular-value decomposition, any "collinearity" is reduced to a single small coefficient on a single variable in the new coordinate system. Thus, there is nothing to be gained, and substantial unification to be lost, by insisting that collinearity be used *only* to describe a combination involving two or more of the candidate variables.

The singular values d_i of X are the square roots of the singular values of $X'X$, and from the singular-value decomposition of X we can write

$$X'X = V\Lambda V', \qquad (3.6.6)$$

where $\Lambda = \mathrm{diag}(\lambda_1, \ldots, \lambda_p)$, and $\lambda_j = d_j^2$. These quantities are *eigenvalues* of $X'X$ corresponding to the *eigenvectors* v_i. That is, $(X'X)v_i = \lambda_i v_i$. Expression (3.6.6) is called the *spectral decomposition* of $X'X$. Clearly, any collinearity measure based upon the singular values of X can be expressed in terms of the eigenvalues of $X'X$, and it is sometimes beneficial to do so for computational or statistical reasons.

The simplest conditioning diagnostic is the *condition number* of X, $\kappa(X) = d_1/d_p$. In section 3.5 we have discussed this measure. It figures prominently in overall measures of the stability of the least-squares problem, and its common logarithm can nearly be interpreted as the number of significant digits lost in obtaining a generalized inverse for X. Moreover, Thisted (1987) notes that the condition number is the ratio of standard deviations between the most- and least-precisely estimated linear combinations of regression coefficients. Stewart (1987) discusses the suitability of the condition number as a collinearity diagnostic.

Since small singular values correspond to near collinearities, it is natural to ask what the effect on X might be of eliminating that particular linear combination of the X's from consideration. The singular-value decomposition makes the answer straightforward—the next smallest linear combination corresponds to the next smallest singular value. This suggests looking at the sequence of condition numbers d_1/d_p, $d_1/d_{p-1}, \ldots, d_1/d_1 = 1$. Belsley, Kuh, and Welsch (1980) call these quantities *condition indices*. Stewart (1987) proposes an alternative set of condition indices which more clearly relate particular variables to near collinearities. Stewart's indices are closely related to the *variance inflation factors* of Marquardt (1970).

If some linear combination of the variables exhibits little variability, it is often useful to know which linear combination that is. The coefficients of v_p provide this information, and it is good practice to inspect the singular vectors corresponding to small singular values. Such an approach to model diagnosis has been discussed in detail in Thisted (1976) and Thisted (1978). A closely related approach, called *latent root regression analysis*, considers the eigenvalues and eigenvectors of $Z'Z$, where $Z = [X\,Y]$; see Webster, Gunst, and Mason (1974) and Mason, Gunst, and Webster (1975).

Belsley, Kuh, and Welsch (1980) note that $Var(\hat{\beta}_k) = \sigma^2 \sum_{i=1}^{p} v_{ki}^2/\lambda_i$. They define the quantity $\phi_{ki} \equiv v_{ki}^2/\lambda_i$, which they call the *kith variance decomposition component*, and they suggest examining the proportion of variance in each $\hat{\beta}_k$ associated with each of the singular vectors v_i. This approach may be helpful when particular regression coefficients are the primary quantities of interest.

Occasionally one is interested in the effects of conditioning of the X matrix on statistical properties of possible estimators. For instance, ridge regression estimators of a certain form can be constructed so as to be minimax (with respect to sum of squared errors loss in the regression coefficients), provided that the X matrix is not too badly conditioned. The relevant measure of collinearity in this case (perhaps better termed a *minimaxity index*, is given by

$$mmi = \sum_{i=1}^{p} \frac{\lambda_p^2}{\lambda_i^2};$$ (3.6.7)

when this number exceeds two, a minimax ridge regression estimator can be constructed (Thisted, 1982). Note that this minimaxity index is the sum of squares of the condition indices for $(X'X)^{-1}$.

3.6.6 Other diagnostic methods

The previous sections have discussed diagnostic methods based upon residuals, leverage, singular values, and combinations of these quantities. Other diagnostic methods address the adequacy of the regression model as a whole, or the plausibility of requisite or desirable assumptions about the structure in the data such as Gaussian errors or constant variances. Because many of the issues involve systematic patterns not reflected in the model, the diagnostic methods which are most successful at indicating when such patterns may be present are graphical displays. For instance, a plot of residuals (possibly Studentized) against fitted values from a regression may give a strong visual indication when the variance about $E(Y|X)$ is increasing as a function of $E(Y|X)$. Similarly, a plot of residuals against sequence number (when the data are gathered sequentially in time) may reveal dependence between adjacent observations. Specific diagnostic measures or test statistics can be constructed to be sensitive to specific sorts of departures from the standard assumptions about a linear model, and these quantities can be used to supplement the visual displays (which may provide more information about the nature of the departure when one exists than a single number could convey). Examples of departure-specific statistics in the regression model are the heteroscedasticity diagnostics of Cook and Weisberg (1983), and the serial-correlation test statistics of Durbin and Watson (1951). Most such diagnostic measures are based on the quantities which we have already computed, and there is little more to say from the computational standpoint, with the possible exception of plotting techniques. The computational aspects of statistical graphics are the topic of Chapters 12, 13, and 14.

Polasek (1984) has extended influence diagnostics such as (3.6.4) to the setting of generalized least squares (see section 3.10), that is, when the variance matrix of the errors is proportional to a known matrix Σ rather than proportional to the identity matrix. Other generalizations to essentially nonlinear settings are discussed in Chapter 4.

It may be thought that the conditional distribution of Y depends in some fashion on a linear combination of the X's, but that it is the conditional mean of $f(Y)$ rather than Y that has this linear behavior. This suggests embedding the linear model $Y = X\beta + \epsilon$ in a larger family of models such as the Box-Cox transformation family of power transformations:

$$f_\lambda(Y) = \frac{Y^\lambda - 1}{\lambda} = X\beta + \epsilon,$$

where $f_0(Y)$ is taken to be $\ln(Y)$ by continuity (Box and Cox, 1964). The computational details of estimating λ and monitoring the likelihood are discussed in Chapter 4, which deals with nonlinear aspects of statistical models.

EXERCISES

3.56. [40] Develop an efficient algorithm for use in conjunction with the Householder computations that produces a vector of residuals obtained by updating the vector of residuals from the previous step in the computation.

3.57. [30] Write an algorithm which computes the residuals from the kth step of the Householder computations by first applying the Householder reflections in reverse order to $(y_p^1, y_p^2, \ldots, y_p^k, 0, \ldots, 0)'$, the first k components of the rotated Y vector at step k, and then subtracting this result from Y.

3.58. [15] Why is the distribution of r_i approximately t rather than *exactly* t when the model is correct?

3.59. [23] Show that, under the assumption of independent Gaussian errors in the linear regression model, that t_i in (3.6.3) has a Student t distribution with $n - p - 1$ degrees of freedom.

3.60. [24] Show that $(n - p - 1)s^2(i) = (n - p)s^2 - \left(e_i^2/[1 - p_{ii}] \right)$.

3.61. [18] Prove expression (3.6.5).

3.62. [24] Prove that the values of $DFFIT_i$ are unaffected by changing the coordinates in which the columns of X are expressed, that is, if X is replaced by XA, where A is any $p \times p$ nonsingular matrix, the resulting values of $DFFIT$ are unchanged.

3.63. [17] Write down what might be called $DFFIT_{ij}$, the effect on the jth fitted value of omitting the ith observation.

3.64. [15] Show that $e_i(i) = e_i/(1 - p_{ii})$.

3.65. [48] Develop an efficient method for computing any regression diagnostic defined on multiple rows of the X matrix, for all possible subsets of rows of X.

3.66. [50] Consider any diagnostic measure D defined on sets of rows of an X matrix. Let $D(I)$ denote the value of D when applied to the rows of X whose indices are in the set I, and let $M_j = \max_{|I|=j} D(I)$, $j = 1, \ldots, n$. Prove that the amount of computation required to compute $\{M_j \mid j = 1, \ldots, n\}$ must grow exponentially in n.

3.7 Regression updating

By *regression updating* we mean computing regression estimates by modifying the current estimates to reflect adding observations (rows) to, or subtracting observations from, the data set. The problem of introducing or deleting additional variables (columns) has been considered in section 3.1.3. It is sometimes useful to distinguish the result of adding to the data set from subtracting data from the data set, in which case the former is referred to as *updating* the regression, and the latter is referred to as *downdating*. Updating methods discussed here are closely related to the ideas and methods of the *Kalman filter*, which is discussed in Chapter 4.

The problem of suitably modifying a regression equation arises in several contexts. Consider four examples. First, the data may arrive sequentially in time, and after each new data point is received, the regression estimates must be revised. As a second example, consider performing regression computations when the number of observations is so large that not all of the observations can be present simultaneously in random-access memory. In this case, an algorithm is required that can perform the regression computations sequentially as data are read in from an external storage device. As a third example, recall that in the previous section we noted that regression diagnostics for the leverage of sets of points can be constructed by examining the changes in the regression coefficients caused by deleting two or more points at a time from the set. To examine all $\binom{n}{2}$ two-point subsets, for instance, it is not necessary to compute $n(n-1)/2$ regressions based on $(n-2)$ points each. Instead, great computational savings can be achieved by repeatedly using downdating and updating formulae to remove observations from, and then to reintroduce observations into, the data set. A final example arises in the context of smoothing using locally linear fits, discussed in detail in Chapter 6. Such smoothing methods, which generalize the simple notion of moving averages, require least-squares lines to be fit to portions of the data which fall within a "moving window." As the window moves from left to right through the data set, some points enter the window on the right, while others depart on the left. Each time the set of points falling within the window changes, the regression equation must be modified by updating or downdating.

Although some relatively simple updating procedures based on updating the $(X'X)^{-1}$ matrix are not numerically stable, it is possible to construct algorithms for updating which are highly stable. Unfortunately the same cannot be said for downdating, which is an inherently unstable procedure, the degree of instability depending upon the relationship of the row to be deleted to the remainder of the data set (through the singular-value decomposition of the latter).

It is relatively straightforward to update the matrix of corrected sums of squares and cross-products; this was discussed at length in section 2.3.4.

Of course, for regression, what is of importance are the Cholesky or QR factorization for this matrix, its inverse, and the regression coefficient estimates. The key idea of this section is that it is more efficient to update the QR computations directly, rather than to recompute the QR decomposition for an updated version of the SSCP matrix each time. Similar comments apply to Cholesky updating, the details of which we shall outline.

3.7.1 Updating the inverse matrix

A most useful formula can be derived from the *Sherman-Morrison-Woodbury formula*

$$(A + uv')^{-1} = A^{-1} - A^{-1}u(I + v'A^{-1}u)^{-1}v'A^{-1}, \qquad (3.7.1)$$

where A is $p \times p$ symmetric and nonsingular, u and v are each $p \times q$ and of rank q, and $(I + v'A^{-1}u)^{-1}$ exists. Let x_i' denote the i-th row of X. By choosing $v = x_i$ and $u = \pm x_i$, the formula above becomes

$$(X'X \pm x_i x_i')^{-1} = (X'X)^{-1} \mp \frac{(X'X)^{-1}x_i x_i'(X'X)^{-1}}{1 \pm x_i'(X'X)^{-1}x_i}. \qquad (3.7.2)$$

The positive choice for x_i corresponds to adding row i to the existing X matrix, and the negative choice corresponds to deleting row i from X. Formula (3.7.1) was reported in an abstract by Sherman and Morrison (1949) for the case $q = 1$; and Woodbury (1950) reported the general case in an unpublished memorandum. The regression updating formula of (3.7.2) was published independently by Plackett (1950) and Bartlett (1951), whose names are often associated with the result. Cook and Weisberg attribute the finding to Gauss in 1821.

This result is highly useful for proving results about residuals and regression diagnostics, as well as about matrix inverses in general. However, as with most regression methods based on the inversion of the $X'X$ matrix, it is not recommended for general use on the basis of relative numerical instability, particularly when used for downdating. A stability analysis of downdating using the Sherman-Morrison-Woodbury formula has been given by Stewart (1974) and by Yip (1986). Direct updating of the Householder QR decomposition or the Cholesky factor is preferable from the numerical standpoint. When it is necessary at the same time to obtain regression diagnostics which depend directly on the inverse matrix, or when it is essential to have the inverse matrix at each step for other purposes, formula (3.7.2) may well be worth the numerical compromise, particularly if the application does not require deletion of data points.

3.7.2 Updating matrix factorizations

The numerical superiority of orthogonal factorizations can be retained in the context of updating by applying special formulædirectly to the matrices derived from the decompositions. These methods depend upon the following theorem.

Theorem 3.7–1. *Let X be an $n \times p$ matrix, and let $X = QR$ be its QR decomposition. Let w be a $1 \times p$ vector. Then*

$$\begin{pmatrix} X \\ w \end{pmatrix} \quad and \quad \begin{pmatrix} R \\ w \end{pmatrix}$$

have the same $p \times p$ Cholesky factor.

Updating the Cholesky factorization directly is straightforward; $\begin{pmatrix} R \\ w \end{pmatrix}$ is already upper triangular except for the last row, which can be made equal to zero (if R has full rank) by applying at most p Givens rotations. The theorem above appears to apply only to the case of updating, since

$$\begin{pmatrix} R \\ w \end{pmatrix}' \begin{pmatrix} R \\ w \end{pmatrix} = R'R + w'w = X'X + w'w = \begin{pmatrix} X \\ w \end{pmatrix}' \begin{pmatrix} X \\ w \end{pmatrix}.$$

However, premultiplication of w by $i = \sqrt{-1}$ produces

$$\begin{pmatrix} R \\ iw \end{pmatrix}' \begin{pmatrix} R \\ iw \end{pmatrix} = R'R - w'w = X'X - w'w = X_{(w)}'X_{(w)},$$

where $X_{(w)}$ represents X with row w deleted. Interestingly enough, all of the computations needed to accomplish the downdating operation this way can be done using real arithmetic only. This method has been set forth by Golub (1969) and by Chambers (1971). The numerical stability of downdating the Cholesky factor has been studied by Stewart (1979). Lawson and Hanson (1974) present several algorithms for updating the Householder computations as each new row is added. Doing the regression computations this way requires approximately twice the number of floating-point operations as would doing the computation once with all n observations at hand. Although this algorithm does not adapt easily to downdating, they present a straightforward alternative algorithm for downdating Q and R simultaneously.

FORTRAN programs for updating and downdating both the triangular factor of the QR decomposition and the Cholesky factorization can be found in LINPACK (1979). Well-tested subroutines in the public domain for updating the orthogonal basis Q in the QR decomposition are not currently available.

3.7.3 Other regression updating

Chambers (1977) gives an interesting interpretation of regression updating in terms of weighted regression, in which adding or deleting a row corresponds to an augmented regression problem with weights ± 1. The augmented regression problem can then be solved by updating the previous solution using any of the methods discussed above.

The singular-value decomposition can be updated using a methods described by Businger (1970).

EXERCISES

3.67. [18] Prove formula (3.7.2).

3.68. [25] Prove Theorem 3.7–1.

3.8 Principal components and eigenproblems

The notion of *principal components* refers to a collection of uncorrelated random variables formed by taking linear combinations of a set of possibly correlated random variables. The original variables can then be represented in terms of this uncorrelated set, which sometimes makes either computation or interpretation easier. The term "principal components" is used to refer both to samples and to populations. We first consider the notion of population principal components in order to fix ideas. In practice, the sample components are of interest. The coefficients in these linear combinations are related to eigenvectors of covariance matrices, and the corresponding eigenvalues have useful statistical interpretations. For this reason, we begin with a brief discussion of eigenproblems.

3.8.1 Eigenvalues and eigenvectors

An *eigensystem* is the collection of eigenvectors and corresponding eigenvalues associated with a square matrix, and we shall use the term *eigenproblem* as shorthand for the problem of computing the eigensystem from a given matrix. (More generally, eigenvalues are defined for linear transformations on finite-dimensional real inner-product spaces taking values in the same space. We treat only the numerical computation of eigensystems here. Symbolic manipulation of eigensystems, while important, is beyond the scope of this volume.) The eigenproblems of interest in statistics are almost always those associated with real positive semidefinite matrices, and we shall outline the definitions and properties of eigensystems in terms of such matrices. The results we state below without proof; these are well-known and can be found in textbooks on linear algebra such as Bentley and Cooke (1973).

Let A be a real, symmetric, positive semidefinite $p \times p$ matrix. A $p \times 1$ vector $v \neq 0$ is said to be an *eigenvector* of A if there exists a real number λ for which $Av = \lambda v$. Since any scalar multiple of v is also an eigenvector, it is helpful to reserve the term for eigenvectors of unit length, and we shall adopt this convention. The value λ corresponding to v is said to be an *eigenvalue* of A. If A has rank k, then there exist k mutually orthogonal (unit) eigenvectors v_i, corresponding to positive eigenvalues λ_i. If the eigenvalues are distinct, then the collection $\{v_1, v_2, \ldots, v_k\}$ is unique. If the eigenvalues are not distinct, then any linear combination of eigenvectors corresponding to the same eigenvalue will also be an eigenvector.

The eigenvalues also satisfy the determinantal equation $|A - \lambda I| = 0$, and any solution to this equation must be an eigenvalue of A. The left-hand side of this equation is known as the *characteristic polynomial of A*. The largest eigenvalue, λ_1, can be characterized as the maximum possible value of $v'Av/v'v$, and the vector v which attains this maximum is the eigenvector corresponding to λ_1.

Eigenvalues and eigenvectors are also known as *latent roots* and *latent vectors*, and as *characteristic roots* and *characteristic vectors*. These terms are interchangeable.

COMMENT. If A is the covariance matrix for a $p \times 1$ random vector X, then the variance of the linear combination $v'X$ is given by $v'Av$. Thus eigenvalues of variance matrices can be interpreted as variances of particular linear combinations of the component random variables. In particular, λ_1 is the largest variance obtainable by taking linear combinations of X with weights having unit length.

The eigenvalues also have a geometrical interpretation. The equation $x'Ax = c$ describes an ellipsoid in $\mathcal{R}^p$. The eigenvalues are proportional to the lengths of the axes of this ellipsoid.

3.8.2 Generalized eigenvalues and eigenvectors

The solutions to many problems in multivariate analysis involve quantities which generalize the usual notion of eigenvalues and eigenvectors. The eigenvalues of a symmetric positive semidefinite matrix A can be thought of as a measure of "size"; if A is a variance matrix, the eigenvalues are proportional to variances of particular linear combinations of the corresponding random variables. Many univariate statistical problems involve comparing a mean square associated with a hypothesis to a residual mean square, in effect, comparing two variances. Naturally, if the problem is rescaled, both variances are scaled by the same amount. We can think of the problem as measuring the size of one variance *relative to* the size of another. In multivariate problems the situation is similar, but we must assess the size of one variance *matrix* relative to another. The eigenvalues of a variance matrix A, then, represent its "size" relative to a particular reference matrix, namely, the identity I. The generalized eigenvalue measures "size" of A relative to another positive definite matrix B.

Let B be a real, symmetric, positive definite matrix, and let A be real, symmetric, and positive semidefinite. The *eigenvalues of* A *with respect to* B are given by the solutions λ to the determinantal equation $|A - \lambda B| = 0$. A value λ is an eigenvalue of A with respect to B if, and only if, there exists a vector v for which $Av = \lambda Bv$; such a vector is said to be the eigenvector of A with respect to B corresponding to λ. The largest generalized eigenvalue is the maximum of $v'Av/v'Bv$, and by convention, the generalized eigenvectors are normalized so that $v'Bv = 1$.

Clearly, the eigenvalues of A with respect to B, of AB^{-1}, of $B^{-1}A$, and of $B^{-1/2}AB^{-1/2}$ are all the same; however the corresponding eigenvectors differ, being respectively v, Bv, v, and $B^{1/2}v$.

3.8.3 Population principal components

Let x be a $p \times 1$ random vector, with mean $E(x) = \mu$ and variance matrix $V(x) = \Sigma$. The *(population) principal components* of x are the p linear combinations $z = \Gamma x$ whose elements are uncorrelated with one another. The coefficients of these linear combinations are given by the rows of the $p \times p$ matrix Γ. By virtue of this definition,

$$V(z) = V(\Gamma x) = \Gamma \Sigma \Gamma' = \Lambda, \qquad (3.8.1)$$

where $\Lambda = \operatorname{diag}(\lambda_1, \lambda_2, \ldots, \lambda_p)$ is a diagonal matrix. Since Λ is a variance matrix, it follows immediately that all of the $\lambda_i \geq 0$. Since Γ can be premultiplied by a permutation matrix so as to rearrange the z_i's, we can take $\lambda_1 \geq \lambda_2 \geq \cdots \geq \lambda_p \geq 0$ without loss of generality. It follows from the singular value decomposition of Σ that matrices Λ and Γ satisfying (3.8.1) exist; furthermore Γ is orthogonal and, if the λ_i's are distinct, it is unique as well. Using this result, we may rewrite (3.8.1) as

$$\Sigma = \Gamma' \Lambda \Gamma. \qquad (3.8.2)$$

This is called the *spectral decomposition* of Σ. The ith row of Γ is the *eigenvector* of Σ which corresponds to the ith *eigenvalue* λ_i. Note that, since Λ is diagonal, the elements of z are uncorrelated with one another, and if z has a normal distribution then they are independent as well. It is this property of the principal components that makes them play a central role in multivariate analysis.

3.8.4 Principal components of data

Consider now an $n \times p$ data matrix X, and assume (solely for ease of exposition) that the column means of X have been removed, so that each column of X has mean zero. If the rows of X are independent observations from the p-variate population of the previous section, the covariance matrix Σ can be estimated by $\hat{\Sigma} = X'X/n = CSSCP/n$. This matrix has a spectral decomposition, say, $\hat{\Sigma} = \hat{\Gamma}'\hat{\Lambda}\hat{\Gamma}$ The *(sample) principal components* of X are the columns of $Z = X\hat{\Gamma}$. It is clear from this definition that any pair of columns of Z are orthogonal; hence their sample correlation is zero. Note that, if X has the singular-value decomposition $X = UDV'$, where D is $p \times p$, then $Z = UD$, $\hat{\Gamma} = V$, and $\hat{\Lambda} = D^2$.

In the previous paragraph we assumed that the columns of X had their means removed. If the means were not removed, then $X'X/n$ would be an estimate for $\Sigma + \mu\mu'$ which, in the absence of strong prior information that μ is near zero, is rarely of interest.

If, in addition to removing the means, the columns of X are each rescaled to have squared length n, then $X'X/n$ is the sample correlation matrix R. The principal components (eigenvalues and eigenvectors) of this matrix can be extracted as well. Because the standardized variables are unit-free, the principal components of R are sometimes more easily interpreted than are the principal components of $\hat{\Sigma}$. There is no direct relationship between the eigenvalues and eigenvectors of R and $\hat{\Sigma}$.

From the computational standpoint, if the SVD of X has already been computed, then no further computations are required to obtain the principal components. If the principal components are to be examined in the context of a regression problem in which all of the other computations are accomplished using the SSCP matrix, then it is more efficient both in terms of storage and computing time to take advantage of the positive definiteness and symmetry of covariance matrices by using a specialized eigensystem program. Efficient methods for extracting the eigenvalues and eigenvectors of a real, symmetric, positive semi-definite matrix are discussed in Stewart (1974). Well-tested and numerically stable FORTRAN subroutines for this purpose can be found in EISPACK (Smith, *et. al.*, 1976); FORTRAN subroutines for extracting the SVD are available in LINPACK.

EXERCISES

3.69. [15] Let A be a real, symmetric, positive semi-definite $p \times p$ matrix. Prove that $|A - \lambda I| = 0$ if, and only if, λ is an eigenvalue of A.

3.70. [12] Let A be a real, symmetric, positive semi-definite $p \times p$ matrix. Prove that $v'Av = 0$ implies that $Av = 0$.

3.71. [05] Show that the result in the previous exercise fails to hold if the requirement that A be positive semidefinite is omitted.

3.72. [14] Let A be a real, symmetric, positive semi-definite $p \times p$ matrix. Prove that $\max v'Av/v'v$ is attained when v is the eigenvector of A corresponding to the largest eigenvalue λ_1, and that the maximum value is precisely λ_1.

3.73. [12] Show that the eigenvalues of A with respect to B, of AB^{-1}, of $B^{-1}A$, and of $B^{-1/2}AB^{-1/2}$ are all the same; however the corresponding eigenvectors differ, being respectively v, Bv, v, and $B^{1/2}v$.

3.9 Solving eigenproblems

The algorithms for obtaining solutions to eigenproblems and for the closely related problem of finding a singular value decomposition are among the most beautiful yet least known of all algorithms used in statistics. Although the computations themselves shed very little light on statistical aspects of the problem, their simplicity and elegance make them worth study. In addition, they illustrate in fundamental ways the use of more basic techniques of matrix linear algebra. They are models of the clever use of Householder and Givens transformations, which are such important elements of other statistical computations. For these reasons, and to give the interested reader some background for delving into the rich literature in numerical analysis on these topics, the rest of this section is devoted to an overview of the computational aspects of eigensystems as they arise in statistics.

The last proviso is an important one, since nearly all of the eigenproblems encountered in statistical applications are instances of the *symmetric eigenvalue problem,* which is substantially less complicated, and numerically much more stable, than its asymmetric sibling. The symmetry here refers to the symmetry of the $p \times p$ matrix A whose eigenvalues and eigenvectors are to be extracted. Much of the time statistical interest focuses on matrices A which are not only symmetric, but positive semidefinite as well. Symmetry ensures that the eigenvalues of a real matrix are real, and the condition ensures as well that there is an orthonormal basis for $\mathcal{R}^p$ composed of eigenvectors of A. In the asymmetric case there are no real eigenvectors corresponding to complex eigenvalues, and while the sum of the (algebraic) multiplicities of the eigenvalues is always p, this may exceed the sum of the dimensions of the corresponding eigenspaces. (Complex eigenproblems do arise in some statistical problems; for such problems the algorithms here can be extended to Hermitian matrices.) There are substantial computational advantages to the symmetric problem as well, reducing both computer storage and the number of operations required to obtain the result.

The modern theory of numerical computation and analysis of eigensystems begins with the 1961 papers of Francis, in which the QR algorithm is developed. The landmark reference in the field is *The Algebraic Eigenvalue Problem,* by J. H. Wilkinson (1965). The development that follows relies heavily on the synthesis of subsequent work compiled by Golub and Van Loan (1983). The remainder of this section is devoted to the symmetric QR algorithm for finding eigenvalues, the Golub-Reinsch algorithm for the singular value decomposition, which is based upon the symmetric QR procedure, and finally a discussion of the symmetric version of the generalized eigenproblem, which arises frequently in multivariate analysis.

3.9.1 The symmetric QR algorithm

The basic QR algorithm for eigenanalysis is a generalization of the *power method*. We begin our discussion with the latter procedure, which is interesting in its own right. This leads to the QR method proper, and a dramatic enhancement obtained by incorporating the idea of *origin shifts*.

3.9.1.1 The power method

Let A be a real symmetric matrix, with largest eigenvalue λ_1 and a unique unit eigenvector v_1 corresponding to this eigenvalue. Let $x_0 \in \mathcal{R}^p$ be any vector which is not orthogonal to v_1. (A randomly selected vector x_0 will be orthogonal to v_1 with zero probability, so this is rarely a practical concern.) Finally, define $x_i = Ax_{i-1}/|x_{i-1}|$ for $i \geq 1$. The x_i sequence will converge to v_1. The method as stated here fails if $|\lambda_1|$ has multiplicity greater than one, and convergence to the solution is very slow if the magnitude of λ_1 is not well-separated from that of the next largest eigenvalue. It is easy to see why this is so. Represent x_0 as a linear combination of the ordered eigenvectors:

$$x_0 = w_1 v_1 + w_2 v_2 + \ldots + w_p v_p, \quad w_1 \neq 0,$$

since x_0 and v_1 are not orthogonal. After n iterations, x_n will be a multiple of $A^n x_0$, which can be written as

$$A^n x_0 = w_1 \lambda_1^n v_1 + w_2 \lambda_2^n v_2 + \ldots + w_p \lambda_p^n v_p$$

$$= w_1 \lambda_1^n \left(v_1 + \left(\frac{\lambda_2}{\lambda_1} \right)^n \left(\frac{w_2}{w_1} \right) v_2 + \ldots + \left(\frac{\lambda_p}{\lambda_1} \right)^n \left(\frac{w_p}{w_1} \right) v_p \right).$$

$$(3.9.1)$$

The parenthesized quantity is going to v_1 at a rate determined by $|\lambda_2/\lambda_1|^n$. What is more, since the algorithm renormalizes x_{i-1} at each step, it is clear that $\pm|x_i|$ is converging to λ_1, where the sign is chosen to be negative when the elements of x_i alternate in sign from one iteration to the next, and positive otherwise.

The power method often converges quite slowly due to lack of separation of λ_1 from λ_2, but has the advantage of being easy to program correctly. It has the added advantage that one need only be able to compute Ax; when A is large and sparse, the entire matrix need not be maintained in memory. An obvious disadvantage is that the algorithm produces only the largest eigenvalue and eigenvector.

The power method described above can be generalized in the following way to extract k eigenvalues and eigenvectors. Begin with a $p \times k$ matrix X_0

having orthogonal columns of unit length. The matrix AX_0 is the result of applying the first step of the power iteration to each of the columns of X_0. If this iteration were continued, each column of the result would converge to a multiple of v_1. To avoid this, at each step we re-orthogonalize the columns of X_i before proceeding. Intuitively, if the first column is converging to v_1, then the second (and subsequent) columns must be converging to other vectors orthogonal to v_1. Under mild conditions this will be the case. The multi-dimensional power iteration, then, can be described as follows: Let X_0 be a matrix with k orthonormal columns. For $i = 1, 2, \ldots$, let $T = AX_{i-1}$, and then compute the QR decomposition $T = X_i R_i$.

When A is symmetric with distinct eigenvalues, X_i converges to the k eigenvectors corresponding to the k largest eigenvalues, and R_i converges to a diagonal matrix containing these eigenvalues. Under general conditions, convergence of this algorithm occurs at the rate $(\lambda_{k+1}/\lambda_k)^n$. The proof is beyond the scope of this book; Golub and Van Loan (1983) give a thorough treatment of the general case.

COMMENT. To what will X_i and R_i converge in the asymmetric case, or in the case that A has non-distinct eigenvalues? When A is not symmetric, X_i will not converge to a constant real matrix, nor will R_i converge to a diagonal matrix (since the eigenvalues may have complex values). Instead, the k-dimensional space spanned by the columns of X_i will converge to a space S which is *invariant* under A, that is, for $x \in S, Ax \in S$. What is more, S can be written as the direct sum of invariant subspaces $S_1 \oplus S_2 \oplus \cdots \oplus S_r$, where $x \in S_j$ implies that $Ax \in S_j$ and $|Ax|/|x| = |\lambda_j|$, and where λ_j is a possibly complex eigenvalue. The sequence of matrices R_i will converge *in form* to a block upper triangular matrix, with each diagonal block being either a multiple of the identity matrix or a 2×2 upper-triangular matrix. The multiples of the identity matrices converge to eigenvalues of A. The diagonal elements of the remaining blocks have diagonal elements whose product converges to $|\lambda_j|^2$, for some eigenvalue λ_j. Each of the blocks corresponds to one of the invariant subspaces S_j.

3.9.1.2 The QR algorithm with origin shifts

We turn now to a special form of the iteration described above in the symmetric case. This specializes the QR method for the general case proposed by Francis (1961), which is treated in depth in Chapter 7 of Golub and Van Loan (1983). Let A be $p \times p$ and symmetric, and let the QR decomposition of A be $A = QR$. Consider now the matrix $A_1 = Q'AQ = RQ$. It is clear that A_1 is symmetric, that A_1 has the same eigenvalues as A, and that if the columns of Γ are the eigenvectors of A, then the columns of $Q'\Gamma$ are the eigenvectors of A_1. Notice that A_1 can be formed in two steps: form the QR decomposition of A, then post-multiply the R part by Q. This process can be repeated indefinitely, to produce a sequence of matrices whose eigenvalues are the same as those of A, and whose eigenvectors are simply related to those of A. Note, too, that this is simply the power iteration applied with $k = p$ and $X_0 = Q$. Thus, this repeated iteration will converge to a diagonal matrix of eigenvalues, together with a matrix of corresponding eigenvectors. (When the eigenvalues are not distinct, the space spanned by the columns corresponding to multiple eigenvalues will converge to an invariant subspace, although these columns need not converge to a particular basis for this subspace. The numerical sensitivity of an invariant subspace depends upon the extent to which its eigenvalue is separated from the others. Again, in statistical computations this is rarely an issue.)

Although the power iteration just described will produce the desired result, it is possible to improve the algorithm in three major respects. The first reduces the number of arithmetic calculations that need to be done at each step, and the second dramatically accelerates the rate at which convergence occurs from step to step. These two improvements constitute the heart of the symmetric QR algorithm. The third improvement makes use of a remarkable theorem to avoid explicitly computing most of the matrices in terms of which the QR method is described!

First, instead of starting with A, begin with $A_0 = U'AU$, where U is chosen to be a $p \times p$ orthogonal matrix such that A_0 is *tridiagonal* in form. The construction of such a matrix U is left as an exercise. A matrix is said to be tridiagonal if the only non-zero elements are either on the diagonal, or on the super- or sub-diagonal. Put another way, A is tridiagonal if $a_{ij} = 0$ whenever $|i - j| > 1$. (When A is zero below the subdiagonal— but possibly nonzero to the right of the superdiagonal—A is said to be in upper *Hessenberg form*. Upper Hessenberg matrices have $a_{ij} = 0$ whenever $i - j > 1$.) Of course, A_0 is symmetric, since A is. What makes this initial step useful is that after each subsequent step of the QR-power iteration, the resulting matrix A_i remains tridiagonal! Since most of the elements of a tridiagonal matrix are zero, the number of actual arithmetic computations required to perform the matrix multiplications at each step is dramatically reduced. What is more, the QR decomposition of a tridiagonal matrix requires only $O(p)$ flops, since the subdiagonal can be eliminated with just $p - 1$ Givens rotations.

Second, we know that convergence of the power iteration depends upon the ratio of the small eigenvalues. The notion of *origin shifts* makes it possible to compute the solution to a related problem for which this ratio is markedly smaller (and convergence is commensurately faster). Note that A and $A - \mu I$ have the same eigenvectors, and the two sets of eigenvalues differ by μ. If $|\lambda_{k+1}/\lambda_k|$ is close to one, then convergence of these eigenvalues and their corresponding eigenvectors will be quite slow. However, if μ is chosen to be close to λ_{k+1}, then the ratio $|(\lambda_{k+1} - \mu)/(\lambda_k - \mu)|$ can be quite small, and convergence will be much more rapid. To implement this, then, we obtain an approximate eigenvalue μ at each step of the QR-power iteration, and compute the QR decomposition of $A_i - \mu I$, then let $A_{i+1} = RQ + \mu I$. An obvious choice for the approximate eigenvalue μ is the current value of the lower-right element a_{pp}; a more generally satisfactory choice is the *Wilkinson shift*, equal to the eigenvalue of the lower right 2×2 block closer to a_{pp}. Either choice produces cubic convergence in practice. Lawson and Hanson (1974) give a proof of global quadratic convergence. Without shifts the general QR method exhibits only linear convergence.

3.9.1.3 The implicit QR algorithm

A final bonus is that neither the shifted matrix $A_i - \mu I$ nor its QR factorization need be computed explicitly. Instead, exploiting the symmetry and the simplicity of the symmetric tridiagonal structure, together with the remarkable *implicit Q theorem*, an entire iteration can be computed *implicitly* using a sequence of $p - 1$ symmetrically applied Givens rotations (see Section 3.1.6.2). One complete iteration of the implicit QR algorithm with Wilkinson shift requires something under $14p$ flops plus p square roots.

Theorem 3.9–1 (Implicit Q Theorem). *Let Q and V be orthogonal matrices such that $Q'AQ = H$ and $V'AV = K$, where A is a $p \times p$ matrix, H and K are zero everywhere below the subdiagonal, and all of the subdiagonal elements of H are nonzero. If the first column of Q and V are the same, then the remaining columns of Q and V agree (except for sign changes), and $|h_{i,i-1}| = |k_{i,i-1}|$, for $i = 2, \ldots, p$.*

The theorem asserts that any two orthogonal matrices which reduce the same matrix A to upper Hessenberg form and which agree in the first column must be essentially the same. How can this be applied to the symmetric eigenproblem? A single step of the QR iteration (with shift) comprises the transition from A_i to A_{i+1}, where both matrices are tridiagonal (and hence, Hessenberg). Note that all symmetric Hessenberg matrices are tridiagonal. Moreover, $A_{i+1} = RQ + \mu I = Q'(A_i - \mu I)Q + \mu I = Q'A_iQ$. Thus, if we can find *any* Q for which $Q'A_iQ$ is tridiagonal and which agrees in the first column with the orthogonal matrix which converts A_i to upper triangular form, the implicit Q theorem tells us that we are done (without having formed R or $A_i - \mu I$ at all).

Fortunately, getting the first column of Q is easy. Note that $Q'(A_i - \mu I) = R$, so that one way to compute Q is to express it as the product of $p - 1$ Givens transformations which zero the subdiagonal elements of $A_i - \mu I$. Let $Q = J_1 \cdots J_{p-1}$, where J'_r zeros the rth subdiagonal of $(J_1 J_2 \cdots J_{r-1})'(A_i - \mu I)$. Note that the first row of Q' is just that of

J_1', since the remaining J_r's leave the first row untouched. Thus, the first column of Q is the first column of J_1. Now $J_1' A_i J_1$ is not tridiagonal, indeed, it has the form

$$
J_1' A_i J_1 = \begin{pmatrix}
* & * & + & & \\
* & * & * & & \\
+ & * & * & * & \\
& & * & * & * \\
& & & * & *
\end{pmatrix}.
$$

Here we have used "+" to denote the new elements off the tridiagonal introduced by the transformation. It is easy to zero the element below the subdiagonal with a Givens transformation G_2' which, by symmetry, will also zero the element to the right of the superdiagonal when applied on the right. Unfortunately, this transformation introduces new children off the tridiagonal:

$$
G_2' J_1' A_i J_1 G_2 = \begin{pmatrix}
* & * & & & \\
* & * & * & + & \\
& * & * & * & \\
& + & * & * & * \\
& & & * & *
\end{pmatrix}.
$$

But a new Givens rotation G_3 can get rid of these:

$$
G_3' G_2' J_1' A_i J_1 G_2 G_3 = \begin{pmatrix}
* & * & & & \\
* & * & * & & \\
& * & * & * & + \\
& & * & * & * \\
& & + & * & *
\end{pmatrix}.
$$

And finally, one more rotation will get rid of the annoying off-diagonals once and for all:

$$
G_4' G_3' G_2' J_1' A_i J_1 G_2 G_3 G_4 = \begin{pmatrix}
* & * & & & \\
* & * & * & & \\
& * & * & * & \\
& & * & * & * \\
& & & * & *
\end{pmatrix} \equiv Q' A_i Q \equiv A_{i+1}.
$$

The implicit Q theorem guarantees that the orthogonal matrix Q that we have constructed is precisely the one needed in the QR algorithm, and we have made the transition to A_{i+1} without ever explicitly having computed a QR decomposition!

In the algorithm as we have described it, Givens rotations are used to triangularize A implicitly. This will be possible only if all of the subdiagonal elements are nonzero. When one of the subdiagonal elements becomes zero (to a given small precision), the resulting matrix contains two submatrices, each of which is tridiagonal. The general method can then be applied to these smaller subproblems. When this occurs, the problem is said to *decouple*. The problem decouples at the next iteration whenever the shift μ exactly equals one of the eigenvalues.

The algorithm essentially as we have described it is detailed in Golub and Van Loan (1983) and in Lawson and Hanson (1974). Computer programs to compute the symmetric eigenproblem can be found in EISPACK, which contains routines for the more general problems as well.

3.9.2 The Golub-Reinsch singular value decomposition algorithm

At first glance, the problem of computing the singular value decomposition of an $n \times p$ matrix $X = UDV'$ would seem more difficult than that of the symmetric eigenproblem. It turns out, however, that the two problems are intimately related, and that the singular value decomposition can be easily and stably computed using a variant of the QR method described in the preceding subsection.

Recall the notation of section 3.5.3, in which $n \geq p$, U is an $n \times n$ orthogonal matrix, V is a $p \times p$ orthogonal matrix, and D is an $n \times p$ "diagonal" matrix, that is $D_{ij} = 0$ for $i \neq j$. We denoted the upper $p \times p$ submatrix of D by D. The diagonal entries are taken to be nonincreasing and positive; these are the *singular values* of X. The columns of U and V are, respectively, the left and right *singular vectors* of X. Notice that $X'X = VD^2V'$, and that $XX' = UDDU'$, each of which is a symmetric positive semidefinite matrix. Thus the singular values could be computed as the square roots of the eigenvalues of $X'X$, the right singular vectors as the eigenvectors of $X'X$, and the left singular vectors as the eigenvectors of XX'.

The Golub-Reinsch (1970) algorithm in effect computes the symmetric QR reduction, but does so in factored form, eliminating the need to form either $X'X$ or XX' explicitly. The algorithm starts by reducing X to upper bidiagonal form, that is, to an $n \times p$ matrix B which is zero except on the diagonal and superdiagonal. The superdiagonal elements are then systematically reduced in magnitude by an implicit form of the QR iteration.

3.9.2.1 Householder-Golub-Kahan bidiagonalization

The reduction of an $n \times p$ matrix X to upper bidiagonal form $B = U_{bd}XV'_{bd}$ can be accomplished using a sequence of Householder reflections in a manner first described by Golub and Kahan (1965). Clearly, B has the same set of singular values as X. First, the sub-diagonal elements of the first column of X are made zero by an $n \times n$ Householder transformation U_1. This makes $n - 1$ elements of the first column equal to zero. Next, the last $p - 2$ elements of the first row are made zero by a $p \times p$ Householder matrix V_1 postmultiplying U_1X. This operation leaves the first column unchanged, but modifies the remaining columns. At this point, the matrix $U_1 X V_1'$ is zero below the diagonal in column 1, and zero to the right of the superdiagonal in row 1. A Householder transformation U_2 which zeros the $n-2$ positions below the second diagonal element is now applied on the left, after which a Householder transformation V_2 is applied on the right to zero the $p-3$ elements to the right of the superdiagonal in row 2. These transformations, by construction, do not alter the first row or column, so at the end of step 2, $U_2 U_1 X V_1' V_2'$ has both the first two rows and the first two columns in the required form. Repeating this process a total of p times, the matrix $U_{bd} X V_{bd} \equiv (U_p U_{p-1} \cdots U_1) X (V_p V_{p-1} \cdots V_1)'$ will be in upper bidiagonal form. (Note that $V_{p-1} = V_p = I$.) Note that the "off-bidiagonal" storage space can be used at each step to store the coded form of the Householder transformation which zeroed that portion of a column or row.

3.9.2.2 The SVD of a bidiagonal matrix

The procedure that we describe here is essentially that embodied in the algorithm of Golub and Reinsch (1971), which in turn represents an improvement over its predecessor developed by Golub and Kahan (1965) and implemented in slightly different form in a 1967 computer program due to Businger and Golub. Let B denote the bidiagonalized matrix obtained by the Householder-Golub-Kahan procedure.

Recall that the right-hand matrix V' in the SVD of B is the same as the matrix of eigenvectors of $B'B$, which in turn we know from the development of the implicit QR algorithm can be represented as a product of Givens transformations. In particular, at each step of the QR iteration, we implicitly shift by an approximate eigenvalue μ and begin a series of Givens transformations which re-triangularize $B'B$. The first of these Givens transformations, say V_1, can be computed easily enough; what if we simply apply it directly to B without forming the product $B'B$? In schematic form,

$$BV_1 = \begin{pmatrix} * & * & & & \\ & * & * & & \\ & & * & * & \\ & & & * & \end{pmatrix} \quad V_1 = \begin{pmatrix} * & 0 & & & \\ + & * & * & & \\ & & * & * & \\ & & & * & \end{pmatrix}.$$

Now apply a Givens transformation U_1' on the left to eliminate the newly-introduced subdiagonal element; this reintroduces a superdiagonal element in the first row, and a new element in the first row to its right:

$$
U_1'BV_1 = \begin{pmatrix} * & * & + & \\ 0 & * & * & \\ & & * & * \\ & & & * \end{pmatrix}.
$$

The next step removes the unwanted element to the right of the super-diagonal using a Givens rotation V_2 involving only the second and third columns, which introduces a new element into the subdiagonal of the third row, which is then removed with a Givens rotation U_2' on the left operating on rows two and three which zeros this element:

$$
U_2'U_1'BV_1V_2 = U_2' \begin{pmatrix} * & * & 0 & \\ & * & * & \\ & + & * & * \\ & & & * \end{pmatrix} = \begin{pmatrix} * & * & & \\ & * & * & + \\ & 0 & * & * \\ & & & * \end{pmatrix}.
$$

The next pair of rotations returns the matrix to bidiagonal form:

$$
U_3'U_2'U_1'BV_1V_2V_3 = U_3' \begin{pmatrix} * & * & & \\ & * & * & 0 \\ & & * & * \\ & & + & * \end{pmatrix} = \begin{pmatrix} * & * & & \\ & * & * & \\ & & * & * \\ & & 0 & * \end{pmatrix} \equiv U'BV \equiv B^*.
$$

Since $(B^*)'B^* = V'B'BV$, and since the first column of V is that required to triangularize the shifted matrix $B'B - \mu I$, the implicit Q theorem guarantees that V is the matrix required in the QR iteration, and that $(B^*)'B^*$ is the next iterate in the decomposition of $B'B$. The analogous statement is true for $B^*(B^*)'$ and U. Thus, we have *implicitly* computed the correct left and right matrices U' and V, and we have made the transition to the next iterate, B^* without ever explicitly having computed a triangular factor R *or* the matrix product $B'B$.

Computer programs which implement the SVD can be found in LIN-PACK and EISPACK, as well as in Lawson and Hanson (1974).

3.9.2.3 The Chan modification to the SVD

In 1982, Chan proposed a modification to the Golub-Reinsch procedure outlined above which has some computational advantages when $n > 5p/3$ (Chan, 1982a,b). Since this will often be the case in regression problems in statistics, the idea behind the Chan modification and its relative merits are worth some discussion.

The approach taken by Chan is designed to speed up the bidiagonalization step of the SVD. This is done by first converting X to upper triangular form, and the $n \times p$ matrix $\begin{pmatrix} R \\ 0 \end{pmatrix}$ can be converted to upper bidiagonal form ignoring the $(n - p) \times p$ block of zeros. Unless n is much larger than p, the extra work involved in computing the triangular decomposition is more than the savings from creating a block of zeros. Rough flop counts done by Golub and Van Loan (1983) suggest that the break-even point is when $n \approx (5/3)p$.

Chan also introduces some modifications in the implicit iteration which speed things up somewhat when the matrices U and V must be accumulated (as they often must for use in statistical applications). The matrices U_r are the product of Givens transformations, but in accumulating them on the left, each one affects two columns of the matrix being built, each of which is of length n. However, the matrix to which the Givens rotations are being applied is effectively $p \times p$. Thus, these left-hand Givens rotations can be accumulated in a $p \times p$ matrix, and then applied once at the end to obtain the $n \times n$ result.

A FORTRAN program which uses the Chan modification when $n \geq 2p$ and the Golub-Reinsch method otherwise is described in Chan (1982b), and is available through the ACM Algorithms Distribution Service. The program involves almost 3000 lines of code. A speedup by almost a factor of two is possible over the unmodified version for very large problems, although for problems of modest size the complexity of the Chan approach will outweigh the increased speed obtainable.

3.9.3 The generalized eigenproblem

We introduced the generalized eigenproblem in section 3.8.2; for real $p \times p$ matrices A and B with A symmetric and B positive definite, we consider the solutions in λ and v of $(A - \lambda B)v = 0$. With these restrictions the problem is referred to as the *symmetric-definite generalized eigenproblem*. The solutions in λ are called the eigenvalues of A with respect to B, and corresponding to them are eigenvectors v. There will be exactly p eigenvalues of this system (counting multiplicities), and due to the symmetry of A and B, they will be real.

COMMENT. More generally, if A and B are any two complex-valued $p \times p$ matrices, then the set of all matrices of the form $A - \lambda B$, $\lambda \in \mathcal{C}$, is said to be a *matrix pencil*. The set of solutions to the determinantal equation $|A - \lambda B| = 0$ is the set of eigenvalues of the pencil, and is denoted $\lambda(A, B)$. The number of elements in $\lambda(A, B)$ is p, provided that B is nonsingular. Otherwise, $\lambda(A, B)$ may be empty, may have a finite number of elements, or may even be the entire complex plane. A general algorithm for computing $\lambda(A, B)$ is called the QZ algorithm (Moler and Stewart, 1973); it is similar in spirit to the QR iteration described above.

Unfortunately, in the common symmetric-definite case, the QZ method destroys symmetry and definiteness.

Since B is nonsingular, the eigenvalues we seek are identical to those of $C \equiv B^{-1/2}AB^{-1/2}$, so that one approach to the problem is to apply the QR method to the latter matrix. A suitable method for obtaining a matrix square root for B in this scheme is to use the Cholesky factorization. An efficient implementation requires about $7p$ flops (Martin and Wilkinson, 1968). Unfortunately, if B is ill-conditioned then C will have some very large entries, causing the computed eigenvalues to be contaminated with rounding error even if C itself is not badly conditioned. Golub and Van Loan (1983, Chapter 8) describe heuristic measures which can be taken to minimize this difficulty, at the expense of first computing the spectral decomposition of B. Fix and Heiberger (1972) present an algorithm for the ill-conditioned problem.

In multivariate analysis, the matrices A and B which arise are *both* positive semidefinite, since each is a matrix sum of squares associated either with a (multivariate) hypothesis or with the residual from the model. Consequently, A can be written as the inner product matrix $A = C'C$ for some $n \times p$ matrix C—and in practice A is often computed in this way! Similarly, B may be written as $D'D$ for some $m \times p$ matrix D. The pencil of interest in these problems, then, has the form $C'C - \lambda D'D$. Remarkably, there is a generalization of the singular value decomposition which makes it possible to diagonalize C and D simultaneously, and to obtain each λ as the square of the ratio of a pair of these diagonal elements, without explicitly forming the products $C'C$ and $D'D$. A numerically stable algorithm for computing the GSVD is given in Stewart (1983). A useful discussion of the GSVD can be found in Paige and Saunders (1983).

3.9.4 Jacobi methods

The Jacobi method for eigenvalue extraction is interesting for two reasons: it provides an alternative approach to diagonalizing a symmetric matrix, and it is an example of a basic computational technique that, like many others, becomes popular and then recedes in importance. All of the algorithms described earlier in this section have come into existence since 1960, so that the modern theory of solving eigensystems is less than thirty years old. Before that time, the Jacobi method (1846) and its variants were essentially the only ones available.

The basic method is elegant in its simplicity. Consider a $p \times p$ symmetric matrix A. Let $S(A)$ denote the sum of squares of the off-diagonal elements of A, that is,

$$S(A) = \sum_{i=1}^{p} \sum_{\substack{j=1 \\ j \neq i}}^{p} a_{ij}^2.$$

If $S(A) = 0$, then A must be diagonal, otherwise $S(A) > 0$. For any orthogonal matrix Q, $Q'AQ$ and A have the same eigenvalues. Given A, the Jacobi method finds a Q for which $S(Q'AQ) < S(A)$; thus $Q'AQ$ is closer to a diagonal matrix than is A.

Some hundred years before Givens, Jacobi noted that orthogonal rotations (Givens transformations) could be used to reduce $S(A)$. First note that the sum of squares of all of the elements of A (called the *Frobenius norm* of A, and denoted by $\|A\|$) is unchanged by orthogonal transformation. Second, let a_{ij} be the off-diagonal element with largest magnitude, and let $Q = G[i, j, \theta]$ be the elementary rotation which zeros the i, j-element of $Q'AQ$. This choice minimizes $S(Q'AQ)$ over all elementary rotations Q, and it is easy to show that $S(Q'AQ) < S(A)$ whenever A has non-zero elements off the diagonal. The new matrix then has a largest off-diagonal element, which can then be made zero with a new elementary rotation, and the iteration repeated. This method, called the *classical Jacobi algorithm*, ultimately has quadratic convergence, and is very simple to program.

A variant of this method, called the *serial Jacobi algorithm*, avoids the search for the largest element of A at each step (which requires $O(p^2)$ computations) by simply rotating on each of the non-zero off-diagonal elements in turn, starting with the superdiagonal elements in the first row, and continuing to subsequent rows. This method also has quadratic convergence, and is much faster, since it avoids the search. This can be improved somewhat by skipping the rotation when the corresponding $|a_{ij}|$ is very small (since $S(A)$ will only be reduced by $2a_{ij}^2$). Another variant, due to Nash (1975), uses the Jacobi technique to compute the singular value decomposition; Maindonald (1984) contains an implementation of this algorithm in BASIC.

The Jacobi methods are not currently used, except in special circumstances. Except for very small matrices, the QR iteration has usually converged in the time it takes the serial Jacobi algorithm to make one pass through the off-diagonal elements of A. In the late 1970's and early 1980's, Jacobi methods became attractive once again because they could be written using very small computer programs, making it possible to implement them on mini- and microcomputers with small random access memories (and essentially unlimited computing time). Advances in semiconductor technology have made limited memory a thing of the past, so general interest in Jacobi methods has once again begun to wane. But the Jacobi idea is not dead, or even dying. The technique is admirably suited to *parallel computation*, in which different central processors simultaneously work on different parts of the problem at the same time. Since the rotation $G[i, j, \theta]$ affects only rows i and j and columns i and j, while one CPU is doing the rotation on position i, j, another can be rotating the k, l position, provided that i, j, k, and l are distinct. The QR algorithm, on the other hand, does not lend itself to parallel computation in any obvious way. Stewart (1985), describes an algorithm which, in the symmetric case, is a variant of the Jacobi method, and which has a parallel implementation that can compute the eigenvalues in time proportional to the order of the matrix.

EXERCISES

3.74. [19] Prove (3.9.1).

3.75. [17] Prove that the eigenvalues of $A_1 = Q'AQ$ are the same as those of A, and that the eigenvectors of A_1 are the columns of $Q'\Gamma$, where the columns of Γ are orthogonal unit-length eigenvectors of A.

3.76. [22] Let A be a $p \times p$ symmetric matrix. Let Q be a Householder matrix that puts zeros in the last $p - 2$ positions of the first column of A. Show that QAQ' is symmetric, and that the elements of the first row and column of this matrix are zero except on the diagonal, super-diagonal, and sub-diagonal.

3.77. [20] Show that the matrix A of the previous exercise can be reduced to tridiagonal form by $p - 2$ Householder transformations. This process is called *Householder tridiagonalization*.

3.78. [24] Show that the process of Householder tridiagonalization can be accomplished in $2p^3/3$ flops.

3.79. [25] Show that, if the matrix $A_0 = U'AU$ is the Householder tridiagonalization of A, then the orthogonal matrix U can be obtained at the same time, at the cost of an additional $2p^3/3$ flops.

3.80. [24] Prove that, if A is a symmetric, tridiagonal matrix with QR decomposition $A = QR$, then the matrix RQ is also symmetric and tridiagonal.

3.81. [30] Let A be a $p \times p$ symmetric tridiagonal matrix, with all subdiagonals nonzero. Show that when μ is exactly equal to one of the eigenvalues of A, the QR iteration with shift μ decouples at the next step. In particular, $a_{p-1,p} = 0$ at the next iteration, and $a_{pp} = \mu$. (Lawson and Hanson, 1974).

3.82. [28] Prove Theorem 3.9-1.

3.83. [20] Show that, if A is a $p \times p$ symmetric matrix, and Q is orthogonal, that $\|A\| = \|Q'AQ\|$.

3.84. [25] Find the Givens rotation Q of section 3.9.4 which makes the i,jth element of $Q'AQ$ equal to zero.

3.85. [21] Show that $S(Q'AQ) - S(A) = 2a_{ij}^2$, where Q is defined as the Givens rotation that makes the i,j-element of $Q'AQ$ equal to zero.

3.10 Generalizations of least-squares regression

Although the linear model introduced in section 3.1 is quite generally applicable, its mathematical structure is not sufficiently general to encompass all aspects of statistical modeling. We have focused so far on the linear model, in which

$$Y = X\beta + \epsilon,$$
$$\epsilon \sim (0, \sigma^2 I),$$

(3.10.1)

and X is an $n \times p$ matrix of full rank. Because it is both generally useful in its own right, and a natural starting point for other analyses, this basic linear regression model and its least-squares solution have been extended, refined, and elaborated over the course of several decades. Any elaboration, of course, must be given a new name to distinguish it from the standard framework, and it is natural to call a more general approach something like "generalized least squares (GLS)," or "the general linear model (GLM)," or "generalized linear models (GLIM)." Indeed, each of these terms has been applied to some extension of (3.10.1), but each has been applied to a *different* extension. In this section we discuss these different generalizations and examine computational aspects of GLS, and we touch briefly on computation associated with the other extensions of ordinary least squares (OLS).

3.10.1 GLM: The general linear model

The term *general linear model (GLM)* refers to linear models that have come to be associated with the analysis of variance and analysis of covariance as opposed to regression. GLM is *mathematically* little more than (3.10.1), but *statistically* there are substantial differences between it and the regression models we have discussed up to now. That X has full rank in (3.10.1) implies that its columns are linearly independent, and this greatly simplifies the computational details that comprise the preceding sections. What is more, the use of generalized inverses as outlined in section 3.5 makes it possible in principle to deal with the rank-deficient case, although in regression problems the need rarely arises. Another way of writing (3.10.1), however, focuses on the expected value of Y as a linear combination of columns of X:

$$E(Y) = X\beta,$$
$$V(Y) = \sigma^2 I.$$

(3.10.2)

When the data are sampled under k different circumstances, such as from k distinct populations or under k different experimental conditions, the mean of Y is often modeled as having k different values, one for each of the k sampled groups. We can express this model in terms of (3.10.2) or (3.10.1) by using an X matrix with k columns, each entry of which is either a one or a zero, position i of column j having the value one if the ith observation came from group j and zero otherwise. Such a column is called an *indicator variable* or *dummy variable*. A dummy variable can in general be used to distinguish observations in a regression according to a categorical variable. The number of dummy variables required is equal to the number of different categories.

While in principle nothing has changed, in practice, a great deal has taken place. The matrix X no longer has a column consisting entirely of ones, hence our discussion of regression computations based on a centered SSCP matrix, or a regressor matrix adjusted for the constant column, does not apply to this X matrix. Second, if we introduce a column of ones, the new X matrix becomes rank-deficient, which means either that we must resort to generalized-inverse methods, we must impose restrictions on the parameter vector β, or we must delete a column of X in order to obtain parameter estimates. Third, if more than one categorical variable is included, additional columns must be dropped, or restrictions applied, since the sum of the columns of a dummy variable representing a single categorization is always equal to a column of ones. Fourth, the new matrix is sparse, so there is additional structure that we have not explicitly taken advantage of in our earlier discussion of computing. Fifth, a categorical variable with k different categories produces k (or $k-1$) dummy variables, so that even including a small number of categorical variables may produce an X matrix with a very large number of columns.

The computations associated with the general linear model are primarily those which deal with these facets of the regression problem with categories, with or without single columns representing continuous variables. In contrast to the computational methods for linear regression, the additional methods required to deal with the analysis of variance models are not of more general statistical interest, nor are they likely to be required in computer programs that statisticians may need to write for themselves. Consequently, we shall not discuss them at any greater length here. Programs such as SAS (SAS Institute, 1982a, 1982b) have excellent capabilities for carrying out the GLM computations on data, and are generally quite adequate. An account of the computational details associated with the general linear model can be found in Chapter 9 of Kennedy and Gentle (1980).

3.10.2 WLS: Weighted least squares

A second elaboration of (3.10.1) does not change the mean structure of Y, but allows Y a more general variance structure. This is often appropriate. For instance, the observations in a regression problem may come from instruments whose precision is not equal, and it is sensible to want to give more weight to the data coming from the more precise measurements. Similarly, the observations on which the regression analysis is performed may be means of other measurements; unless the number of measurements going into each of these means is the same, the variances in the regression observations will differ. In the latter case, we at least know the relative magnitudes of the variances of the observations; they are merely the ratios of sample sizes associated with the means. In the former case, it is often reasonable to assume that we know the relative precisions of the measuring instruments. In either case, we can write a linear model for Y in terms of X as

$$Y = X\beta + \epsilon,$$
$$\epsilon \sim (0, \sigma^2 V),$$
(3.10.3)

where V is a diagonal matrix whose ith diagonal element $v_{ii} = 1/w_i$ represents the relative variance of the ith observation to the others. It is important to note that V is known, although σ^2 may not be. Under these circumstances, we can premultiply both sides of the equation in (3.10.3) by $V^{-1/2}$, to obtain

$$Y^* \equiv V^{-1/2}Y = V^{-1/2}X\beta + V^{-1/2}\epsilon$$
$$\equiv X^*\beta + \epsilon^*,$$
where
$$\epsilon^* \sim (0, \sigma^2 I).$$
(3.10.4)

Note that this is simply (3.10.1) with an asterisk on each symbol, except that β in (3.10.3) and (3.10.4) is unchanged. Thus, if we should use least squares to estimate β in (3.10.1), then we should be equally willing to use it in (3.10.4), giving

$$\hat{\beta} = (X^{*\prime}X^*)^{-1}X^{*\prime}Y^*$$
$$= (X'V^{-1}X)^{-1}X'V^{-1}Y$$
$$= \sum w_i x_i' y_i / \sum w_i x_i' x_i,$$
(3.10.5)

where in the last expression x_i represents the ith row of X. This estimator is called the *weighted least-squares* (WLS) estimator, as each observation gets weight w_i. Note that the weights are simply the reciprocals of the variances associated with each row. This weighted estimator is the best linear unbiased estimator of β, and in the normal-theory case, the maximum-likelihood estimator as well. The estimate for σ^2 is based on a weighted residual sum of squares, the derivation of which is left as an exercise. In econometrics and business statistics, the basic least-squares estimator $\hat{\beta} = (X'X)^{-1}X'Y$ is referred to as the *ordinary least-squares estimator*, or *OLS* estimator. Clearly, OLS is the special case of WLS with all $w_i = 1$. The ordinary least-squares estimator is the solution β which minimizes $(Y - X\beta)'(Y - X\beta)$. The weighted least-squares estimator is also the solution to a minimization problem; it is the solution which minimizes $(Y - X\beta)'V^{-1}(Y - X\beta)$.

Computationally, little changes when the arbitrary weights of WLS are introduced. If the SSCP matrix or a corrected SSCP matrix is accumulated, the weight corresponding to each row is applied as the terms in that row are accumulated into the total. Thus, the SSCP matrix S of (3.4.8) becomes

$$s_{00} = \sum_{1 \le k \le n} w_k,$$

$$s_{0j} = \sum_{1 \le k \le n} w_k X_{kj}, \qquad (3.10.6)$$

$$s_{ij} = \sum_{1 \le k \le n} w_k X_{ki} X_{kj}.$$

This is simply $X^{*'}X^*$ in the notation of this section. A corrected version can also be constructed, provided that the correction is based on deviations from the *weighted mean* of each column. The first step of the SWEEP algorithm will replace the weighted column sums s_{0j} by an appropriately weighted mean, and the cross-product terms s_{ij} will be replaced by an appropriately weighted corrected sum of squares and cross products.

When the least-squares computations are to be done using the QR decomposition and Householder transformations, the introduction of weights complicates matters a bit. An obvious solution is to begin by transforming Y and X explicitly by premultiplication by the inverse square-root of V; after having done this, the Householder algorithm can be followed without further modification. When V is ill-conditioned, that is, when w_{max}/w_{min} is large, the transformed problem may be numerically unstable; more stable methods are described in section 3.10.3.

3.10.3 GLS: Generalized least squares

A third generalization of the linear least-squares problem, called *generalized least squares (GLS)* in the econometrics literature, is a minor elaboration on weighted least squares. Suppose that, instead of $\text{Var}(Y) = \sigma^2 V$, where V is a known *diagonal* matrix with positive elements, V is a known *general* positive-definite matrix. Then following exactly the argument in section 3.10.2, the appropriate estimator for β is $\hat{\beta}_{GLS} = (X'V^{-1}X)^{-1}X'V^{-1}Y$.

Similarly, this estimator minimizes $(Y - X\beta)'V^{-1}(Y - X\beta)$, and this latter quantity is the residual sum of squares when $\beta = \hat{\beta}_{GLS}$. The only difference, then, between WLS and GLS is that in the former, V is taken to be diagonal, and in the latter V is taken to be a general variance matrix.

The variance structure V of ϵ is known only in a few special circumstances which, nonetheless, are important. The two most frequent circumstances are when V has intraclass form ($v_{ii} = 1$, and $v_{ij} = \rho$, for $i \neq j$), and when V has autoregressive form ($v_{ij} = \rho^{|i-j|}$). In these cases, the variance structure is determined by a single additional parameter ρ, which occasionally is known, or for which a good estimate exists. A third circumstance is in Monte Carlo computations, in which the variance structure of some random variables is known as a consequence of the structure of the generated random numbers in the simulation. A number of variance-reduction methods are based on generalized least squares.

As in the WLS case, we can compute a matrix square root of V, say $V^{-1/2}$ and premultiply both X and Y. After having done this, we can proceed with the least-squares computations as usual. What distinguishes GLS from WLS computationally is that there is no easy route to forming the appropriate SSCP matrix when the V matrix is not diagonal, so there is no advantage to forming the SSCP matrix and using Cholesky or SWEEP operations to obtain the estimator. Second, ill-conditioning of the V matrix is more likely to occur in practice for the GLS problem than for the WLS problem. Third, there is the matter of numerical stability in computing $V^{-1/2}$, and in premultiplying X and Y by $V^{-1/2}$, that arises in only a trivial way in WLS. A numerically stable method for computing the quantities required in GLS regression can be found in Paige (1979a, 1979b) and in Kourouklis and Paige (1981). The latter article describes the computation of relevant statistical quantities in terms of orthogonal reductions to triangular form.

3.10.4 GLIM: Generalized linear interactive models

A final way of rewriting (3.10.1) when Y has a normal distribution is

$$Y = \mathcal{N}(X\beta, \sigma^2 I). \tag{3.10.7}$$

When written in this way, the notation emphasizes that Y has a mean that is linear in X, and a distribution that is Gaussian, with given mean and variance. The last generalization of the linear model which we shall discuss is based on (3.10.7) and includes GLM with Gaussian errors as a special case.

Instead of writing the mean of Y as a linear function of X as in (3.10.7), we could say that some *function* of the mean of Y is linear in X, that is, $E(Y) = \mu$, and $f(\mu) = X\beta$. Such a function linking the mean of Y to a linear function of X is called a *link function*. Moreover, Y could have some distribution other than the Gaussian distribution, say F. When F is Gaussian and $f(\cdot)$ is the identity function, we have the normal-theory linear model. As an example of the utility of such a generalization, the *logistic regression model* can be expressed in these terms: Y has a binomial distribution $B(1, p)$, with $E(Y) = p$ and $f(p) = \log(p/(1 - p))$. The link function is the *logit transformation*, and it is on the logit scale that the dependency of Y on X appears linear.

This generalization is called the *generalized linear (interactive) model*, or *GLIM*. The model itself was first introduced in a systematic way by Nelder and Wedderburn (1972), and implemented in a computer program called GLIM (Baker and Nelder, 1978) using an interactive approach to model building. In this approach, terms could be added to or deleted from the model one at a time, link functions and error terms changed, and their effects examined. A comprehensive treatment of the model, its theoretical underpinnings, and computational details are contained in McCullagh and Nelder (1983). The "I" in the acronym is usually retained when referring to these models in order to distinguish "generalized linear models" (GLIM) from the (generally) less general "general linear model" (GLM).

Note that in this example, the variance of Y is not constant from one observation to the next, but depends in a systematic fashion on the mean of Y. Thus, the computational methods associated with GLS, which assume that the variance matrix is known up to a scalar multiple, cannot be used. Computing estimates for the parameters in GLIM models requires solving nonlinear systems of equations. General approaches to solving these systems are of broad interest, and are discussed in Chapter 4, with particular attention paid to GLIM models in Section 4.5.6. The main computational tool used for such models is iteratively reweighted least squares, which is described briefly in the next section.

3.10.5 Iteratively reweighted least squares

When the variances of the Y_i's are unequal, the best linear unbiased estimate of β is given by weighted least squares as outlined in section 3.10.2. It is often the case that these variances (and hence, the appropriate weights) are unknown. However, a common situation is for the variance of Y_i to be a known function $v(\cdot)$ of the mean $\mu_i = E(Y_i)$. The linear model asserts that $\mu = X\beta$. If only μ were known, the appropriate weights $w_i = 1/v(\mu_i)$ could be computed and then used to obtain the best estimate for β. But β itself is of interest only because it determines μ! That is, the best estimate for μ based on a data set depends upon μ itself.

This suggests the following estimation algorithm. Begin with a provisional set of relative precisions, say $w_i^{(0)} \equiv 1$, and then estimate β using WLS with this set of weights. Call this estimate $\hat{\beta}^{(1)}$. With this estimate in hand, compute $\hat{\mu}^{(1)} = X\hat{\beta}^{(1)}$, and then obtain a new estimate of the inverse variances $w_i^{(1)} = v(\hat{\mu}_i^{(1)})^{-1}$. Use this new set of weights and WLS to obtain $\hat{\mu}^{(2)}$, and repeat this process until it converges. More generally, the form of the variance function $v(\cdot)$ may also change from iteration to iteration, and may represent an asymptotic variance rather than exact small-sample variance. At each stage a new set of weights is computed and then used in weighted least squares, hence the name for this process *iteratively reweighted least squares (IRLS)*. The method is quite generally useful in solving nonlinear estimation problems; detailed discussion is deferred to Section 4.5.6. IRLS methods are used in the GLIM computer package, for instance. Under fairly mild conditions, the procedure will converge to the maximum likelihood estimator for β, and if $\hat{V}$ is the estimated variance matrix based on the final iterate, the matrix $\hat{\sigma}^2(X'\hat{V}^{-1}X)^{-1}$ will be asymptotically equivalent to the variance matrix of $\hat{\beta}$.

We have assumed that the weights at each iteration are computed from estimated variances of the observations. There is nothing in the above development which relies on this, however, and the weights computed at each stage may be constructed from other criteria. Some robust regression methods, for instance, can be viewed as iterative reweighting schemes in which each observation has a weight in the range $0 \le w_i \le 1$, and in which observations corresponding to very large residuals are assigned weights closer to zero in the next iteration.

EXERCISES

3.86. [12] Show that equations (3.10.6) do in fact produce $X^{*\prime}X^*$.

3.87. [19] Show how to estimate σ^2 in the weighted least-squares problem.

3.88. [05] Show that the weighted least-squares estimator is the vector β which minimizes $(Y - X\beta)'V^{-1}(Y - X\beta)$.

3.89. [23] Write down the matrix SWEEP[0] S, where S is the weighted SSCP matrix of (3.10.6).

3.90. [36] Show how to modify the Householder computations to perform weighted least-squares implicitly, that is, without explicitly premultiplying by $V^{-1/2}$.

3.91. [29] Let $V^{-1}(\delta) \equiv W(\delta) = \text{diag}(w_1, w_2, \ldots, w_{k-1}, w_k(1 + \delta), w_{k+1}, \ldots, w_n)$, with $\delta > -1$. Prove that the kth residual is a monotone decreasing function of δ.

3.92. [48] Characterize the relative size of residuals as a function of the elements of the diagonal weighting matrix W in WLS.

3.11 Iterative methods

In section 3.2 we introduced the problem of solving linear systems, and the methods we developed led directly to a solution which, if performed in infinite-precision arithmetic, would be the exact solution to the linear system. Moreover, this solution could be computed with a fixed number of arithmetic computations determined solely by the size of the input problem. The methods we discuss in this section produce a sequence of approximate solutions which converge on the exact solution (again, under the assumption of infinite-precision arithmetic). They are highly useful in solving large sparse systems. Although these methods have long been studied and employed in the field of numerical analysis, they have rarely been applied in statistical applications, simply because the direct methods have been entirely adequate for most needs in statistical data analysis. The coefficient matrix in the linear systems which arise in statistics are generally matrices of sums of squares and cross-products, and as a result are generally non-sparse. The importance of these iterative methods stems from the fact that they form the building blocks for some of the nonlinear computationally-intensive statistical methods discussed in Chapter 4. The basic ideas, however, are most easily discussed in the context of the linear model, which explains their presence in this chapter.

3.11.1 The Jacobi iteration

Consider once again the linear system of section 3.2 given by

$$A\beta = c, \tag{3.11.1}$$

where A is a $p \times p$ matrix of known coefficients of full rank, c is a given $p \times 1$ vector, and the object is to solve for the $p \times 1$ solution vector β. The first row of this system can be written as

$$a_{11}\beta_1 + a_{12}\beta_2 + \ldots + a_{1p}\beta_p = c_1,$$

which can be rearranged to produce

$$\beta_1 = \left(c_1 - (a_{12}\beta_2 + \ldots + a_{1p}\beta_p)\right)/a_{11},$$

provided that $a_{11} \neq 0$. Similarly, we can write

$$\beta_i = \frac{1}{a_{ii}} \left(c_i - sum_{j \neq i} a_{ij}\beta_j\right), \tag{3.11.2}$$

for $i = 1,\ldots,p$, again provided that all of the diagonal elements a_{ii} are nonzero. Equation (3.11.2) suggests an iterative algorithm for estimating β. Starting from an initial guess $\beta^{(0)}$, compute

$$\beta_i^{(k)} = \frac{1}{a_{ii}} \left(c_i - \sum_{j \neq i} a_{ij}\beta_j^{(k-1)}\right), \quad i = 1,\ldots,p, \tag{3.11.3}$$

for $k = 1, 2, \ldots$. This procedure is called the *Jacobi iteration,* and has also been called the *method of simultaneous displacements.*

Note that if $\beta^{(k-1)}$ ever equals the solution β, subsequent iterations leave the estimate unchanged. This suggests that, if $\beta^{(0)}$ is sufficiently close to β, then the procedure will converge. Note, too, that if A is a diagonal matrix with nonzero entries, then the correct solution is produced on the first iteration. This suggests that the Jacobi iteration will work well when A is "nearly" a diagonal matrix. Finally, note that if A is sparse with nonzero diagonals, the amount of computational effort required at each step is very small. Since the LU decomposition of A will generally be nonsparse, in this case the saved computational effort using the Jacobi method may compensate for the additional work entailed by iteration, making this approach competitive with or superior to direct methods of solution. What is more, only the nonzero elements of A need be stored when using the Jacobi iteration, which makes it attractive for solving very large problems.

Under what circumstances will the Jacobi method converge to the solution of (3.11.1)? If we write $A = D + N$, where D contains the diagonal elements of A and N the off-diagonal ("nondiagonal") elements, then we can write the iteration of (3.11.3) as

$$D\beta^{(k)} + N\beta^{(k-1)} = c \tag{3.11.4}$$

or

$$\begin{aligned} \beta^{(k)} &= D^{-1}\left(c - N\beta^{(k-1)}\right) \\ &\equiv D^{-1}c + H\beta^{(k-1)}. \end{aligned} \tag{3.11.5}$$

To examine the error at the kth step, write $\epsilon^{(k)} = \beta^{(k)} - \beta$. Since $\beta = D^{-1}c + H\beta$, subtracting this from both sides of (3.11.5) gives

$$\begin{aligned} \epsilon^{(k)} &= H\epsilon^{(k-1)} \\ &= H^k \epsilon^{(0)}. \end{aligned} \tag{3.11.6}$$

As $k \to \infty$, the right-hand side of (3.11.6) will go to the zero vector, provided that each eigenvalue of $H = -D^{-1}N$ is less than one in magnitude (see, for instance, Forsythe and Wasow, 1960). Note that this is a global statement; if H satisfies the condition, then convergence takes place *regardless* of the initial starting value.

3.11.2 Gauss-Seidel iteration

In equations (3.11.3) in computing $\beta_i^{(k)}$ we use $\beta_j^{(k-1)}$ for $j < i$, even though we have already computed the (presumably) better estimate $\beta_j^{(k)}$. An alternative to the Jacobi iteration which always uses the most recently updated values for β_j is called the *Gauss-Seidel iteration,* also sometimes called the *method of successive displacements* (Forsythe and Moler, 1967). Replacing equations (3.11.3), we have

$$\beta_i^{(k)} = \frac{1}{a_{ii}} \left(c_i - \left(\sum_{j<i} a_{ij}\beta_j^{(k)} + \sum_{j>i} a_{ij}\beta_j^{(k-1)} \right) \right), \quad i = 1,\ldots,p,$$
(3.11.7)

for $k = 1, 2, \ldots$. The analysis of this iteration is analogous to that of the Jacobi procedure, only instead of D and N containing the diagonals and off-diagonals of A, respectively, D now contains the lower triangular part of A including the diagonal, and N holds the remaining upper triangular piece (excluding the diagonal). If we again define $H = -D^{-1}N$, all of the analysis of section 3.11.1 goes through unchanged. In particular, convergence will occur for any starting guess $\beta^{(0)}$ if and only if H has all eigenvalues less than one in magnitude.

As with the Jacobi method, the Gauss-Seidel iteration converges quickly when the diagonal elements dominate the off-diagonals. A remarkable fact about the Gauss-Seidel algorithm is that convergence is guaranteed whenever A is symmetric positive definite (as it will generally be in regression computations, for instance); a proof can be found in Chapter 10 of Golub and Van Loan (1983).

COMMENT. Householder (1964, page 115) has an interesting historical commentary on the Gauss-Seidel iteration, which begins, "Forsythe has remarked that the Gauss-Seidel method was not known to Gauss and not recommended by Seidel. Gauss did, however, use a method of relaxation, as the term is used here, annihilating at each step the largest residual.... For hand computation this is natural and more efficient than the straight cyclic process; for machine computation the search is time-consuming."

3.11.3 Other iterative methods

As equation (3.11.6) indicates, convergence depends upon $|\lambda_1|^k$, and can be quite slow if the largest eigenvalue of H is close to one in magnitude. When this is the case, convergence of the Gauss-Seidel iteration can be accelerated using various schemes. Among these is the *method of successive over-relaxation (SOR)*. To describe this approach, define

$$\gamma_i^{(k)} = \frac{1}{a_{ii}} \left(c_i - \left(\sum_{j<i} a_{ij}\beta_j^{(k)} + \sum_{j>i} a_{ij}\beta_j^{(k-1)} \right) \right), \quad i = 1,\ldots,p,$$
(3.11.8)

so that the kth Gauss-Seidel iterate is simply $\beta_i^{(k)} = \gamma_i^{(k)}$. The kth SOR iterate is given by

$$\beta_i^{(k)} = \omega\gamma_i^{(k)} + (1 - \omega)\beta_i^{(k-1)}. \qquad (3.11.9)$$

When $0 \le \omega \le 1$, $\beta^{(k)}$ lies on the line segment connecting the previous iterate to what would be the next Gauss-Seidel iterate. Typically in SOR, the appropriate value of ω is *greater* than one, producing an estimate which takes a bigger step away from the current estimate than the Gauss-Seidel method would, but in the same direction. In general, the relaxation parameter ω must be estimated. There is a method due to von Mises which bears the same relation to the Jacobi iteration as Gauss-Seidel does to SOR; the method is little used.

The basic iterations in the Gauss-Seidel and Jacobi methods in effect use one row at a time. The matrices involved, however, can be partitioned into submatrices of order greater than 1×1, and the same sequence of iterations can be performed, solving one block at a time. There are block-matrix versions of both of the basic iterative schemes.

The discussion above makes no direct use of the symmetry often encountered in statistical problems, and one might expect that either efficiency or accuracy (or both) might benefit from taking advantage of this additional structure when it is present. This is indeed the case, and symmetric variants of the Gauss-Seidel iteration, for instance, have been developed. The fundamental feature of such methods is a "double sweep" through the estimated parameters. After updating β_1 through β_p, the order is then reversed, estimating β_p through β_1 on the next sweep. This double sweep ensures an exponential decrease in the size of the error vectors. In addition, extrapolation and acceleration methods can be applied to speed convergence. These topics, as well as the single-point and block methods, are discussed in Martin and Tee (1961).

EXERCISES

3.93. [01] Why is it true that $\beta = D^{-1}c + H\beta$ in the Jacobi iteration?

3.94. [22] For symmetric A with nonzero diagonals, show that the right-hand side of (3.11.6) goes to zero for every starting vector $\epsilon^{(0)}$ if and only if $H = -D^{-1}N$ has no eigenvalue greater than one in absolute value.

3.95. [28] (Continuation) Repeat the preceding exercise under the assumption that A is a general real matrix with nonzero diagonal elements.

3.96. [17] Show that, if the largest eigenvalue of H in magnitude is less than one and has a unique eigenvector v associated with it, then $\epsilon^{(k)}/|\epsilon^{(k)}| \to v$, provided that $v'\epsilon^{(0)} \neq 0$.

3.12 Additional topics and further reading

This chapter has focused on least-squares regression and the computations which surround it. There are a number of topics which from a statistical standpoint are closely related, but which present quite different computational facets. Our purpose here is to indicate what those facets are, and to provide an entry into the appropriate literature for those who wish to pursue them.

3.12.1 L_p regression

The least-squares solution to the regression problem minimizes $\sum(y_i - \hat{y}_i)^2$. When the error vector ϵ has a Gaussian distribution, this produces the maximum likelihood estimator for the coefficients β, but for other error distributions the least-squares criterion for estimation is not so obviously appealing. What is more, a single aberrant data point can exert tremendous influence on the estimated regression coefficients; holding all other y values fixed, the estimates for β are simply linear functions of y_i. This sensitivity to outliers is a consequence of the large penalty that least-squares imposes for large residuals. When the error distribution has heavier tails than the Gaussian—and in practice this is often the case—large residuals should be less heavily penalized (since it is more likely that they represent large actual deviations from the regression surface). One natural approach is to choose β so as to minimize an alternative measure of overall lack of fit, such as

$$\tilde{\beta} : \min_{\beta} \sum |y_i - \hat{y}_i| = \sum |y_i - x_i\beta|. \qquad (3.12.1)$$

The estimate $\tilde{\beta}$ is called the *least absolute-value (LAV)*, or *minimum absolute deviation (MAD)* estimator. The minimization problem (3.12.1) also arises in the decision-theoretic formulation of the Gaussian case, when the coefficients are to be estimated with respect to absolute-error loss.

A computational disadvantage to the LAV formulation is that the solution vector $\tilde{\beta}$ cannot be represented in closed form. Indeed, even if X has full rank, the LAV solution need not be unique. A statistical disadvantage is that there are no readily available estimates for asymptotic standard errors to associate with the estimates. In addition, when the errors do follow the Gaussian distribution, the LAV estimates sacrifice considerable efficiency. Nonetheless, the LAV regression plane is much less affected by outliers in the data than is the least-squares estimate. A slightly more general formulation is to compute

$$\tilde{\beta}^{(p)} : \min_{\beta} \sum |y_i - \hat{y}_i|^p, \qquad (3.12.2)$$

where $1 \leq p \leq 2$. This solution is referred to as the L_p regression estimate. L_1 is LAV; L_2 is least squares. Although the L_p regression problem presents some interesting computational problems for $1 < p < 2$, such problems are seldom of interest in practical situations. The interested reader should consult the survey of L_p regression methods found in Chapter 11 of Kennedy and Gentle (1980). The remainder of our discussion of L_p regression will deal with the common L_1 problem, that is, with least absolute-value regression.

COMMENT. There is some room for confusion in notation here. The term L_p regression is used due to the fact that the minimization criterion minimizes the ℓ_p norm $\|y - \hat{y}\|_p \equiv (\sum |y_i - \hat{y}_i|^p)^{1/p}$. When $1 \le p \le 2$, the outer exponent $1/p$ is irrelevant, and we have omitted it from our definition of L_p regression, as other values for the exponent are rarely of practical interest in statistics. We have also used p to indicate the column order of the regressor matrix X; there is obviously no relationship between the number of columns of X and an appropriate exponent for the minimization criterion. In the remainder of this section, we shall revert to our earlier practice of having p represent the number of columns in X, except when we use the generic term "L_p regression."

In the LAV regression problem, we can write

$$\sum |y_i - \hat{y}_i| = \sum \frac{1}{|y_i - \hat{y}_i|} (y_i - \hat{y}_i)^2$$
$$= \sum w_i (y_i - \hat{y}_i)^2,$$

(3.12.3)

provided that none of the residuals is exactly zero. Thus, in principle, LAV regression can be computed using iteratively reweighted least squares, starting from the least-squares solution. The difficulty arises in that, for some solution, at least p of the residuals *will* be zero, so that the weights will converge to infinity. As they do so, numerical instabilities can be introduced into the solution.

Two possibilities suggest themselves. First, whenever a residual became very small, we could replace it by a small positive quantity $\epsilon > 0$. While this will retard the instability, it will also in general impede convergence and will introduce some inaccuracy in the final solution. Second, we can note that if a particular residual (say r_i) is zero, that implies a constraint on the solution vector β, namely, $y_i = x_i\beta$, where x_i is the ith row of X. When the weight w_i associated with any single observation becomes too large, one could set the corresponding residual to zero, apply the constraint to reduce the dimension of the problem by one, and then omit that observation in computing the solution to the reduced problem. Although this seems straightforward, at a later stage the sum of absolute residuals can often be reduced still further by setting some new residual r_j to zero and by allowing an old residual such as r_i to become non-zero. Without applying the constraint, this method is not guaranteed to converge; with the constraint, it is not clear how to allow old observations to reenter.

An alternative approach for LAV regression—and the one on which the most efficient algorithms are based— is to note that the problem is equivalent to a *linear programming problem*. The linear-programming formulation

$$
\begin{array}{ll}
\text{minimizes} & \sum e_i^+ + \sum e_i^- \\
\text{subject to} & X\beta + e^+ - e^- = y \\
& e^+ \geq 0 \\
& e^- \geq 0.
\end{array}
\tag{3.12.4}
$$

Here there are $2n + p$ unknowns, namely the components of β, e^+, and e^-. Since $y_i = x_i\beta + (e_i^+ - e_i^-)$, it is clear that at the solution either $e_i^+ = 0$ or $e_i^- = 0$, or both. Solving the linear program in this form is inefficient, and efficient implementations make use of the relationships among the $2n + p$ variables to reduce both the amount of storage and the computing time required to find the solution. A full discussion of linear programming methods is beyond the scope of this book; Kennedy and Gentle (1980) discuss some aspects of linear programming in the context of LAV regression, and Gentle (1977) gives an excellent survey of the topic.

An efficient FORTRAN program for computing LAV solutions using the linear programming approach is given by Barrodale and Roberts (1974). A survey of such methods, incorporating linear equality or inequality constraints, can be found in Barrodale and Roberts (1977). The ROSEPACK (Coleman, et al, 1980) package includes a similar routine for obtaining the L_1 solution. The most efficient programs for computing LAV regression estimates appear to be those of Armstrong, Frome, and Kung (1979) and of Bloomfield and Steiger (1980). The former program is a modification of that of Barrodale and Roberts which, in turn, is a specially-tuned variant of the linear programming approach. The latter program is based on an algorithm for finding the p variables which, when made to have zero residuals, will minimize the sum of residual magnitudes. The Bloomfield-Steiger algorithm is markedly superior to the Barrodale-Roberts algorithm for $n > 100$ and all $p \geq 2$; its performance is somewhat inferior to Barrodale-Roberts on small data sets. A useful review paper on LAV estimation is Dielman and Pfaffenberger (1982). Narula (1982) gives an excellent review of the literature on L_1, L_2, and L_∞ regression, including discussion of multiple regression, subset selection, and constrained estimation in this context.

3.12.2 Robust regression

Least absolute-value regression is one method for giving large residuals less weight in regression computations. Related methods either limit the influence of individual observations ("bounded-influence regression") or assign nearly zero weight to those observations with very large residuals, in effect removing markedly aberrant points from the fitting process ("robust regression"). The computational details associated with these methods are quite similar to one another. We shall illustrate them in the context of M-estimation for robust regression; the details of bounded-influence regression and R-estimation are similar.

The process of assigning small or zero weights to observations with large residuals is called *downweighting*. The *M-estimators* are generalizations of maximum-likelihood estimators (hence, the name), which downweight large residuals; they estimate β so as to minimize

$$\sum_{i=1}^{n} \phi\left(\frac{y_i - x_i\beta}{s}\right) \equiv \phi\left(\frac{r_i}{s}\right), \tag{3.12.5}$$

where $\phi(\cdot)$ is a convex function and s is a scaling factor. When $\phi(z) = z^2$ we have least-squares, which is non-robust; when $\phi(z) = |z|$ we have LAV regression, which is highly non-efficient in the Gaussian case. Many M-estimators are a compromise between OLS and LAV, behaving like the former for moderate size residuals, and like the latter for large residuals. Thus, for example, one choice for ϕ (due to Huber) is given by

$$\phi(z) = \begin{cases} z^2/2, & \text{if } |z| \le c; \\ c|z| - c^2/2, & \text{if } |z| > c, \end{cases} \tag{3.12.6}$$

where c is a fixed positive constant. The tuning constant c controls the trade off between robustness and Gaussian efficiency loss.

If ϕ is a convex differentiable function and s is fixed, then the minimization problem can be solved by setting to zero the partial derivatives of $\phi(r_i/s)$ with respect to β, which yields two instructive expressions. To obtain the first of these expressions, set

$$\begin{aligned} -\frac{s}{2}\frac{\partial}{\partial \beta_j} \sum_{i=1}^{n} \phi\left(\frac{r_i}{s}\right) = 0 &= \sum_{i=1}^{n} x_{ij}\phi'\left(\frac{r_i}{s}\right) \\ &\equiv \sum x_{ij}\psi\left(\frac{r_i}{s}\right) \\ &= X_j'\psi, \end{aligned} \tag{3.12.7}$$

where X_j represents the jth column of X, and $\psi_i \equiv \psi(r_i/s)$. In matrix terms, we can write $X'\psi = 0$, which is analogous to $X'r = 0$, which is satisfied by the least-squares estimator. Thus, we can view M-estimators in regression as similar to OLS, but with the orthogonality conditions applied to a *modified* set of residuals. To compute such a solution one can proceed iteratively by approximating $\psi(r_i/s)$ by a linear function in β, say $\hat{\psi}_i$. If $\hat{\psi}_i$ is now used to estimate ψ_i in (3.12.7), the resulting equation is exactly that of an (ordinary) linear regression. Solving this equation for β gives a new value which can then be used as the starting point for the next iteration. This scheme is called the *method of modified residuals*.

Alternatively, we can obtain a second computationally useful expression by continuing from the second line in (3.12.7), writing

$$0 = \sum_i x_{ij}\psi\left(\frac{r_i}{s}\right) = \sum x_{ij}\left(\frac{r_i}{s}\right)\frac{\psi(r_i/s)}{r_i/s}$$

$$= \sum x_{ij}r_i\frac{\psi(r_i/s)}{r_i} \qquad (3.12.8)$$

$$= \sum w_i x_{ij} r_i.$$

This can be written in matrix terms as $X'Wr = 0$, or $X'WY = X'WX\beta$, so that the solution is given by $\hat{\beta} = (X'WX)^{-1}X'WY$, which is a weighted least-squares estimator. Since the weights themselves depend upon β, the natural approach to computing this solution is to use iteratively reweighted least squares (IRLS), starting from an initial estimate such as the OLS estimator or the LAV estimator. Huber (1981) notes that the method based on IRLS generally converges in fewer steps than does the method of modified residuals, but the computational effort per step in the latter is often considerably less than that in the former.

Provided that $\psi(u)/u$ is bounded and monotone decreasing for $u > 0$ and s is fixed, each step of the IRLS computation of (3.12.8) decreases (3.12.5), unless the previous value for β achieved the minimum, in which case, (3.12.5) is unchanged (Dutter, 1975).

Huber (1981) gives a detailed analysis of the computational aspects of robust regression. His discussion covers other approaches to robustness than M-estimation, as well as more general problems in which the scaling factor s must be estimated as well. Huber's Chapter 6 examines the relative merits of IRLS and modified residuals, and gives a proof of Dutter's IRLS result stated above. Coleman, et al (1980) describe a package of FORTRAN subroutines which provide a quite general package for robust regression based on IRLS; their software includes regression diagnostics and plotting routines as well as estimation programs for a variety of M-estimates based on eight different weighting functions $\psi(\cdot)$. The paper by Holland and Welsch (1977) is devoted to the topic of robust regression computation using iteratively reweighted least squares.

3.12.3 Subset selection

We have treated the regression problem as if the X matrix of predictor variables were known, and not subject to dispute. In many practical situations, however, measurements are available on a large number of similar or related variables, and the precise collection of them on which a regression model for Y should be based is not at all clear *a priori*. It is natural in such situations to try to identify a small number of X variables in terms of which most of the observed variability in Y can be described. It is this subset of the potential variables which then make up the X matrix on which regression analysis is performed. Methods for finding subsets of a set of candidate predictors which are in some sense optimal (or nearly so), are called *subset selection methods,* of which we shall briefly discuss two approaches in the least-squares context.

3.12.3.1 All-subsets regression

Suppose that there are k candidate variables from which to choose. A natural first crack at the selection problem is to compute all regressions with $p = 1$, all regressions with $p = 2$, and so on, up to the single regression with $p = k$ predictors. For subsets of a given size p, we can identify the one with highest R^2, say, and such a subset is a candidate for being called the "best subset of size p." From the computational standpoint, this requires us to compute $2^k - 1$ different regressions; even for moderate k the cost rapidly becomes prohibitive, roughly doubling for each variable added to the candidate set. Fortunately, all $2^k - 1$ regressions need not be computed explicitly, as there are algorithms which need only compute a fraction of all of the regressions in order to identify the best subsets of each size. The news is not all good, however, since even the best of known algorithms cannot stem the exponential growth of the work required as the number of input variables increases. Whether it is possible to do so remains an open question in theoretical computer science. The computing aspects of interest in the all-subsets problem are combinatorial rather than numerical, and will be deferred to Chapter 9. Interested readers can find an excellent tutorial on such methods in Hocking (1977). An algorithm to find the best subset of a given size in LAV regression is given by Armstrong and Kung (1982).

3.12.3.2 Stepwise regression

A second feature of the all-subsets solution is that it is often the case that several subsets of regressors of the same size will all have values of R^2 within a few fractions of a percent of one another. From the statistical standpoint, these subsets are essentially equivalent to one another, but some of them may be more easily interpreted than others of them. What this suggests is that it may not be worth the computational price to obtain the *best* subset of each size, provided that we can obtain a *very good* one. The methods collectively known as *stepwise regression* require an amount of work proportional to k^2, which grows much more slowly than 2^k, making such methods feasible even on data sets with very large numbers of predictors. We describe a variant of stepwise regression called *forward stepping*.

The basic idea behind stepwise regression is to begin with an initial set of predictors (say, for instance, the empty set), and then to add one variable at a time so as to produce at each step the largest possible value of R^2 consistent with adding just a single variable to the set. This produces an increasing sequence of nested subsets of predictors. Although in general this algorithm need not produce the best subset of each size—it is possible for the best sequence of subsets not to be nested—in practice the subsets produced have R^2 values which invariably are within a percent or two of the maximum possible R^2 values. Whether this must be the case is an open question.

From the computational standpoint, forward stepping is easily accomplished using the SWEEP algorithm of section 3.4. Jennrich (1977) gives a detailed description of one such implementation of a general stepwise regression program, based on the SWEEP method. Indeed, the computational advantages for stepwise methods in OLS derive from the straightforward way in which a single variable can be added to or deleted from the model without recomputing the entire regression.

COMMENT. From the statistical standpoint, stepwise regression procedures are problematical because they do not fit easily into the theoretical paradigms of mathematical statistics. They are widely used because scientists and scholars with large numbers of variables find them genuinely useful in suggesting relationships and potential models to consider *and* because they are computationally inexpensive to use. Stepwise regression is an excellent example of statistical practice leading statistical theory as a consequence of computational progress.

A few theoretical difficulties associated with subset-selection methods are worth mentioning. Although it is natural to rank subsets of a given size according to some measure of goodness-of-fit, it is by no means clear that maximizing R^2 without paying attention to other structural details of the problem produces good solutions, much less optimal ones in any sense; the R^2 criterion is used because it is both plausible and easy to compute with. When the goal is actually to identify a "correct" subset of variables, then the particular subset selected is of interest. There is currently no theory which describes the probability of subset methods finding the correct subset, in the sense of finding the subset with highest population R^2; such a theory would correspond to studying the operating characteristics of a statistical test. The problem of choosing between subsets of different size when the subsets have been determined by an optimizing procedure such as stepwise regression has been little studied. If the emphasis is on the regression coefficient estimates, there is little known about the sampling distribution of regression coefficients obtained using least-squares estimation after subset selection.

The comments of the previous paragraph should not be construed as reasons not to use stepwise methods (or best-subset methods); rather, they indicate the wide range of possibilities in providing an adequate theoretical understanding of the behavior of such methods.

3.12.3.3 Other stepwise methods

The problem of subset selection arises whenever there is a large set of candidate variables for use in prediction or classification; this includes such areas as nonlinear regression, loglinear models, GLIM models, analysis of covariance, and discriminant analysis. Many multivariate linear methods are based on the assumption of Gaussian errors, which leads directly to linear least-squares estimates. Whenever this is the case (as it is, for instance, in linear discriminant analysis), stepwise computations based on the SWEEP algorithm can fruitfully be employed.

EXERCISES

3.97. [05] In the linear-programming formulation of LAV regression, why must either $e_i^+ = 0$ or $e_i^- = 0$, or both, at the solution?

3.98. [48] Investigate methods for computing the best subset of regressor variables in the robust regression setting using M-estimation.

3.99. [50] Consider the regression of a variable Y on subsets S of the columns of an $n \times k$ matrix X with columns X_i. Investigate the minimum amount of computation required to find the sequence $S_1, S_2, \ldots, S_k$, where S_i is that subset of $X_1, X_2, \ldots, X_k$ of size i which has maximum R^2 in the corresponding regression problem. In particular, derive lower bounds on the amount of computation required as a function of k.

3.100. [50] Let $S_1, S_2, \ldots, S_k$ be defined as in the preceding exercise. Let $R_s^2(i)$ denote the value of R^2 attained when Y is regressed on the variables in S_i. Consider a stepwise regression algorithm such as forward stepwise regression which produces a nested collection of subsets, denoted by T_i, for $i = 1 \ldots k$, with corresponding values of R^2 given by $R_t^2(i)$. Derive a nontrivial lower bound for $\alpha_i = R_t^2(i)/R_s^2(i)$.

3.101. [48] Suppose that $Y = X\beta + \epsilon$, where X is $n \times k$, but that only p of the elements of β are nonzero; let S denote the set of variables corresponding to these nonzero elements. Let S_p denote the p-variable subset of the X variables determined by forward stepwise regression. Describe the probability that $S_p = S$ in terms of β. Does backward stepping change this result?

4 NONLINEAR STATISTICAL METHODS

The fundamental computational building blocks for the linear models developed in Chapter 3 are those of numerical linear algebra. The linear algebraic structure arises naturally as a consequence of fitting models for data whose mean exhibits a structure linear in the parameters to be estimated. In the case of independent Gaussian errors, the likelihood surface is quadratic, so that an explicit maximum can be obtained in closed-form, which in turn can be computed by solving a system of linear equations in the parameters. In this chapter we consider the computational requirements that arise in treating two kinds of nonlinear models. The first kind involves models for data whose means are smooth but nonlinear functions of the parameters; this is the classic example of nonlinear regression. The second kind are models for data which, despite possible linear structure, have likelihoods that are not quadratic in the parameters due to such factors as non-Gaussian errors, missing data, or dependence. The basic computational tools required for most such models are those involving optimization and solution of simultaneous nonlinear equations. These two areas of computation shall be the major focus of this chapter.

These are not the only kinds of nonlinear statistical methods, of course. Nonlinear methods whose computational flavor is primarily discrete or combinatorial include isotonic regression, all-subsets regression, dynamic programming, multiple classification analysis ("regression trees"), and clustering. Although the primary emphasis of this chapter is on unconstrained optimization, the added wrinkles associated with introducing constraints are also combinatorial in nature. We touch on some of the difficulties arising in constrained optimization in Section 4.5.7, but we defer more detailed discussion. The computational methods associated with these and other related statistical problems are discussed in Chapter 9.

Maximum-likelihood estimation and likelihood-based inference are of central importance in statistics and data analysis. For that reason, we begin the chapter with computational maximum likelihood as the context in which we discuss function maximization. When the likelihood is not quadratic, setting its derivative to zero produces a nonlinear system of equations in the parameters. After a brief general discussion of maximum likelihood estimation, we then discuss general methods for optimization and for solving nonlinear equations which can be applied in this context, first in the single-parameter case, and then in the vector-parameter case. After discussing the general methods, in each instance we show how they can then be applied in the context of important statistical problems. These sections are followed by a general discussion of function optimization methods which, although perhaps unsuitable for the maximum likelihood problem, nonetheless are indispensable in computing other nonlinear statistical methods. The chapter concludes by discussing several classes of statistical problems in which these computational methods play a prominent role: the so-called "computer-intensive" fitting techniques, estimation of missing data, and modeling time-series dependence.

As in Chapter 3, the emphasis in this chapter will be on numerical methods which are generally suited to *statistical* problems. Nonlinear optimization is exceedingly difficult in general, however, statistical problems almost always exhibit special structure that can be exploited to advantage. Our discussion will examine in greatest detail those methods which rely on such structure and will omit more general approaches.

4.1 Maximum likelihood estimation

In this section we shall examine the basic features of maximum likelihood estimation. Although the fundamental statistical ideas are the same whether the parameter of interest is a scalar or a vector, the computational consequences are rather different, and to some extent the notation must be modified slightly in the latter case. For this reason, we treat the scalar and vector cases separately.

4.1.1 General notions

Consider random variables X_1, X_2, ..., X_n which are independent and identically distributed (*i.i.d.*) according to a distribution with density $P(x \mid \theta)$, where the parameter θ is possibly a $p \times 1$ vector, and write $X = (X_1, \ldots, X_n)$. The joint density of X_1, X_2, ..., X_n is given by

$$P(x_1, \ldots, x_n \mid \theta) = \prod_{i=1}^{n} P(x_i \mid \theta) \equiv L_x(\theta). \qquad (4.1.1)$$

Each realization $x = (x_1, \ldots, x_n)$ produces a different function $L_x(\cdot)$; this function is a sufficient statistic for θ, and is called the *likelihood function* for θ. Sir Ronald Fisher proposed using the value of θ that maximizes the likelihood as a single quantity summarizing the information about θ contained in the data. This maximizing value is called, naturally enough, the *maximum likelihood estimate (MLE)* for θ, and is often denoted by $\hat{\theta}$.

From a theoretical standpoint, it is usually more convenient to focus on the logarithm of the likelihood, which we shall denote by $\ell_x(\cdot) = \ln L_x(\cdot)$. Clearly, the maxima of ℓ and L coincide. When the likelihood is formed from independent, identically distributed components, the log-likelihood is a sum of n *i.i.d.* components. If the observations have a Gaussian distribution with known variance, then the log-likelihood function is a quadratic. This is a particularly simple case with which to deal, and a standard to which other log-likelihoods can be compared. Since we examine likelihoods at fixed values of x, it is usually unnecessary to maintain the subscript x in the notation, so unless the omission would cause confusion, we write simply $\ell(\theta)$ for $\ell_x(\theta)$.

COMMENT. We have defined the likelihood in terms of a sample of independent, identically distributed random variables $X_1, \ldots, X_n$. More generally, the likelihood for a random vector X—measure-theoretic niceties aside—is simply its joint density $P(x \mid \theta)$ considered as a function of θ for each fixed x, that is, $L_x(\theta) = P(X \mid \theta) = P(X_1, \ldots, X_n \mid \theta)$.

COMMENT. If $\ell_x(\cdot)$ is the log likelihood for a parameter θ, then any function of the form $g_x(\cdot) = c(x) + \ell_x(\cdot)$ carries the same information about θ, and can be considered equivalent to the log likelihood. All of the derivatives of g coincide with those of ℓ. What is more, since x is considered fixed here, g is simply ℓ shifted by a fixed scalar. When comparing log likelihoods at different parameter values it is the difference in log likelihoods that matters, so that any added constant simply cancels (see Chapter 1 of Cox and Hinkley, 1974). For this reason, we shall identify all such functions, and we may add constants $c(x)$ depending on the observed x whenever it is convenient to do so.

4.1.2 The MLE and its standard error: The scalar case

Consider now the case in which the parameter of interest θ is a scalar parameter. The value of θ, say $\hat{\theta}$, at which the log likelihood is maximized satisfies the condition

$$\dot{\ell}(\theta) \equiv \frac{d\ell(\theta)}{d\theta} = 0, \qquad (4.1.2)$$

where here and throughout we shall denote (partial) differentiation with respect to θ by a raised dot (˙). This equation is called the *likelihood equation,* and the log-likelihood $\dot{\ell}(\theta)$ is also called the *score function.* Thus solutions of the likelihood equation coincide with roots of the score function. In the Gaussian case, $\ell(\cdot)$ is quadratic, and hence has a unique maximum. In this case, finding a solution to the likelihood equation is particularly simple, since (4.1.2) is simply a single linear equation in θ.

When $\ell(\cdot)$ is not quadratic, then the likelihood equation is a nonlinear equation, and the MLE $\hat{\theta}$ is a root of that equation. Thus we are led immediately to a study of numerical techniques for finding roots of non-linear equations in order to solve the problem of finding maximum values of nonquadratic functions. In general, $\ell(\cdot)$ may have multiple maxima, or may even have minima or points of inflection. At each such point, equation (4.1.2) is satisfied. Consequently, we must also be alert to the possibility of finding a spurious solution to the likelihood equation.

As one might expect, the second derivative of the log-likelihood function $\ddot{\ell}(\cdot)$ is of importance. Of course, a solution to the likelihood equation is a local maximum of $\ell(\cdot)$ if the second derivative at that point is negative. But $\ddot{\ell}(\cdot)$ plays a fundamental role both in statistical inference and in computation. The *Fisher information* in X about θ is defined by

$$ \mathcal{I}(\theta) = \left[-E_\theta \ddot{\ell}_X(\cdot) \right] \Big|_\theta = E_\theta \left[\dot{\ell}_X(\cdot) \right]^2 \Big|_\theta , \qquad (4.1.3) $$

where the notation E_θ denotes taking expected values over the distribution of X at the parameter value θ. The Fisher information can be thought of as the result of a two-step process. First, the average of the random functions $\ddot{\ell}_X(\cdot)$ is computed under the assumption that the true parameter governing the distribution of X is θ. Then this function average is *evaluated* at the same value of θ.

Under general conditions on the form of the likelihood, when the true value of θ is θ_0, the asymptotic variance of $\hat{\theta}$ is given by

$$ \text{asy var}_{\theta_0}(\hat{\theta}) = \left[-E\,\ddot{\ell}_X(\cdot) \right]^{-1} \Big|_{\theta_0} = \mathcal{I}(\theta_0)^{-1}, \qquad (4.1.4) $$

where the expectations are taken with respect to the probability distribution of X at the true value of θ, namely θ_0, and the resulting functions are evaluated at the true value of θ as well. The notation of (4.1.4) is somewhat cumbersome but precise; note that the middle term is a *function* which is then *evaluated* at θ_0, emphasizing the two-step process. It is convenient, if somewhat imprecise, to write the function as $E\ddot{\ell}(\theta)$, so that we could then write $\mathcal{I}(\theta) = -E\ddot{\ell}(\theta)$.

Were θ_0 known, a ready-made indicator of the precision of the MLE $\hat{\theta}$ would be its asymptotic standard deviation $1/\sqrt{\mathcal{I}(\theta_0)}$. There are two candidates to use in practical terms as a standard error for $\hat{\theta}$, of which the more obvious is based on the *estimated Fisher information:* $[\mathcal{I}(\hat{\theta})]^{-1/2}$. Note, however, that the asymptotic variance of the MLE is related to an average of the curvature of the likelihood function at θ_0. It may well be that the particular likelihood we have observed gives a better indication of the precision of $\hat{\theta}$ for the particular data we have observed, which suggests a standard error based on the *observed information:* $[-\ddot{\ell}(\hat{\theta})]^{-1/2}$. Kendall and Stuart (1979) have an extended discussion of the statistical properties of these quantities. Methods for finding solutions of the likelihood equations numerically also depend upon the curvature of the likelihood, and both the estimated Fisher information and the observed information play roles in these numerical methods.

4.1.3 The MLE and the information matrix: The vector case

We now consider the more common case in which the parameter of interest θ is a $p \times 1$ vector $(\theta_1, \ldots, \theta_p)$. Once again, when the log likelihood achieves its maximum value, the likelihood equation of (4.1.2) is satisfied, provided that we interpret $\dot{\ell}(\theta)$ to be the vector of partial derivative functions of $\ell(\cdot)$ with respect to the components of θ, that is,

$$\dot{\ell}(\theta) \equiv \left(\frac{\partial \ell(\theta)}{\partial \theta_1}, \frac{\partial \ell(\theta)}{\partial \theta_2}, \ldots, \frac{\partial \ell(\theta)}{\partial \theta_p} \right)'. \tag{4.1.5}$$

The likelihood equation then is a vector equation representing a system of nonlinear equations in the p unknowns θ_k. In the case of a Gaussian mean, $\ell(\cdot)$ is a quadratic form in the parameters, and has a unique maximum, which can be obtained by solving the system of likelihood equations, which would then be linear in θ.

As in the scalar case, the second derivatives are important both statistically and numerically. We shall continue to use the notation $\ddot{\ell}(\cdot)$ when θ is a vector; in this case it denotes the matrix of mixed second partial derivatives with i, j position given by

$$\ddot{\ell}(\theta)_{ij} = \frac{\partial^2 \ell(\theta)}{\partial \theta_i \partial \theta_j}. \tag{4.1.6}$$

This is a matrix-valued function of θ. For those points at which $\ell(\theta)$ achieves a local maximum, $\dot{\ell}(\theta) = 0$, and $\ddot{\ell}(\theta)$ is negative definite. In the vector case, matters are complicated by the fact that a zero of the likelihood may be neither a local maximum nor minimum, but a saddle point—a local maximum with respect to one direction in $\mathcal{R}^p$ and a local minimum with respect to another. In these cases the matrix $\ddot{\ell}(\theta)$ is other than negative definite.

The Fisher information, or *information matrix*, for θ is defined as in (4.1.3) to be

$$I(\theta) = \left[-E_\theta \ddot{\ell}_X(\cdot) \right]\Big|_\theta = E_\theta \left[\dot{\ell}_X(\cdot) \right] \left[\dot{\ell}_X(\cdot) \right]'\Big|_\theta. \tag{4.1.7}$$

Under general conditions, when the true value of θ is θ_0, the asymptotic variance matrix of the MLE $\hat{\theta}$ is given by the inverse of the information matrix; so that (4.1.4) holds in the vector case as well. As in the scalar case, standard errors are obtained by taking square roots of the diagonal elements of the inverse of an estimate for $I(\theta_0)$, either the estimated Fisher information $I(\hat{\theta})$ or the observed information $-\ddot{\ell}(\hat{\theta})$.

EXERCISES

4.1. [10] Show that $L_x(\cdot)$ is a sufficient statistic for θ.

4.2. [05] Show that, if $X = (X_1, \ldots, X_n)$, where the components of X are independent, and identically distributed, that $\ell_X(\theta)$ can be written as the sum of n i.i.d. components.

4.3. [14] When $X_1, \ldots, X_n$ are i.i.d. random variables, show that $\mathcal{I}(\theta)$ is equal to $n \cdot i(\theta)$, where $i(\theta)$ is the Fisher information for θ in a single observation X_i.

4.4. [17] The density for the Cauchy distribution location family is given by $P(x \mid \theta) = 1/[\pi\{1 + (x - \theta)^2\}]$. Let $X = (X_1, \ldots, X_n)$. What are $\ell_X(\theta)$, $\ddot{\ell}_X(\theta)$, and $\mathcal{I}(\theta)$?

4.2 Solving $f(x) = 0$: x scalar

As the previous section indicates, maximum likelihood estimation of a single parameter involves either directly maximizing $\ell(\cdot)$ (discussion of which we defer to section 4.5), or finding a solution to the likelihood equation (4.1.2), which in general is nonlinear in θ. This latter approach leads us to consider numerical methods for finding the roots of nonlinear equations. Although single parameter problems occasionally arise in statistical practice, simultaneous estimation of several parameters is much more common. The single-parameter case is of interest, however, for three reasons. First, maximum likelihood estimation of a single parameter does arise frequently in investigations of statistical theory and in Monte Carlo studies. Second, an important method for finding maximum likelihood estimates of several variables—variously called *cyclic ascent, Gauss-Seidel, or backfitting*—involves maximizing a conditional likelihood in one parameter at a time. Third, several multivariable root-finding methods are generalizations of the simpler single-variable techniques.

We begin this section with an investigation of several common methods for solving a one-parameter nonlinear equation. Rice (1983) gives a comprehensive survey of such methods. For consistency with the numerical literature, and to emphasize that these are general methods quite apart from the statistical contexts in which they arise, we shall denote by $f(x) = 0$ a nonlinear equation in x whose roots we wish to find. All of the methods we examine are iterative, that is, they produce a sequence of values x_1, x_2, $\ldots$, which converges (one hopes!) to the solution. After discussing the basic algorithms themselves, therefore, we discuss the collateral issues of choosing a starting value for this sequence, and of deciding when to stop the iteration and to accept the current value x_i. Following the discussion of general methods, we shall then apply them to some important statistical examples.

A note on the computations. In this section we provide numerical examples to illustrate the methods discussed. The computations were done using double-precision arithmetic conforming to the IEEE p754 Binary Floating Point standard, discussed in section 2.4. The convergence criterion used in each example is the same—a relative error in the solution of less than 10^{-6}.

4.2.1 Simple iteration

The simplest and often easiest method for solving $f(x) = 0$ is to solve a related fixed-point equation. The method to be described here can be used when $f(x)$ is continuous. First, reformulate $f(x) = 0$ in the form $g(x) = x$. From a starting value x_0, compute $x_i = g(x_{i-1})$ until convergence is attained. By "reformulate" here, we mean to rearrange the form of the basic likelihood equation until it has the desired form $g(x) = x$; this can always be done in several different ways. The iteration is defined by Algorithm 4.2.1.

Algorithm 4.2.1 (Simple iteration):

Find a function $g(x)$ for which $g(x) = x \implies f(x) = 0$
$x_0 :=$ initial-value
for $i := 1$ **to** ∞ **until** convergence **do**
    $x_i := g(x_{i-1})$.

The most obvious reformulation is to set $g(x) = x + f(x)$, so that $x_i = x_{i-1} + f(x_{i-1})$. It is instructive to examine this iteration in detail. Written this way, the adjustment to the current iterate x is simply $f(x)$. Under what conditions will successive application of this adjustment lead to convergence? A necessary condition is that f be defined when evaluated at each iterate. Thus, we must require that for some possibly infinite interval $I = [a, b]$, $x \in I$ implies that $g(x) = x + f(x) \in I$. What is more, $f(x)$ cannot be changing too rapidly near the solution, otherwise, the adjustment will move us *away* from the true solution. The definition of a Lipschitz condition captures this notion. A function $g(x)$ is said to satisfy a *Lipschitz condition* on I with index L if, for any $s, t \in I$, $|g(s) - g(t)| \leq L|s - t|$.

The following theorem guarantees existence and uniqueness of a solution, as well as convergence, of the simple iteration scheme. The proof is left as an exercise.

Theorem 4.2.1. *Let $g(x)$ be a continuous function defined on the interval $I = [a, b]$ such that $g(x) \in I$ whenever $x \in I$, and satisfying a Lipschitz condition with $L < 1$. Then for any $x_0 \in I$, the sequence defined by $x_i = g(x_{i-1})$ converges to the solution s of the equation $g(x) = x$, and the solution is unique.*

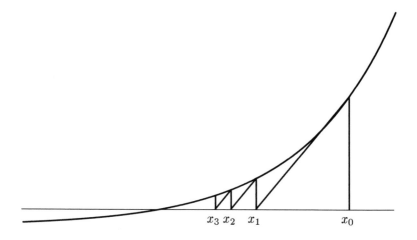

FIGURE 4.2.1 *Rescaled simple iteration, or the method of parallel chords. At each step, the next iterate is obtained by drawing a chord having a given constant slope through the point $(x, f(x))$ and then computing the point of intersection with the horizontal axis.*

A necessary and sufficient condition for a differentiable function g to satisfy the necessary Lipschitz condition is that $|g'(x)| \leq L$ on I for some constant $L < 1$. There is no guarantee that the function f whose root we seek will satisfy $|1 + f'(x)| < 1$ on $[a, b]$. Note, however, that if f' is bounded and does not change sign, we can rescale the problem. If $f(x) = 0$ is satisfied then so is $\alpha f(x) = 0$, and the latter equation can be solved using simple iteration applied to $g(x) = x + \alpha f(x)$, provided only that α is chosen so that $|1 + \alpha f'(x)| < 1$. We refer to this method as *rescaled simple iteration*, or the *parallel chord* method. Figure 4.2.1 illustrates the way in which a convergent iteration proceeds. Successful rescaling may require some knowledge about the function and its derivative.

The basic equation $f(x) = 0$ can sometimes be reformulated using other functions for $g(x)$ than $x + \alpha f(x)$. It is usually the case that $f(x)$ can be rewritten by focusing on one occurrence of the variable x in the expression for $f(x)$ and "solving" the equation for this particular occurrence of the symbol x. An example will make the method clearer. Let

$$f(x) = -3062(1 - \xi)e^{-x}/[\xi + (1 - \xi)e^{-x}] - 1013 + 1628/x, \qquad (4.2.1)$$

where ξ is considered here to be a fixed constant. This is the derivative of a log-likelihood function to be examined in detail later.

In addition to performing simple iteration with $g_1(x) = x + f(x)$, by "solving" for the x in the right-most term of (4.2.1) we can rewrite the equation $f(x) = 0$ as

$$x = \frac{1628[\xi + (1 - \xi)e^{-x}]}{3062(1 - \xi)e^{-x} + 1013[\xi + (1 - \xi)e^{-x}]} \equiv g_2(x), \qquad (4.2.2)$$

and can use this expression for the fixed-point iteration. Similarly, we could solve for e^{-x} in the expressions above and then (after taking logarithms and negating) use the resulting expression $g_3(x)$ in yet another alternative iteration. In fact, of these three possibilities, only the iteration using g_2 defined by (4.2.2) converges. The simple iteration using $g(x) = x + f(x)$ fails to converge because the Lipschitz condition fails; this can be remedied by rescaling. With $\alpha = 1/1000$, convergence is moderately rapid; this is illustrated in the table below. Solving for e^{-x} and iterating produces an inherently unstable iteration which diverges rapidly. Starting with an initial value of $x_0 = 2$, and using the value $\xi = 0.61489$, the iterations produce the results in Table 4.2.1.

	$g_2(x) =$ Equation (4.2.2)		$g_4(x) = x + f(x)/1000$	
i	x_i	$f(x_i)$	x_i	$f(x_i)$
0	2.00000000	-438.259603329	2.00000000	-438.259603329
1	1.30004992	-207.166772208	1.56174040	-326.146017453
2	1.11550677	-75.067503354	1.23559438	-166.986673341
3	1.06093613	-24.037994642	1.06860771	-31.623205997
4	1.04457280	-7.374992947	1.03698450	0.583456921
5	1.03965315	-2.232441655	1.03756796	-0.033788786
6	1.03817307	-0.673001091	1.03753417	0.001932360
7	1.03772771	-0.202634180	1.03753610	-0.000110591
8	1.03759369	-0.060988412	1.03753599	0.000006329
9	1.03755336	-0.018354099	1.03753600	-0.000000362
10	1.03754122	-0.005523370	1.03753600	0.000000021
11	1.03753757	-0.001662152	1.03753600	-0.000000001
12	1.03753647	-0.000500191	1.03753600	0.000000000
13	1.03753614	-0.000150522	1.03753600	-0.000000000
14	1.03753604	-0.000045297	1.03753600	0.000000000

TABLE 4.2.1 *Convergence of two methods for fixed-point iteration. The first pair of columns gives the result after reformulating the basic equation (4.2.1) in the form of equation (4.2.2), with $\xi = 0.61489$. The second pair gives the result for a scaled version of simple iteration on the basic equation. Both of these versions are extremely stable, although convergence is not rapid when x_0 is far from the solution s.*

Simple iteration using the basic formulation $g(x) = x + f(x)$ often works in the contexts that arise in computational statistics, and it is generally the easiest method to implement in a computer program. When $g(x)$

is appropriately formulated, either through scaling or rewriting the basic equation, fixed-point iteration can be extremely stable. For the equation in Table 4.2.1, for instance, iteration using both g_2 and g_4 converge when the starting value is $x_0 = 50$ or more which, in this example, corresponds to a starting value several thousand standard errors away from the correct value. (Convergence is rather slow for simple iteration with rescaling, however, with g_4 requiring 62 iterations to achieve eight-digit accuracy in the result. Iteration using g_2 still requires only 15 iterations in this case.)

Because it is easy to program and relatively stable with rescaling, fixed-point iteration is worth knowing about and trying; if convergence fails, no more than five minutes may have been lost. It has many difficulties, however, which make it unsuitable for general use. Primary among these difficulties are the Lipschitz condition required for convergence, the lack of information concerning the second derivative, and a rate of convergence which can be painfully slow.

4.2.2 Newton-Raphson

The simple iteration method discussed in the last section requires $f(x)$ to be continuous and to satisfy a Lipschitz condition restricting its variability so that near the solution to $f(x) = 0$, f must be fairly flat, having slope between zero and -2. The *Newton-Raphson* iteration, or *Newton's method*, also makes some requirements on $f(x)$ which seem at first to be more restrictive, but which are commonly satisfied in practice. What is more, under general conditions the Newton-Raphson iteration converges quadratically. These two facts make the Newton-Raphson method one of the most widely known and used procedures.

The order of convergence of an iteration is defined as follows. Denote the solution being sought by s, and denote the error at step i by $\epsilon_i = |x_i - s|$. Then the sequence $x_1, x_2, \ldots$ is said to exhibit *convergence of order β* if $\lim_{i \to \infty} \epsilon_{i+1} = c\epsilon_i^\beta$, for some non-zero constant c. When $\beta = 2$ the method is said to exhibit quadratic convergence, since the error at step $i + 1$ is proportional to the square of the error at the previous step. The order of convergence for simple iteration is generally linear ($\beta = 1$).

Newton's method requires $f(x)$ to be twice continuously differentiable and $f' \neq 0$ at the solution s. At the ith stage in the iteration, the function is approximated by the line tangent to it at x_i. The point at which the tangent line crosses the horizontal axis is taken to be the next value in the iteration, as illustrated in Figure 4.2.2. We derive the method by expanding $f(x)$ in a Taylor series about the current iterate, writing

$$0 = f(s) = f(x_i) + (s - x_i)f'(x_i) + \frac{(s - x_i)^2}{2}f''(x^*),$$

where x^* is a point between s and x_i. When x_i is sufficiently close to

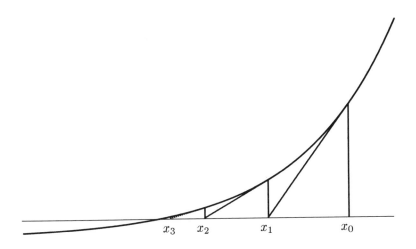

FIGURE 4.2.2 *Newton's method for solving a nonlinear equation. The next iterate at each step is obtained from the point of intersection of the horizontal axis with the tangent to $f(\cdot)$ through the point $(x, f(x))$.*

s, the remainder term will be small relative to the other terms (provided that $f'(s) \neq 0$). Dropping the remainder, then, provides an approximation which can be rearranged to obtain

$$s \approx x_i - \frac{f(x_i)}{f'(x_i)},$$

which is the basis for the Newton-Raphson algorithm given below.

Algorithm 4.2.2 (Newton-Raphson iteration):

$x_0 :=$ initial-value
for $i := 0$ **to** ∞ **until** convergence **do**
$\quad x_{i+1} := x_i - f(x_i)/f'(x_i).$

This algorithm requires that the starting value be sufficiently close to the solution to guarantee convergence. We illustrate the method using the example of section 4.2.1, and for contrast, we show the performance of the simple (scaled) iteration as well. The results in Table 4.2.2 use a starting value of $x_0 = 1.5$ instead of the value $x_0 = 2$ used in the earlier example of simple iteration because the Newton-Raphson iteration diverges from that point. This illustrates the dependence of Newton-Raphson on a starting value that is sufficiently close to the solution. Note that the scaled simple iteration does better than Newton-Raphson initially, but the quadratic convergence takes over near the solution, approximately doubling the number of correct digits at each step from iteration 3 onward.

	Newton-Raphson		$g_4(x) = x + f(x)/1000$	
i	x_i	$f(x_i)$	x_i	$f(x_i)$
0	1.50000000	-303.107943850	1.50000000	-303.107943850
1	0.73097994	504.767684105	1.19689206	-140.029381516
2	0.93274493	126.993377592	1.05686267	-19.952291505
3	1.02440139	14.115961892	1.03691038	0.661929931
4	1.03732279	0.225472101	1.03757231	-0.038393959
5	1.03753594	0.000059670	1.03753392	0.002195521
6	1.03753600	0.000000000	1.03753611	-0.000125653
7	1.03753600	0.000000000	1.03753599	0.000007191

TABLE 4.2.2 *Convergence of the Newton-Raphson iteration. The first pair of columns gives the results of Newton-Raphson iteration on equation (4.2.1) with $\xi = 0.61489$. The second pair of columns gives the results of simple iteration with rescaling, for comparison.*

COMMENT. We can consider rescaled simple iteration to be a modification of Newton's method in which the derivative $f'(x)$ is replaced by a constant. The optimal value for this constant is $f'(s)$, corresponding to $\alpha = -1/f'(s)$. This reduces the Lipschitz index L nearly to zero in a neighborhood of the solution. The choice of $\alpha = 1/1000$ is particularly fortuitous in this case, since $f'(s) \approx -1057$. In this case, nearly quadratic convergence can be achieved.

COMMENT. When Newton steps are expensive due to the complexity of evaluating the derivative, the previous comment suggests a hybrid method that is much less expensive. After each Newton step is taken, several rescaled fixed-point steps are taken using $\alpha = -1/f'(x)$, where x is the most recent point at which f' has been evaluated.

The advantage of Newton-Raphson is its rapid convergence. In addition, because it uses information about the derivative of f, that information —which in maximum likelihood estimation can be used to obtain a standard error for a parameter estimate—is available at the end of the computation. It is not always easy to write down the derivative of f in closed form, however, and sometimes it is not possible at all. It is becoming increasingly common to investigate statistical estimators whose likelihood function can be specified *as an algorithm*, but not as a conventional closed-form mathematical expression. In such cases Newton's method is unavailable. Another drawback is that the computational cost per Newton-Raphson iteration can be quite high, particularly if the derivative of f is substantially more complex than is f itself. In Table 4.2.2, for instance, the cost of Newton-Raphson iterations take almost twice as long to compute as do those of rescaled simple iteration. While computational efficiency is rarely an issue when computing the solution to a single equation—the iterations of Table 4.2.2 took less than a second on a personal computer using either method—they can become paramount when the iteration process is repeated many thousands of times as, for instance, they are in a Monte Carlo study of alternative estimators.

4.2.3 The secant method (regula falsi)

When the derivative of $f(x)$ is either hard or impossible to write down (and hence, to program), or when the computational effort required to evaluate $f'(x)$ is very large compared to that for $f(x)$, Newton-Raphson iteration is impossible or costly to carry out. An alternative is to approximate the derivative by a finite difference, that is, to write

$$f'(x_i) \approx \frac{f(x_i) - f(x_{i-1})}{x_i - x_{i-1}}.$$

The approximate Newton-Raphson iteration can then be expressed in the following algorithm.

Algorithm 4.2.3 (Secant method):

Choose x_0, x_1
for $i := 1$ to ∞ until convergence
$x_{i+1} := x_i - f(x_i)(x_i - x_{i-1})/[f(x_i) - f(x_{i-1})].$

This iteration is called the *secant method* because it approximates the function $f(x)$ by the secant line through two successive points in the iteration, rather than the tangent at a single point used in the Newton-Raphson iteration, as illustrated in Figure 4.2.3. This method is also known as *regula falsi*, the "method of false position," although we shall not use this term. The secant method is easy to program. It provides an approximation to the derivative which is quite good, provided that the two iterates on which it is based are close to one another. It has the disadvantage of requiring two starting values rather than one, it is as sensitive as Newton-Raphson to the choice of these values, and it makes the same assumptions on f that Newton's method does. Using the example of the previous sections, the secant iteration produces the results in Table 4.2.3.

COMMENT. The term *regula falsi* is used in many texts on numerical analysis to refer to specific secant-like techniques. Unfortunately, differ-

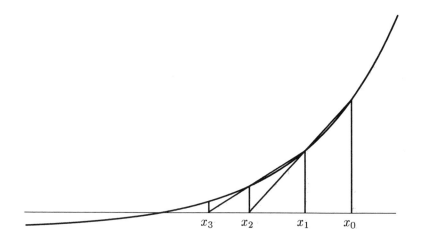

FIGURE 4.2.3 *The secant method for solving a nonlinear equation. At each step the next iterate is obtained from the intersection of the horizontal axis with the secant line connecting the two most recently computed points on the graph of $f(\cdot)$.*

	Newton-Raphson		Secant method		Muller's method	
i	x_i	$f(x_i)$	x_i	$f(x_i)$	x_i	$f(x_i)$
0	1.5000000	-303.1079438	1.5000000	-303.1079438	1.5000000	-303.1079439
1	0.7309799	504.7676841	1.4900000	-299.1303510	1.4900000	-299.1303510
2	0.9327449	126.9933776	0.7379614	487.4959303	1.4800000	-295.0793033
3	1.0244014	14.1159619	1.2040223	-145.1703250	0.9814545	63.6802300
4	1.0373228	0.2254721	1.0970809	-58.5768607	1.0459031	-8.7548953
5	1.0375359	0.0000597	1.0247395	13.7467475	1.0377325	-0.2076831
6	1.0375360	0.0000000	1.0384896	-1.0070151	1.0375361	-0.0001350
7	1.0375360	0.0000000	1.0375511	-0.0159851	1.0375360	0.0000000
8	1.0375360	0.0000000	1.0375360	0.0000189	1.0375360	0.0000000
9	1.0375360	0.0000000	1.0375360	-0.0000000	1.0375360	0.0000000

TABLE 4.2.3 *Convergence of the secant method and Muller's method. Using equation (4.2.1) with $\xi = 0.61489$, the Newton-Raphson, secant, and Muller methods are compared. The secant method requires two starting values. By choosing the second initial value($x_1 = 1.49$) close to the first, x_2 from the secant iteration is very close to x_1 from Newton-Raphson. Because x_1 and x_2 are far apart, however, the secant and the tangent are quite different, so that subsequent iterates are not close. Muller's method requires three starting values. It can be considered a compromise between the secant method and Newton's method.*

ent authors have used this term for at least three related, but slightly different, techniques. The most common usage, and the one that we shall adopt, considers the term *regula falsi* to be a synonym for the secant method. Other algorithms which are occasionally called *regula falsi* include a variant of bisection (discussed in section 4.2.4) in which the two endpoints are known to bracket a root of $f(x)$, and a variant of the secant method in which one end-point of the secant is fixed, say at x_0.

The secant method can be thought of in the following way. Two points lying on the curve $y = f(x)$ are found. These points determine a straight line, and the root of the equation determined by the line is taken to be the next iterate. This suggests a generalization of the secant method which takes three points on $y = f(x)$ determined by the most recent three iterates, fits a quadratic curve, and then uses a root of the resulting quadratic equation as the next iterate. The root chosen is the one closest to the most recent iterate. This procedure is known as *Muller's method,* an example of which is illustrated in Table 4.3.3. Muller's procedure requires continuous third derivatives near s. Muller (1956) shows that the order of convergence for the secant method is 1.618, and for Muller's method is 1.839, which approaches the second-order convergence of Newton-Raphson. What is more, successive iterates can be computed using only the previously computed values of $f(x_i)$, so that each iteration is little more expensive than an iteration of the secant or fixed-point procedures. These features make it sufficiently attractive to mention, but the added complexity of the procedure makes it more difficult to program, and this added complexity rarely compensates for the faster convergence. It has, in addition, the property that the roots of the quadratic equation at some step may be complex rather than real. In most statistical contexts this is an annoyance, although occasionally it is an advantage when a polynomial with possibly complex roots is being solved. Muller's algorithm is given below in the real case.

4.2.4 Bracketing methods

The methods for solving $f(x) = 0$ discussed up to this point have essentially required that $f'(x)$ exist and be well-behaved. When the behavior of the first derivative is either unknown or unpleasant, other numerical methods based on *bracketing* can be used. The basic bracketing method requires that two points $x_0 < x_1$ be known for which $f(x_0)f(x_1) < 0$, and that f be continuous on $[x_0, x_1]$. These conditions imply that f has a root in the interval $[x_0, x_1]$, and the various algorithms proceed by systematically reducing the length of the interval in which the root is known to lie. The simplest such method is *bisection* (Algorithm 4.2.5).

The bisection method divides the current interval at its midpoint, and takes as the next interval the half which continues to bracket a root. Bi-

Algorithm 4.2.4 (Muller's method):

Choose x_0, x_1, x_2
$f_0 := f(x_0)$
$f_1 := f(x_1)$
$d_1 := f_1 - f_2$
for $i := 2$ **to** ∞ **until** convergence
  $f_i := f(x_i)$
  $d_i := f_i - f_{i-1}$
  $dd := (d_i - d_{i-1})/(x_i - x_{i-2})$
  $s := d_i + (x_i - x_{i-1})/dd$
  $x_{i+1} := x_i - 2f_i/(s + \text{sgn}(s)\sqrt{s^2 - 4f_i \cdot dd})$

section has the discouraging feature that convergence, while guaranteed, is often excruciatingly slow. A modification applies the secant method (*regula falsi*) to the bracketing interval, drawing the secant connecting $(L, f(L))$ and $(U, f(U))$, and then splitting the interval at the point where the secant intersects the horizontal axis. (Some authors use the term *regula falsi* only in the context of reducing an interval which brackets a root.) This method, which we shall refer to as the *secant-bracket method* (Algorithm 4.2.6), speeds convergence at the initial steps. However, as soon as the interval is reduced to one on which $f(x)$ is either concave or convex, one of the two endpoints of the interval becomes fixed in subsequent iterations, producing only linear convergence.

Algorithm 4.2.5 (Bisection):

Choose $x_0 < x_1$ such that $f(x_0)f(x_1) < 0$
$L := x_0$
$U := x_1$
for $i := 1$ **to** ∞ **until** convergence
  $x_{i+1} := (L + U)/2$
  if $f(x_{i+1})f(L) < 0$ **then** $U := x_{i+1}$
  else $L := x_{i+1}$

A further refinement has become known as the *Illinois method* (Algorithm 4.2.7), its origin attributed to the staff of the computation center at the University of Illinois in the early 1950's (Jarratt, 1970). If the unmodified endpoint (say U) is the same one as in the previous iteration, the Illinois method uses a line segment connecting the new endpoint $(L, f(L))$ with $(U, f(U)/2)$.

Algorithm 4.2.6 (Secant-bracket method):

Choose $x_0 < x_1$ such that $f(x_0)f(x_1) < 0$
$L := x_0$
$U := x_1$
for $i := 1$ **to** ∞ **until** convergence
   $x_{i+1} := [Lf(U) - Uf(L)]/[f(U) - f(L)]$
   if $f(x_{i+1})f(L) < 0$ **then** $U := x_{i+1}$
   else $L := x_{i+1}$

Algorithm 4.2.7 (Illinois method):

Choose $x_0 < x_1$ such that $f(x_0)f(x_1) < 0$
$L := x_0$
$U := x_1$
$FL := f(L)$
$FU := f(U)$
for $i := 1$ **to** ∞ **until** convergence
   $x_{i+1} := [L \cdot FU - U \cdot FL]/[FU - FL]$
   if $f(x_{i+1}) \cdot FL < 0$ **then**
      $U := x_{i+1}$
      $FU := f(U)$
      if $f(x_i)f(x_{i+1}) > 0$ **then** $FL := FL/2$
   else $L := x_{i+1}$
      $FL := f(L)$
      if $f(x_i)f(x_{i+1}) > 0$ **then** $FU := FU/2$

The Illinois method can be shown to have order of convergence 1.442 (Jarratt, 1970), which is superior to the rate of 1 for either bisection or secant-bracket, and nearly as good as the secant method (at 1.618). Table 4.2.4 compares the three bracketing methods on the example of section 4.2.1.

4.2.5 Starting values and convergence criteria

All of the methods that we have discussed for finding a root of $f(x) = 0$ have been iterative. A fully specified algorithm based on an iterative scheme must have three components: a method for deciding on a starting value for the iteration x_0, a method for obtaining the next iterate from its predecessors, and a method for deciding when to stop the iterative process. The preceding sections examine ways to generate the iterative sequence itself; the starting values and stopping criteria are treated as given. In real

	Bisection		Secant-bracket		Illinois	
i	x_i	$f(x_i)$	x_i	$f(x_i)$	x_i	$f(x_i)$
0	1.5000000	-303.1079439	1.5000000	-303.1079439	1.500000	-303.1079439
1	1.0000000	41.6103765	1.0000000	41.6103765	1.000000	41.61037650
2	1.2500000	-176.4527128	1.0603542	-23.4567728	1.060354	-23.45677280
3	1.1250000	-83.2881067	1.0385964	-1.1196486	1.038596	-1.11964857
4	1.0625000	-25.5957894	1.0375851	-0.0519056	1.036625	0.96379295
5	1.0312500	6.6979564	1.0375383	-0.0024030	1.037537	-0.00126763
6	1.0468750	-9.7601013	1.0375361	-0.0001112	1.037536	-0.00000143
7	1.0390625	-1.6108105	1.0375360	-0.0000052	1.037536	0.00000143
8	1.0351563	2.5233892	1.0375360	-0.0000002	1.037536	0.00000000
9	1.0371094	0.4512749	1.0375360	-0.0000000	1.037536	0.00000000
10	1.0380859	-0.5810175	1.0375360	-0.0000000	1.037536	0.00000000
				etc.		
19	1.0375347	0.0013554	1.0375360	0.0000000	1.037536	0.00000000
20	1.0375357	0.0003471	1.0375360	0.0000000	1.037536	0.00000000

TABLE 4.2.4 *Convergence of three bracketing methods. The example used is $f(x)$ from equation (4.2.1) with $\xi = 0.61489$. Note that the secant-bracket sequence is monotone decreasing from iteration 2 onward. This indicates that the left endpoint is fixed (at 1.00) from the start of the iteration. The Illinois method detects this and makes a correction at iteration 4, leading to improved estimates from iteration 5 onward.*

life, that is simply not the case. This section discusses the practicalities of selecting initial values and determining convergence.

Starting values matter in two ways. First, if the initial value for the iteration is too far away from a solution, the iteration can diverge; an example of this behavior was mentioned in the context of Newton's method. Second, it is possible that the equation $f(x) = 0$ has multiple roots. In this case, the root to which the sequence converges will depend upon the starting value for the iteration. Unfortunately, there is little constructive theory about choosing starting values, however a few words of general advice can be given. Much of the time, the equation being solved is similar to another equation whose roots are easily obtained, so that a root of the latter can be used as a starting point for the equation of actual interest. In statistical problems, particularly when $f(x)$ is the the score function for a parameter, a preliminary estimator such as a moment estimator can be used to get sufficiently close for the iteration to converge rapidly. Sometimes there is no alternative to starting with a guess or two and to observing the progress of the iteration, hoping that it will be possible to adjust the starting value or the iteration method as needed to achieve convergence. A useful approach in general is to graph the function. This can often provide not only good starting values, but also some insight concerning an appropriate form for the iteration.

The problem of detecting multiple roots, and of settling on the appropriate one when more than one root is found, is a difficult one. Probably the most successful general approach for discovering whether there are multiple roots is to start the iteration several times, from the vicinity of possible multiple solutions if enough is known about the function, or using randomly chosen starting points otherwise. In the context of general-purpose statistical software or in that of large-scale Monte-Carlo computations—in which a large number of different equations having the same general structure must be solved—a thorough understanding of the particular class of problems being solved is the only route to a satisfactory scheme for selecting x_0.

There is more to say about stopping an iteration than starting one, but the problem is by no means an easier one to deal with completely. Rice (1984) has a lengthy discussion of alternatives; some of the more common ones are discussed here. There are two reasons for bringing an iteration to a halt—either the iteration has converged, or it has not. Since the solution of the equation is not known explicitly, the decision as to whether an iteration has converged is based on monitoring either the sequence of iterates to see if x_i is sufficiently close to x_{i-1}, or the sequence of function evaluations $f(x_i)$ to see if these become sufficiently close to zero. The two most common definitions for successive iterates to be "sufficiently close" are embodied in the *absolute convergence criterion*, which asserts convergence when $|x_i - x_{i-1}| < tol$, and the *relative convergence criterion*, which asserts convergence when $|(x_i - x_{i-1})/x_{i-1}| < tol$, where *tol* is a preselected tolerance. The absolute convergence criterion is most suitable when the solution s is close to zero; in this case the denominator of the relative criterion can foster numerical difficulties. On the other hand, when s is large the relative criterion is generally more satisfactory. Indeed, if the floating-point representation of $s + tol$ is the same as that of s, then it is possible for the absolute criterion never to be satisfied.

It is also possible for $|x_i - x_{i-1}|$ to be small even though $|f(x_i)|$ is not close to zero, however the reverse situation, in which f is nearly zero despite relatively large differences in x, is more common. Hence, it is advisable to monitor the size of the function as well as the relative size of the iterates in deciding whether convergence has been achieved.

There is a subtle point that arises when deciding how to stop an iterative process. The value for x_i is computed from the values already obtained of x_{i-1}, $f(x_{i-1})$ and possibly an estimate for $f'(x_{i-1})$. We can test for convergence without having computed $f(x_i)$ and $f'(x_i)$. If x_i passes the convergence test, should we evaluate f and f', or should we simply use the most recent values as sufficiently good approximations? While it is often true that $f(x_i)$ is acceptably close to $f(x_{i-1})$, the corresponding statement is not true for f', particularly when f' is estimated from finite differences. It is *always* advisable to evaluate f and f' afresh at the final value of the iterate. Failure to do so in maximum-likelihood computations can produce serious inaccuracies in standard errors, which are obtained from derivative estimates.

When has an iteration failed to converge? A practical answer is to say that convergence fails once some prespecified number of iterations has been computed without satisfying the convergence criterion. This has the effect of placing a maximum computational cost on a single attempt to find a root. The primary disadvantage of this approach is that a rapidly divergent iteration can cause floating-point overflow before the iteration limit is reached, preventing the computer program from gracefully dealing with the problem, say by restarting the iteration at a different point. A divergent iteration is fairly easy to detect by comparing current values for $f(x_i)$ to that of $f(x_0)$. The difficult case to detect is an iteration which *cycles*. This can occur, for instance, if $f''(x)$ is ill-behaved near s when Newton-Raphson iteration or the secant method is used.

4.2.6 Some statistical examples

In this section we examine some nonlinear equations which arise in statistical contexts and aspects of their numerical solution. In each case, we shall pay some attention to choice of starting values, iteration method, and convergence tests.

These examples all involve maximum likelihood estimation of a single parameter θ, so that the purpose of solving the nonlinear equation is to find a zero of the derivative of the log-likelihood function, that is, $\dot{\ell}(\cdot)$ will play the role that $f(\cdot)$ does in the preceding sections. This root $\hat{\theta}$ will correspond to a maximum of the likelihood provided that the second derivative is negative at the solution. In addition $-\ddot{\ell}(\hat{\theta})$ can be used as a standard error for $\hat{\theta}$, so that for both numerical and statistical reasons it will be desirable to gain some information about $\ddot{\ell}(\cdot)$.

In section 4.2.2 we noted that rescaled simple iteration can be viewed as an approximation to Newton's method when $\alpha = -1/f'(s)$ or, in the maximum-likelihood context, $\alpha = -1/\ddot{\ell}(\theta_0)$. The *method of scoring* is a compromise between rescaled fixed-point iteration and Newton's method. At each step, the method of scoring performs a rescaled fixed-point step based on the *expected* Fisher information rather than the *observed* Fisher information of Newton-Raphson, using $\alpha = -1/E(\ddot{\ell}(\hat{\theta})) = 1/I(\hat{\theta})$. The method of scoring is unique to statistical applications in the sense that there is no analog of it which can generally be applied to solve a nonlinear equation.

4.2.6.1 A multinomial problem

A classic example of maximum likelihood estimation is due to Fisher (1925) and arises in a genetics problem. Consider a multinomial observation $X = (m_1, m_2, m_3, m_4)$ with class probabilities given by

$$
\begin{aligned}
p_1 &= (2 + \theta)/4, \\
p_2 &= (1 - \theta)/4, \\
p_3 &= (1 - \theta)/4, \\
p_4 &= \theta/4,
\end{aligned}
\tag{4.2.3}
$$

where $0 < \theta < 1$. The sample size is $n = \sum m_i$. The parameter θ is to be estimated from the observed frequencies $(1997, 906, 904, 32)$ from a sample of size 3839. The log-likelihood function and its derivatives are given by

$$
\ell(\theta) = m_1 \log(2 + \theta) + (m_2 + m_3) \log(1 - \theta) + m_4 \log \theta, \tag{4.2.4}
$$

$$
\dot{\ell}(\theta) = \frac{m_1}{2 + \theta} - \frac{m_2 + m_3}{1 - \theta} + \frac{m_4}{\theta}, \tag{4.2.5}
$$

and

$$
\ddot{\ell}(\theta) = - \left\{ \frac{m_1}{(2 + \theta)^2} + \frac{m_2 + m_3}{(1 - \theta)^2} + \frac{m_4}{\theta^2} \right\}. \tag{4.2.6}
$$

Expression (4.2.5) can be rewritten as a rational function, the numerator of which is quadratic in θ. One of the roots is negative, so the other root is the one we seek. (Note that even though the score function is defined for $\theta < 0$, the log-likelihood function is not.) Although this equation can be solved explicitly, we shall use it to illustrate the iterative methods discussed above.

Because we have an explicit and relatively simple expression (4.2.6) for the derivative of the score function, Newton-Raphson is a logical candidate for the iterative method. For comparison, we shall also show how the method of scoring performs. Choosing a starting value is not difficult in this case; an unbiased estimator for θ is given by $\tilde{\theta} = (m_1 - m_2 - m_3 + m_4)/n = 0.05704611$. We shall start with the convergence criterion of a relative change in $\hat{\theta}$ of less than 10^{-6}, with the view that either the tolerance or the criterion may need to be changed if convergence is not achieved within a few iterations. Table 4.2.5 shows the results of applying the two methods. Using Newton-Raphson, the derivative of the score function at $\hat{\theta}$ is -27519.22288. This should be compared to the negative of the Fisher information at $\hat{\theta}$, which is 29336.52362. The standard errors from the two methods are 0.006028 and 0.005838, respectively.

Once again, the sensitivity of Newton-Raphson to choice of starting values is illustrated in Table 4.2.6, which shows what happens to Newton's method and to the scoring method with $\hat{\theta}_0 = 0.5$. One might be led to such a choice by simply noting that θ must be in $(0, 1)$ and by taking

	Newton-Raphson		Scoring	
i	$\hat{\theta}_i$	$\dot{\ell}(\hat{\theta})$	$\hat{\theta}_i$	$\dot{\ell}(\hat{\theta})$
0	0.05704611	-387.74068038	0.05704611	-387.74068038
1	0.02562679	376.95646890	0.03698326	-33.88267279
2	0.03300085	80.19367817	0.03579085	-2.15720180
3	0.03552250	5.24850707	0.03571717	-0.13386352
4	0.03571138	0.02527096	0.03571260	-0.00829335
5	0.03571230	0.00000059	0.03571232	-0.00051375
6	0.03571230	-0.00000000	0.03571230	-0.00003183

TABLE 4.2.5 *Two methods for maximum likelihood estimation in a multinomial problem. The scoring method converges more rapidly at the start, but Newton-Raphson's quadratic convergence takes over in the last few iterations.*

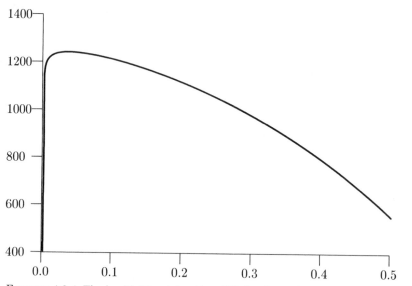

FIGURE 4.2.4 *The log-likelihood function $\ell(\theta)$ for the multinomial example. Note that starting values very close to zero or much larger than 0.1 are unreasonable and likely to cause difficulty for Newton methods.*

the midpoint of that interval. This "easy way out" of the starting-value problem leads to disaster for Newton's method, which converges to the wrong root! This difficulty is easily avoided by plotting the log likelihood before selecting a starting value, as we have done in Figure 4.2.4.

	Newton-Raphson		Scoring	
i	$\hat{\theta}_i$	$\dot{\ell}(\hat{\theta})$	$\hat{\theta}_i$	$\dot{\ell}(\hat{\theta})$
0	0.50000000	-2757.20000000	0.50000000	-2757.20000000
1	0.14134077	-948.94105740	0.05112008	-307.92073569
2	-0.06989831	-1114.89695350	0.03664009	-24.94396888
3	-0.19853655	-562.81128214	0.03576968	-1.57659006
4	-0.40797728	-109.58876500	0.03571586	-0.09778861
5	-0.46586259	-1.74894631	0.03571252	-0.00605820
6	-0.46681399	-0.00030477	0.03571232	-0.00037529
7	-0.46681415	-0.00000000	0.03571230	-0.00002325

TABLE 4.2.6 *Sensitivity of Newton's method to choice of starting values. Newton's method converges to the negative root of the likelihood equation. The scoring method converges to the correct root.*

4.2.6.2 Poisson regression

Thisted and Efron (1986) consider the problem of predicting the number of occurrences of rare words in a new passage of given length written by an author whose distribution of word frequencies is known. Thisted and Efron consider, in particular, whether a nine-stanza poem attributed to Shakespeare discovered in late 1985 by Gary Taylor, an American Shakespearean scholar, is consistent in patterns of word usage with that of the Bard. One approach to this question employs a Poisson regression model discussed at length by Cox and Hinkley (1974). The model discussed in this section has the drawback that fitted values of the Poisson parameter need not be positive; it has the virtue, however, of illustrating the solution of a single nonlinear equation, which is why we employ it here. A more suitable Poisson regression model is discussed in Section 4.3.5.2.

Let m_x be the number of words in the new passage used exactly x times in the known Shakespearean literature, and denote by M_j the jth "decade count," defined by $M_j = \sum_{x=10j}^{10j+9} m_x$. Under plausible assumptions, one can consider M_j to be a Poisson variate with mean $\alpha+\theta(X_j-\overline{X})$, where the X_j's are known constants with $\overline{X} = \sum X_j/n$, where α is a known constant which we shall take to equal $\overline{M} = \sum M_j/n = 11.8$, and where n is the number of decades being examined. Under the hypothesis of Shakespearean authorship, $\theta = 1$. We seek to estimate θ, and to obtain a standard error for the estimate as well. Table 4.2.7 gives the observed counts, and the regressors X_j derived from the Shakespearean literature, for the first $n = 10$ decades.

Let $x_j = X_j - \overline{X}$. Under the model given above, the log-likelihood

$j =$	0	1	2	3	4	5	6	7	8	9
M_j	40	10	21	16	11	7	4	4	2	3
X_j	30.11	13.91	10.63	8.68	7.14	5.99	5.23	4.37	4.09	3.54

TABLE 4.2.7 *Observed word frequencies and predictors for Poisson means, by decade (j), for a passage of disputed authorship attributed to Shakespeare.*

function is given by

$$\ell(\theta) = -n\alpha + \sum M_j \log(\alpha + \theta x_j) - \sum \log(M_j!), \qquad (4.2.7)$$

so that the score function and its derivative are given by

$$\dot{\ell}(\theta) = \sum \frac{M_j x_j}{\alpha + \theta x_j}, \qquad (4.2.8)$$

and

$$\ddot{\ell}(\theta) = -\sum \frac{M_j x_j^2}{(\alpha + \theta x_j)^2}. \qquad (4.2.9)$$

Once again, we have a ready expression for the derivative of $\dot{\ell}(\theta)$, the function whose zero we seek. Thus Newton-Raphson is a good choice. The method of scoring can also be used, since, from (4.2.9) we have $I(\theta) = \sum x_j^2/(\alpha + \theta x_j)$. We also illustrate the results of using the secant method.

A sensible starting value for the iteration naturally suggests itself—the null hypothesis value of $\theta = 1$ corresponding to Shakespearean authorship. For the secant method, we choose the second starting value of 1.1 to be a point nearby the first. The choice of starting value ought not to be critical here, however, since equation (4.2.9) shows that the second derivative of the log-likelihood is always negative, indicating that the function is concave, and that any root found is a local maximum of the likelihood. Consequently, there is at most one finite root. A relative convergence criterion -10^{-6} is used here— is also a plausible one from which to start, as a value of θ near zero is *a priori* unlikely.

Newton-Raphson converges in four iterations, the secant method in six, and scoring in eight, as Table 4.2.8 shows. The secant method approximates $\ddot{\ell}(\cdot)$ using finite differences. By iteration eight, the successive iterates are so close that $\hat{\theta}_i \approx \hat{\theta}_{i-1}$ to machine precision, resulting division by very small numbers. This causes the derivative estimate to become unstable, as can be seen in the table.

EXERCISES

4.5. [09] Let I be the closed interval $[a, b]$. Prove that, if $g(x)$ is con-

	Newton-Raphson		Secant		Scoring	
i	$\hat{\theta}_i$	$\ddot{\ell}(\hat{\theta}_i)$	$\hat{\theta}_i$	$\ddot{\ell}(\hat{\theta}_i)$	$\hat{\theta}_i$	$-\mathcal{I}(\hat{\theta}_i)$
0	1.0000000	-26.501744	1.0000000	0.000000	1.0000000	-32.395811
1	1.5392046	-37.571555	1.1000000	-26.229910	1.4411021	-38.843866
2	1.5072874	-35.365164	1.5447926	-29.400659	1.4972872	-40.649545
3	1.5062607	-35.300786	1.4968234	-36.287938	1.5050183	-40.928028
4	1.5062598	-35.300727	1.5059278	-35.000837	1.5060879	-40.967188
5	1.5062598	-35.300727	1.5062626	-35.290440	1.5062360	-40.972621
6	1.5062598	-35.300727	1.5062598	-35.300813	1.5062565	-40.973374
7	1.5062598	-35.300727	1.5062598	-35.300727	1.5062593	-40.973478
8	1.5062598	-35.300727	1.5062598	-36.500000	1.5062597	-40.973492

TABLE 4.2.8 *Poisson regression estimates using three methods. Note that the secant method fails to compute the derivative correctly when taken two iterations beyond convergence, due to numerical instability. The standard errors for $\hat{\theta}$ are 0.1683 for Newton-Raphson and the secant method, and 0.1562 for scoring. The method of scoring, based on expected rather than observed curvature of the score function, gives a slightly smaller standard error.*

tinuous and if $g(x) \in I$ whenever $x \in I$, then the equation $g(x) = x$ has a solution on I.

4.6. [15] Show that the solution of the previous exercise is unique if $g(x)$ satisfies a Lipschitz condition on I with index $L < 1$.

4.7. [14] Prove Theorem 4.2.1. [*Hint:* Show that $(x_i - s) \to 0$.]

4.8. [11] Prove that, if $g(x)$ is differentiable on $I = [a, b]$, and if $|g'(x)| \leq L$ on that interval, then g satisfies a Lipschitz condition with index L on I.

4.9. [12] Prove that, if $g(x)$ is differentiable on $I = [a, b]$, and if g satisfies a Lipschitz condition with index L on I, then $|g'(x)| \leq L$ for all $x \in I$.

4.10. [08] Show that if $|f'(x)|$ is bounded and does not change sign on $[a, b]$, then there exists a constant α for which $|1 + \alpha f'(x)| < 1$.

4.11. [22] Show that if $L < U$, $f(L)f(U) < 0$ and $f(x)$ is either concave or convex on the interval $[L, U]$, then either L or U remains fixed in all iterations of the secant-bracket method.

4.12. [23] Give an example in which Newton's method cycles, even though f and f' are continuous.

4.13. [02] Under what conditions is (4.2.6) less than zero for positive values of θ? (Solutions to $\dot{\ell}(\theta) = 0$ in this range correspond to maxima of the likelihood function.)

4.14. [12] Show that $I(\theta) = n(1 + 2\theta)/[2\theta(1 - \theta)(2 + \theta)]$ for the log likelihood given in expression (4.2.4).

4.15. [09] Show that $\tilde{\theta} = (m_1 - m_2 - m_3 + m_4)/n$ is an unbiased estimate for θ for the multinomial problem of (4.2.3).

4.16. [39] The Cauchy location problem has likelihood function proportional to $\prod_i^n 1/(1 + (x_i - \theta)^2)$. Conduct a numerical study of methods for maximum likelihood estimation of θ. What iterative methods are reasonable to implement? To what extent are the iterations affected by n and by choice of starting values?

4.17. [10] Show that, for estimating the Cauchy location parameter θ from a random sample of size n, the method of scoring is equivalent to solving the likelihood equation using simple iteration with $g(\theta) = 2\dot{\ell}(\theta)/n$.

4.3 Solving $f(x) = 0$: The vector case

The previous section dealt with solving a single nonlinear equation in a single unknown. In most statistical problems, several parameters must be estimated simultaneously. The derivative of the log-likelihood function when θ is a vector produces a *vector* of partial derivatives, one for each component of θ, so that the score function is vector-valued. Setting the score function to zero produces a system of simultaneous equations, usually nonlinear, which must be satisfied at the maximum likelihood estimate $\hat{\theta}$. This section discusses methods for solving such systems numerically.

The motivating statistical example for this section is multivariate maximum likelihood estimation. Consequently, the emphasis here will be on functions $f : \mathcal{R}^p \to \mathcal{R}^p$. This means, in particular, that we shall not discuss overdetermined systems; we defer treatment of such problems (which arise in areas such as nonlinear regression) until Section 4.5.

There are innumerable methods for solving nonlinear systems of equations, which means that we must exercise considerable selectivity in the choice of methods to be discussed in this section. The methods discussed here fall into two categories, those commonly used in computational statistics, and those which deserve to be more widely known. As in Chapter 3, we emphasize those methods of particular utility in statistical research and data analysis.

4.3.1 Generalizations of univariate methods

For the most part, methods for solving nonlinear systems are generalizations of methods for single equations. Each of the methods (except for those based on bracketing) has at least one extension to the multivariate case. The problem of solving nonlinear systems of equations arises most frequently in statistics in the context of optimizing a scalar objective function $F(\cdot)$ in several variables x, either maximizing (in the case of a log-likelihood function) or minimizing (as in nonlinear least-squares or minimum chi-squared fitting). In either case, the system to be solved is obtained by setting the gradient vector of the F to zero. In our notation, we set $f(x) = F'(x) = \nabla F(x)$. The function $f(x)$ is a nonlinear transformation that maps $\mathcal{R}^p$ into itself. The derivative of $f(x)$—the Jacobian of the nonlinear transformation—is the *Hessian matrix* of the objective function. Unlike the general situation of solving nonlinear equations, this Hessian matrix is necessarily symmetric, and is generally nonnegative definite as well. As a consequence, the problem of solving systems of nonlinear equations as they arise in statistical investigation is considerably simpler than is the general problem, and specialized methods which rely on these properties can be employed.

The methods discussed below will for the most part be based on locally linear approximations to the vector function whose root is sought, so that generally the algorithms discussed here will employ a mix of techniques from linear systems and from univariate nonlinear equations. As in the univariate case, we begin by expanding $f(\cdot)$ in a Taylor series about the current estimate of the solution x_k, and then evaluating the series at the solution s.

Unfortunately, the notation we shall need is a bit cumbersome given the simplicity of the ideas represented. A more unified, if less familiar, notation is commonly used in conjunction with tensor computations (McCullagh, 1987). Since $f(\cdot)$ is vector-valued, taking partial derivatives with respect to the components of x produces a matrix of partial derivatives. This matrix can then be evaluated at any point in $\mathcal{R}^p$, so that the derivative of f, denoted by $f'(\cdot)$, can be thought of as a mapping from $\mathcal{R}^p$ into the space of linear functions $\mathcal{L}(\mathcal{R}^p, \mathcal{R}^p)$. We shall denote the particular linear transformation obtained by evaluating the derivative function at x by $f'(x)$; when this linear transformation is then applied to $y \in \mathcal{R}^p$, we write $f(x)y$. The second derivative f'' of f similarly is a mapping $\mathcal{R}^p \to \mathcal{L}(\mathcal{R}^p, \mathcal{L}(\mathcal{R}^p, \mathcal{R}^p))$. Thus, $f''(x)$ can be thought of as a three-dimensional array; $f''(x)w$ is a linear transformation on $\mathcal{R}^p$. This linear transformation evaluated at $z \in \mathcal{R}^p$ is written $f''(x)wz$. With this notation, we can write the Taylor expansion as

$$0 = f(x_k) + f'(x_k)(s - x_k) + \frac{1}{2}f''(x^*)(s - x_k)(s - x_k), \qquad (4.3.1)$$

where $x^* = ts + (1 - t)x_k$, for some $0 \le t \le 1$. Dropping the last term of (4.3.1) leads to the approximate solution

$$s \approx x_k - [f'(x_k)]^{-1}f(x_k) \qquad (4.3.2)$$

This approximate solution can serve as the basis for iterative methods such as Newton-Raphson, in which $f'(x_k)$ is evaluated at each iteration; simple rescaled iteration schemes, in which $f'(x_k)$ is replaced by a constant matrix; and scoring methods, in which $f'(x_k)$ is replaced by an expectation matrix. Since, for each x_k, equation (4.3.2) is simply a linear system, all of the technology of Chapter 3 can be brought to bear in solving it. In particular, there is rarely any need to obtain the inverse of $f'(x_k)$ explicitly during the course if the iteration. Rather, the solution to the linear system $f'(x_k)(s - x_k) = -f(x_k)$ is solved instead.

A word or two about starting values and convergence is in order. Practically everything we had to say about choosing starting values in the univariate case applies as well in the multivariate case, and there is little to add. Poor convergence and convergence to critical points which are not global optima in maximum likelihood problems occur much more frequently in multivariate problems than in univariate ones, so that correspondingly more caution is necessary in trustingly accepting the results of a single iteration. Although deciding when a multivariate iteration has converged is somewhat more problematic, it is usually satisfactory to define relative and absolute convergence in terms of some norm on x (or f) such as the ℓ_2 norm (sum of squares) or the sup-norm (maximum component magnitude).

We turn now to the details of some of these methods. Illustrations of the methods discussed are contained in Section 4.3.5.

4.3.2 Newton-Raphson

The Newton-Raphson iteration, or the multivariate Newton's method, applies (4.3.2) repeatedly until convergence is achieved. This iteration is a straightforward generalization of the univariate procedure, approximating the curves $f_i(x)$ by tangent planes at each iterate. These tangent planes intersect in a line, which in turn intersects the horizontal plane in a single point whose ordinate becomes the next iterate. An alternative interpretation of the multivariate Newton method is that at each iterate, a *direction* is computed $([f'(x_k)]^{-1}f(x_k))$, and the next iterate is obtained from the current one by taking a step in this direction. As in the univariate case, Newton's method converges quadratically in a neighborhood of the solution but the multivariate iteration is even more highly sensitive to starting values.

An even greater drawback than the sensitivity to choice of starting value is the necessity for computing $p(p + 1)/2$ derivatives in the symmetric case, and p^2 derivatives in the general case. Fortunately, in statistical problems, $f(x)$ is usually the gradient function of a scalar function, say $F(x)$, so that the ijth element of $f'(x)$ is a matrix of mixed-partial derivatives $\partial^2 F/\partial x_i \partial x_j$, which is necessarily symmetric. In this case, the matrix $f'(x)$ is called the *Hessian matrix* of F. There are three aspects to the derivative problem. First, there is the problem of actually obtaining the analytic form for the set of derivative functions. Even when this is possible to do by hand, it is often difficult to do so correctly—the first time at least! Second, the derivatives must then be programmed in FORTRAN, C, or some other programming language. Third, the cost of performing so many function evaluations at each step of the iteration may make the algorithm too costly to employ.

When the derivatives can be expressed analytically, it is sometimes helpful to perform or to check the computations using a symbolic algebra package such as MACSYMA or REDUCE; such packages can also produce FORTRAN code as a byproduct of their work, reducing the burden of programming, as well. The code so produced may not be the most efficient, but it has the virtue of being correct!

Newton's method is most useful when derivatives are easy to compute and when good starting values are available. Some statistical problems satisfy these conditions.

4.3.3 Newton-like methods

The univariate secant method can be interpreted in two ways. First, we can consider it an approximation to Newton's method, in which the derivative $f'(x)$ is approximated by a finite difference, so that the secant line is viewed as an approximation to the tangent line at x. A second interpretation of the method is that the function $f(x)$ is approximated by a line which interpolates f at two of its values. To obtain a multivariate version of the secant method, we can generalize either of these approaches. Doing so produces two different, but related, classes of algorithms in the multivariate case, either of which might legitimately be called secant methods. Following general practice, we shall refer to the former as a *discrete Newton* method, and to the latter as a *(multivariate) secant* method. Either approach is costly in terms of the number of function evaluations required at each iteration (on the order of p^2). To reduce the computational cost, we can compute approximations for $f'(x)$ at only some of the iterations, say after every fifth iteration. Rescaled simple iteration is a special case in which the matrix of derivatives f' is approximated but once. Finally, we can follow an intermediate strategy in which a rough—but cheap—approximation to the derivative matrix is made at each step; such methods are called *quasi-Newton* methods. Detailed development of these and related methods, together with a thorough analysis of their convergence properties, is contained in Ortega and Rheinboldt (1970). Only the broadest overview is given here.

4.3.3.1 Discrete Newton methods

Adopting the first of these two views for a moment, the discrete Newton method replaces the matrix $f'(x)$ by an approximation matrix of difference quotients. Denote the ij-th element of $f'(x)$ by J_{ij} (for Jacobian), and denote the unit vector along the nth coordinate axis by e_n. The two most frequently used difference approximations use

$$J_{ij}(x) \approx \frac{f_i(x + h_{ij}e_j) - f_i(x)}{h_{ij}} \qquad (4.3.3)$$

and

$$J_{ij}(x) \approx \frac{f_i(x + \sum_{n=1}^{j} h_{in}e_n) - f_i(x + \sum_{n=1}^{j-1} h_{in}e_n)}{h_{ij}}. \qquad (4.3.4)$$

It is easiest to take $h_{ij} = h_j$ in either form above, and we shall restrict our discussion to this case. In statistical problems such as maximum likelihood, the Jacobian of the score function is symmetric, so that $J_{ij} = J_{ji}$, which suggests that we may choose $h_{ij} = h_{ji}$; this leads in turn to the common choice of $h_{ij} = h_j \equiv h$. Even better is to use a set of discretization parameters $\{h_j\}$. , Since in general the resulting matrices produced using (4.3.3) or (4.3.4) will not be symmetric, when J is known to be so one should use $J_{ij}^* \equiv \frac{1}{2}(J_{ij} + J_{ij}')$.

The matrix J must be recomputed at each iteration. If h is a fixed constant for all iterations, then we forfeit the quadratic convergence of Newton's method in favor of a linearly convergent sequence. The hope is that the computational savings obtained by trading p^2 derivative evaluations for p^2 additional function evaluations more than compensate for the extra iterations needed to reach convergence. If $h \to 0$ as the iteration proceeds, then more rapid rates of convergence can be obtained. A natural choice for the discretization parameter at the kth step is to set $h_j = x_{k,j} - x_{k-1,j}$, where $x_{k,j}$ denotes the jth component of the kth iterate in the sequence. In general, this procedure converges at a rate between linear and quadratic.

COMMENT. The literature on nonlinear optimization contains many different proposals for changing the discretization parameters h_j at each step of the iteration. A noteworthy method which is seldom used in computational statistics bases h_j on the current value of f_j thus, in effect, combining a secant-like method with fixed-point iteration. Such approaches are known as *Steffensen methods*.

4.3.3.2 Generalized secant methods

Taking the second approach to generalizing the one-variable secant method, we can view the secant in the univariate case as an approximating linear function to the function f of interest, which interpolates f at two points.

When $f(x)$ has p components $f_i(x)$, each a function of $x \in \mathcal{R}^p$, the natural generalization is to approximate each f_i by a linear interpolating function. To do so requires evaluating each $f_i(x)$ at $p+1$ points, say at x_{i-p} through x_i, and then finding the hyperplane determined by these points. Each hyperplane can be expressed as a linear equation in x; setting these p linear equations to zero simultaneously and solving the resulting linear system of equations produces the next iterate.

Note that the secant method as described here requires f to be evaluated at $p+1$ points, so that one requires $p+1$ starting values to begin the iteration. It is hard to choose so many starting values automatically, but it is almost always necessary to do so when one is writing either a general-purpose package or a large-scale Monte Carlo study. In either case, the same algorithm will be performed on a large number of similar problems, but different sets of starting values will be appropriate for each problem. This fact alone makes such methods impractical for most statistical problems.

4.3.3.3 Rescaled simple iteration

The simplest way to reduce the cost of approximating $f'(x)$ is simply not to recompute it at all, but rather to use an initial approximation J throughout the iteration, that is,

$$x_{k+1} = x_k - J^{-1}f(x_k). \tag{4.3.5}$$

This is a straightforward generalization of the rescaled simple iteration algorithm discussed in Section 4.2.1. If $J^{-1} = -\alpha I$ or if J^{-1} is the diagonal matrix $-\text{diag}(\alpha_1, \ldots, \alpha_p)$, then this amounts to applying the simple iteration algorithm to each component of $f(x)$ separately. Our earlier discussion suggests that the choice $J = f'(x_0)$ is a good one.

Such methods will converge from a starting value sufficiently close to the solution s provided that J^{-1} is sufficiently close to $f'(s)^{-1}$. More concretely, we must have $\rho(I - J^{-1}f'(s)) < 1$, where $\rho(A) = |\lambda_{max}(A)|$ is the spectral radius of A. The convergence is generally linear for such methods.

It is difficult to guarantee in advance that this condition will be met. A compromise approach, then, is to take one Newton step followed by several additional iteration steps without updating the matrix J. At that point the derivative matrix can be computed afresh, and several more steps taken.

4.3.3.4 Quasi-Newton methods

As the discussion of rescaled iteration shows, the approximation to $f'(x)$ need not be a good one in order to obtain convergence. What is more, in discrete-Newton methods, we must perform p^2 function evaluations, to approximate the Jacobian, which can be quite expensive. This suggests using a crude approximation to f' that is easily computed at each step of the iteration. Consider the iteration defined by

$$x_{k+1} = x_k - J_k^{-1} f(x_k).$$

To reduce the computational effort in obtaining J_{k+1}^{-1}, we can also require that J_k and J_{k+1} differ only by a matrix of rank one or two. If the difference is of rank one or two then J_{k+1}^{-1} can easily be obtained from J_k^{-1} using the Sherman-Morrison-Woodbury formula (3.7.1), provided that J_0 and the subsequent updated matrices J_k are all nonsingular.

How might such a low-rank update be selected? The discrete-Newton methods re-evaluate the information about the curvature of f at each point in the iteration. An alternative is to accumulate this information as we go along. At each iteration, we move from the current iterate x_k to the next one $x_{k+1} = x_k + d_k$ by taking a step in the direction d_k. Thus, we can easily gain information about the curvature of f in this direction, and this information can be used to update the current estimate J_k used in the iteration. The details of this procedure are clearly discussed in section 4.5.2 of Gill, Murray, and Wright (1981). Such methods are called *quasi-Newton* methods; they are also sometimes referred to as *variable-metric methods*.

COMMENT. The general secant method can be obtained by specifying a particular sequence of rank-one matrices satisfying the nonsingularity condition. Indeed, the computationally efficient way to compute the general secant iteration is to use an updating formula at each step, rather than refitting all $p + 1$ hyperplanes.

Of the quasi-Newton methods based on rank 2 matrix updating, two are of particular interest for statistical computation, the *Davidon-Fletcher-Powell* algorithm, and the *Broyden-Fletcher-Goldfarb-Shanno (BFGS)* algorithm. These algorithms were proposed for solving the optimization problem, wherein $f(x)$ is a gradient vector of a scalar objective function as in maximum likelihood or nonlinear least-squares estimation. Thus, in particular, J is symmetric and positive definite. Both methods ensure that J_k has this property (if J_0 does). Either can be expressed in terms of J or J^{-1}, however their forms are such that Davidon-Fletcher-Powell is somewhat more cumbersome to express in terms of a direct update to J, and BFGS is more awkward to express in terms of an inverse update (to J^{-1}).

The Davidon-Fletcher-Powell method (Davidon 1959; Fletcher and Powell 1963) is moderately well-known in statistical computing, and has been widely recommended. Let $d_k = x_{k+1} - x_k$ and let $g_k = f(x_{k+1}) - f(x_k)$. Then the Davidon-Fletcher-Powell algorithm written in terms of the inverse matrix uses

$$J_{k+1}^{-1} = J_k^{-1} + \frac{d_k d_k'}{d_k' g_k} - \frac{J_k^{-1} g_k g_k' J_k^{-1}}{g_k' J_k^{-1} g_k}. \qquad (4.3.6)$$

(Here we use x' to denote the transpose of x.) This matrix will be non-singular provided that the denominators of the two terms in (4.3.6) are non-zero, and provided that $g_k' J_k^{-1} f(x_k) \neq 0$. Under mild conditions on the log-likelihood function, the Davidon-Fletcher-Powell algorithm can be shown to be well-defined and convergent. For quadratic functions, it converges in a fixed number of iterations to the exact result.

The BFGS algorithm, proposed independently by Broyden (1970), Fletcher (1970), Goldfarb (1970) and Shanno (1970), is generally considered to be superior to the Davidon-Fletcher-Powell method of updating the Jacobian estimate. Written in terms of J, the updating algorithm is

$$J_{k+1} = J_k - \frac{J_k d_k d_k' J_k}{d_k' J_k d_k} + \frac{g_k g_k'}{d_k' g_k}. \qquad (4.3.7)$$

Both Davidon-Fletcher-Powell and BFGS are special cases of a one-parameter family of symmetric positive-definite update methods. The assertion of superiority for the latter is based on extensive studies of the family of updates carried out by numerical analysts.

4.3.4 Nonlinear Gauss-Seidel iteration

One of the least-known yet most useful iterative methods for solving nonlinear systems of equations is the *nonlinear Gauss-Seidel* iteration. The method is a straightforward generalization of the technique for solving linear systems discussed in Section 3.11.2, and is a special case of nonlinear methods of successive over-relaxation. Simply stated, the ith component of $f(x)$ is considered to be a function only of the ith component of x, all of the other components thought of as being fixed at their most recently computed values. This reduces the ith equation $f_i(x) = 0$ to a single nonlinear equation in a single unknown, which is then solved using any convenient method to obtain the updated value for the ith component. After cycling through the p components of f, a new iterate for x will have been computed. The *Gauss-Seidel-Newton iteration* is obtained by using Newton's method as the univariate root-finding method in the nonlinear Gauss-Seidel approach.

What makes this method attractive is that each iteration is composed of p simpler problems, each of which may require far fewer computations to accomplish. Each of these constituent problems is a univariate one, for which nonlinear methods are better understood, more rapidly convergent, and generally easier to carry out automatically. The method is usually easy to program, and a common root-finding subroutine can be employed for each of the univariate problems.

As in the case of linear Gauss-Seidel, the method can be applied to blocks of components of x. For example, if f decomposes into a linear part in the first $p-1$ variables and a nonlinear part in the pth, then the linear part can be solved in a single step, which will then alternate with a single nonlinear equation solver.

If x_0 is sufficiently close to the solution s, then the nonlinear Gauss-Seidel iteration will converge to the solution provided that $\rho(-D^{-1}(s)N(s))$ < 1, where, following the notation of Section 3.11.2, $D(x)$ is the lower triangular part including the diagonal of the matrix $f'(x)$, $N(x)$ is the strictly upper triangular part of $f'(x)$, and ρ denotes the spectral radius of the matrix. This condition will be satisfied when $f'(s)$ is positive definite, as it will be in nondegenerate maximum likelihood and nonlinear regression problems. A remarkable result is the *Global SOR Theorem*, which asserts that nonlinear Gauss-Seidel converges *globally* to a unique solution in $\mathcal{R}^n$, provided that $f'(x)$ is symmetric positive definite for all x, and that the smallest characteristic root of $f'(x)$ is bounded away from zero. A proof can be found in Ortega and Rheinboldt (1970). As in the linear case, symmetry can be exploited to improve convergence by employing a *reverse sweep*, that is, after solving successively for x_1, x_2, ... x_p, one then solves for x_{p-1}, x_{p-2}, ... x_1.

4.3.5 Some statististical examples

We now turn to some problems of statistical interest in which the problem of solving nonlinear systems arises. For each we examine some of the alternative methods, and the results of the tradeoffs involved.

4.3.5.1 Example: Binomial/Poisson mixture

The Annual Report of the pension fund S. P. P. for 1952 (cited in Cramér, 1955) reported that some 4,075 widows received pensions from the fund. A table was given showing the number of children of these widows who were entitled to support from the fund, reproduced here in Table 4.3.1.

Number of children	0	1	2	3	4	5	6
Number of widows	3,062	587	284	103	33	4	2

TABLE 4.3.1 *Distribution of the number of children of 4,075 widows entitled to support from a certain pension fund. The mean number of children in this group is 0.3995.*

The data are not consistent with being a random sample from a Poisson distribution; the number of widows in category 0 is too large. A plausible alternative model for these data is to assume that we are observing a mixture of two populations, population A which is always zero (observed with probability ξ), and population B, which follows a Poisson distribution with parameter λ (observed with probability $1 - \xi$).The problem then is to estimate the parameters ξ and λ.

Let n_i denote the number of individuals observed in category i, and let $N = \sum n_i$. The log-likelihood function is then

$$\ell(\xi, \lambda) = n_0 \log \left(\xi + (1 - \xi)e^{-\lambda} \right)$$
$$+ (N - n_0) \left[\log(1 - \xi) - \lambda \right] + \sum_{i=1}^{\infty} i \, n_i \log \lambda. \qquad (4.3.8)$$

In our example, $N - n_0 = 1013$, $n_0 = 3062$, and $\sum i \, n_i = 1628$. The score functions in the general case are

$$\dot{\ell}_\xi(\xi, \lambda) = \frac{n_0(1 - e^{-\lambda})}{(1 - \xi)e^{-\lambda} + p} - \frac{N - n_0}{1 - \xi} \qquad (4.3.9)$$

and

$$\dot{\ell}_\lambda(\xi, \lambda) = -\frac{n_0(1 - \xi)e^{-\lambda}}{(1 - \xi)e^{-\lambda} + \xi} + \frac{\sum i \, n_i}{\lambda} - (N - n_0). \qquad (4.3.10)$$

(The latter of these is equation (4.2.1), which serves as the basis for the numerical examples of Section 4.2. This equation was solved for λ using a fixed value of ξ close to the maximum likelihood value.) The matrix of second derivatives is not hard to obtain in this case, being

$$\ddot{\ell}_{\xi\xi}(\xi, \lambda) = -\frac{n_0 \left(1 - e^{-\lambda} \right)^2}{\left((1 - \xi)e^{-\lambda} + \xi \right)^2} - \frac{N - n_0}{(1 - \xi)^2}, \qquad (4.3.11)$$

$$\ddot{\ell}_{\xi\lambda}(\xi, \lambda) = \frac{n_0 e^{-\lambda}}{(1 - \xi)e^{-\lambda} + \xi} + \frac{n_0(1 - \xi)(1 - e^{-\lambda})e^{-\lambda}}{\left((1 - \xi)e^{-\lambda} + \xi \right)^2}, \qquad (4.3.12)$$

and

$$\ddot{\ell}_{\lambda\lambda}(\xi, \lambda) = \frac{n_0(1 - \xi)e^{-\lambda}}{(1 - \xi)e^{-\lambda} + \xi} - \frac{n_0(1 - \xi)^2 e^{-2\lambda}}{\left((1 - \xi)e^{-\lambda} + \xi \right)^2} - \frac{\sum i \, n_i}{\lambda^2}. \; (4.3.13)$$

A plausible initial guess for ξ is $n_0/N = 0.7514$ (too large), and an initial guess for λ is $\overline{X} = 0.3995$ (too small). We round these to 0.75 and 0.40, respectively, in the examples below. In each iteration, we use the convergence criterion that relative change in the solution, measured in the l_1 norm, be less than 10^{-6}.

Newton-Raphson is the obvious candidate, given the relative simplicity of $\ddot{\ell}(\cdot)$. This iteration converges in seven iterations from the starting value (0.75,0.40), as given in Table 4.3.2. The iteration converges in six steps, and provides the observed information matrix as a by-product of the computation. Discrete Newton methods produce virtually identical results on this problem. Using the same starting value and a constant difference parameter $h_{ij} = 0.01$, the iteration converged to the same solution in seven iterations. Using starting values of (0.75,0.40) and (0.76,0.41) and taking $h = x_k - x_{k-1}$, where $x' = (\xi, \lambda)$, again produces convergence in seven iterations, and in both cases, the estimated asymptotic covariance matrix agrees with that from Newton-Raphson to five places past the decimal point.

	$\begin{pmatrix}\xi\\\lambda\end{pmatrix}$	$\begin{pmatrix}\dot{\ell}_\xi\\\dot{\ell}_\lambda\end{pmatrix}$	$-\ddot{\ell}^{-1}$	
0	0.7500000	-2951.84522	0.00006	0.00002
	0.4000000	2497.77896	0.00002	0.00011
1	0.6045962	-829.49502	0.00017	0.00012
	0.6205528	813.89802	0.00012	0.00035
2	0.5565734	-25.43959	0.00027	0.00034
	0.8117168	192.38352	0.00034	0.00096
3	0.6148940	-95.07397	0.00018	0.00029
	0.9883717	55.33331	0.00029	0.00132
4	0.6142832	-1.17228	0.00018	0.00032
	1.0335820	3.03125	0.00032	0.00152
5	0.6150562	-0.03124	0.00018	0.00032
	1.0378204	0.01878	0.00032	0.00154
6	0.6150567	-0.00000	0.00018	0.00032
	1.0378391	0.00000	0.00032	0.00154
7	0.6150567	0.00000	0.00018	0.00032
	1.0378391	0.00000	0.00032	0.00154

TABLE 4.3.2 *The Newton-Raphson algorithm applied to the Binomial/Poisson mixture maximum-likelihood estimation problem. Given are the estimates, values of the score function, and the inverse of the observed information matrix at each iteration.*

Simple rescaled iteration does not fare so well in this problem. Using the same starting value as for Newton's method, and using the inverse of the Jacobian matrix at (ξ_0, λ_0) throughout the iteration, produces convergence after 136 iterations. That each iteration is almost 80% faster than the corresponding Newton step obviously does not compensate for the slow convergence. A hybrid method, in which one gets close to the solution with a few (expensive) Newton steps, followed by many (inexpensive) fixed-point iteration steps does not produce practical gains in this problem, but does illustrate the basic idea. After a single Newton step, only 43 additional fixed steps were needed for convergence; after two Newton steps, only 21 more fixed steps were required, and after three Newton steps only 7 more fixed steps were needed.

The method of scoring is no more attractive in this problem than is Newton-Raphson, the expected information (Fisher information) matrix being only slightly simpler to compute by machine (but requiring somewhat more human computation up front to find the form of the information matrix). Indeed, scoring behaves much like the rescaled iteration, requiring 49 iterations to converge. For comparison purposes, the upper triangular part of $I(\hat{\theta})^{-1}$ has entries 0.00012, 0.00007, and 0.00033.

An interesting solution to the problem is obtained by noting that the likelihood equations can be rewritten as

$$\xi = \frac{n_0 e^\lambda - N}{N(e^\lambda - 1)} \tag{4.3.14}$$

and

$$\lambda = \frac{\left(\sum i\, n_i\right)\left(\xi e^\lambda + (1 - \xi)\right)}{n_0 \xi e^\lambda + N(1 - \xi)}. \tag{4.3.15}$$

The nonlinear Gauss-Seidel iteration solves the first of these equations to obtain the next iterate for ξ. Equation (4.3.15), however, gives only an *implicit* solution for λ. This second equation can be applied several times until its successive values have converged. The value for λ so obtained must then satisfy the (4.3.15). This process would complete a single Gauss-Seidel step. This method requires 16 iterations to achieve convergence. Rather than obtaining a solution to equation (4.3.15) by repeated iteration, it is attractive to perform only a single iteration of (4.3.15) at each step; this approach converges after 25 steps, each requiring about 40% of the computational effort of a single Newton-Raphson step. Of course, the Gauss-Seidel method does not provide an estimate of the asymptotic variance matrix, so that some additional computation may be required at the end of the iteration.

COMMENT. In effect, we have just described an algorithm which involves *two* separate iterative processes. The primary iteration involves a Gauss-Seidel step; the secondary iteration is used to obtain λ as part of each primary step. Any single-equation method could be used to obtain λ, given the current value of ξ. If Newton's method were used, for instance, we should say that the iteration is a Gauss-Seidel-Newton procedure. The method actually described using equation (4.3.15), is a Gauss-Seidel-simple iteration algorithm. When only a single step of the secondary iteration is used—as in the last example above—the method is called a Gauss-Seidel one-step iteration method. Such methods generally converge at the same asymptotic rate as the corresponding infinite-step methods.

4.3.5.2 Example: Poisson regression

In Section 4.2.6.2 we discussed a very simple Poisson regression model for word frequencies in a poem attributed to William Shakespeare relative to word frequencies used elsewhere in the Shakespearean literature (Thisted and Efron, 1986). Here we discuss a more plausible model for testing the hypothesis of Shakespearean authorship. Consider the words in the new poem which appeared exactly j times in the Shakespearean canon. Denote the number of such words by y_j. We take each y_j to have a Poisson distribution with parameter λ_k. A model for the Shakespearean canon presented by Thisted and Efron predicts a mean number of such words to be ν_k. For $x = 1,2,\ldots,99$ we take $\log(\lambda_k) = \log(\nu_k) + \alpha + \beta \log(k+1)$. On the hypothesis of Shakespearean authorship, we should have $\beta = 0$. The likelihood equations are

$$\dot{\ell}_\alpha = 0 = \sum_{j=1}^{99} \left\{ y_j - \nu_j e^{\alpha+\beta \log(j+1)} \right\},$$

$$\dot{\ell}_\beta = 0 = \sum_{j=1}^{99} \left\{ y_j - \nu_j e^{\alpha+\beta \log(j+1)} \right\} \log(j+1),$$

(4.3.16)

and the derivatives of the score functions are given by

$$\ddot{\ell}_{\alpha\alpha} = -\sum \nu_j e^{\alpha+\beta \log(j+1)},$$

$$\ddot{\ell}_{\alpha\beta} = -\sum \nu_j \log(j+1) e^{\alpha+\beta \log(j+1)},$$

$$\ddot{\ell}_{\beta\beta} = -\sum \nu_j \log(j+1)^2 e^{\alpha+\beta \log(j+1)}.$$

(4.3.17)

In this problem there is no reason to use anything other than Newton-Raphson. The score functions in (4.3.16) can easily be obtained by adding constants to the values of the first two expressions in (4.3.17), so that there is almost no computational penalty to be paid for computing the derivatives. Indeed, every other method that we have discussed would be more costly than Newton-Raphson, due to the additional number of function evaluations required. Note that the method of scoring and Newton's method coincide in this problem. Table 4.3.3 shows the results of Newton's iteration starting at the initial value $(\alpha, \beta)_0 = (0,0)$, corresponding to the hypothesis of Shakespearean word usage.

4.3.5.3 Example: Logistic regression

A common and useful model for binary response data in the presence of covariates is *logistic regression*, which models the logistic transform of the

	$\binom{\alpha}{\beta}$	$\binom{\dot{\ell}_\alpha}{\dot{\ell}_\beta}$		$-\ddot{\ell}^{-1}$	
0	0.0000000	22.27173	0.10078	-0.02925	
	0.0000000	59.12045	-0.02925	0.00959	
1	0.5152302	-3.61385	0.07004	-0.02076	
	-0.0846989	-8.10592	-0.02076	0.00704	
2	0.4303498	-0.07937	0.07387	-0.02179	
	-0.0667846	-0.14286	-0.02179	0.00734	
3	0.4276000	-0.00006	0.07399	-0.02182	
	-0.0661036	-0.00010	-0.02182	0.00735	
4	0.4275976	-0.00000	0.07399	-0.02182	
	-0.0661030	-0.00000	-0.02182	0.00735	
5	0.4275976	0.00000	0.07399	-0.02182	
	-0.0661030	0.00000	-0.02182	0.00735	

TABLE 4.3.3 *Estimates in a Poisson regression model for word frequencies in a poem attributed to William Shakespeare, using the Newton-Raphson algorithm. Under the hypothesis of Shakespearean authorship, $\beta = 0$. The final estimate is $\hat{\beta} = -0.066$, with a standard error of 0.086.*

probability of response as a linear function of the covariates. More formally, suppose that $y_j \sim \text{Binomial}(n_j, p_j)$, and that corresponding to the jth response, we have a $p \times 1$ vector of covariates x_j. Then the logistic model asserts that

$$\text{logit}(p_j) = \beta' x_j \qquad j = 1, 2, \ldots, k \tag{4.3.18}$$

where

$$\text{logit}(p_j) \equiv \log\left(\frac{p_j}{1 - p_j}\right), \tag{4.3.19}$$

which imply that

$$p_j = \frac{e^{-\beta' x_j}}{1 + e^{-\beta' x_j}} \tag{4.3.20}$$

Typically, the model specified by (4.3.18) includes a constant term, in which case, the first component of each x_j is identically equal to one. The log-likelihood equation is easily written as

$$\ell(\beta) = -\sum_{j=1}^{k} y_j \cdot \beta' x_j - \sum_{j=1}^{k} \log\left(1 + e^{-\beta' x_j}\right), \tag{4.3.21}$$

so that

$$\dot{\ell}(\beta) = \sum (n_j p_j - y_j) x_j, \tag{4.3.22}$$

and

$$\ddot{\ell}(\beta) = -\sum n_j p_j (1 - p_j)(x_j x_j').$$ (4.3.23)

Note that the right-hand side of equation (4.3.23) does not depend upon y_j, so that its expected value is constant. Thus, the method of scoring and Newton-Raphson are equivalent when applied to this model.

As an example, consider the data of Table 4.3.4. The response variable indicates whether an allergic reaction occurred in response to a drug (chymopapain) used in a surgical procedure for removal of herniated lumbar discs. The covariates are the gender of the patient, and whether general or local anesthesia was used during the procedure. A plausible starting value to use for the iteration is to treat gender and anesthesia as unimportant (by setting β_1 and β_2 equal to zero), and then roughly estimating β_0. Since $252/44913 = 0.00561$ is the overall percentage of reactions, this implies that β_0 is approximately $\log(0.00561) \approx -5$. Thus, we select $\beta' = (-5, 0, 0)$ as the initial guess in the iteration. The iteration converges rapidly, as Table 4.3.5 indicates.

The data indicate that there is a significant difference between male and females in terms of the odds of experiencing an allergic reaction, with the odds for females being approximately $\exp(\hat{\beta}_2) = \exp(1.0603) = 2.887$ times as large for females as for males. The standard error associated with $\hat{\beta}_2$ is 0.3626. There is not a significant effect associated with the different types of anesthesia.

j	y_j	n_j	x_j
1	76	20,959	(1,0,0)
2	132	12,937	(1,0,1)
3	15	6,826	(1,1,0)
4	29	4,191	(1,1,1)

TABLE 4.3.4 *Incidence of anaphylactic reactions in 44,913 patients injected with chymopapain during a nineteen-month period. The response variable y_j indicates the number of reactions occurring among n_j patients whose gender and type of anesthesia were recorded. The components of x_j represent, respectively, a constant term, anesthetic route (1 =local, 0 =general), and gender (1 =female, 0 =male).*

EXERCISES

4.18. [24] Show that, if J_k^{-1} is symmetric and positive definite, and if $d_k' g_k > 0$, then J_{k+1}^{-1} defined by equation (4.3.6) is symmetric and positive definite.

	β	$\dot{\ell}_\beta$		$-\ddot{\ell}^{-1}$	
0	-5.0000000	-48.59601	0.05787	-0.03299	-0.04547
	0.0000000	-29.73514	-0.03299	0.13450	0.00015
	0.0000000	46.36485	-0.04547	0.00015	0.11915
1	-5.4351973	-33.69244	0.05940	-0.02889	-0.07347
	-0.3211320	-10.55026	-0.02889	0.15101	0.00005
	0.9209551	-12.17958	-0.07347	0.00005	0.12149
2	-5.6092766	-2.50665	0.06292	-0.02921	-0.08267
	-0.4146997	-1.03179	-0.02921	0.16473	0.00002
	1.0418642	-0.51984	-0.08267	0.00002	0.13055
3	-5.6278903	-0.02642	0.06322	-0.02912	-0.08382
	-0.4306320	-0.01337	-0.02912	0.16638	0.00002
	1.0600526	-0.00373	-0.08382	0.00002	0.13151
4	-5.6280979	-0.00000	0.06323	-0.02912	-0.08384
	-0.4308742	-0.00000	-0.02912	0.16640	0.00002
	1.0602792	-0.00000	-0.08384	0.00002	0.13152
5	-5.6280979	-0.00000	0.06323	-0.02912	-0.08384
	-0.4308742	-0.00000	-0.02912	0.16640	0.00002
	1.0602793	-0.00000	-0.08384	0.00002	0.13152

TABLE 4.3.5 *Newton-Raphson iteration for a logistic regression model. The parameters β_0, β_1, and β_2 represent, respectively, the coefficients of the constant, type of anesthesia, and gender of the patient.*

4.19. [28] Give the inverse update form for the BFGS update algorithm of equation (4.3.7).

4.20. [12] The iterations of Table 4.2.8 concerning Poisson regression can actually be thought of as an illustration of nonlinear Gauss-Seidel. Why?

4.21. [20] Show that the log likelihood for the Poisson/Binomial mixture problem of Section 4.3.5.1 is given by equation (4.3.8).

4.22. [22] Derive equations (4.3.9) and (4.3.10).

4.23. [24] Derive equations (4.3.11), (4.3.12), and (4.3.13).

4.24. [03] Why is $\overline{X} = \sum_i i\, n_i$ too small a guess for λ, and n_0/N too large a guess for ξ in the Binomial/Poison mixture problem?

4.25. [27] The method of scoring requires $I(\hat{\theta})$ at each iteration. What is $I(\theta)$ for the score functions of equations (4.3.9) and (4.3.10)?

4.26. [20] Derive expressions (4.3.16) and (4.3.17).

4.27. [04] Why are scoring and Newton-Raphson equivalent in the Poisson regression problem of Section 4.3.5.2?

4.28. [25] Derive equations (4.3.21), (4.3.22), and (4.3.23).

4.29. [01] In the example of section 4.3.5.3, it is asserted that, if β_1 and β_2 are both zero, then β_0 should be approximately $\log(0.00561) \approx -5$. Why?

4.30. [12] In the chymopapain example of section 4.3.5.3, it was noted that the coefficient of anesthesia is not significantly different from zero in the linear logistic model. On what figures is this conclusion based?

4.4 Obtaining the Hessian matrix

The solutions to maximum-likelihood problems are themselves solutions to linear systems. The inverse of the matrix of partial derivatives of the log-likelihood with respect to the parameter, what we have called $\ddot{\ell}(\hat{\theta})$, has expectation equal to the negative of the asymptotic variance matrix of $\hat{\theta}$. The matrix $H(\hat{\theta}) \equiv \ddot{\ell}(\hat{\theta})$ thus plays a central role in estimation and inference. This matrix is the *Jacobian* of the score function, and the *Hessian matrix* of the log-likelihood function. [Similar comments apply throughout to nonlinear regression problems.] We have seen, however, that even very bad estimates of the Hessian matrix—even data-independent choices—can lead to convergence of the underlying system. What is more, the estimation of derivatives when solving nonlinear systems is often done in a less-accurate and less stable manner than the function evaluations themselves, since the former is much less critical to convergence and accuracy in estimating θ than is the latter. This is particularly true of off-the-shelf nonlinear equations programs, which are designed primarily to obtain numerically accurate solution vectors; the derivatives are generally viewed as a means to this end and of little interest in and of themselves.

In statistical problems, however, the asymptotic variance of $\hat{\theta}$ is often as important for inference as the numerical values for $\hat{\theta}$ itself. Because standard numerical software generally does not address this important statistical need, it is important to recognize that the reported values for $\mathrm{Var}(\hat{\theta})$ may be highly misleading even though $\hat{\theta}$ is estimated numerically quite well. The remainder of this section discusses practical ways to recognize and to deal with this situation.

The quality of the computed Hessian is unlikely to be in doubt when it is computed from explicit formulæ for the entries. The examples of section 4.3 show little instability near the MLE. In this situation, difficulties are most likely to arise from the fact that, when the convergence criterion is satisfied, the final value for $\hat{\theta}$, say $\hat{\theta}_k$, was computed using a Hessian evaluated at $\hat{\theta}_{k-1}$. We might call the difference between $H(\hat{\theta}_{k-1})$ and $H(\hat{\theta}_k)$ the error due to *iteration lag*. This difference is usually minor, and can be eliminated entirely by forcing an additional iteration whose sole purpose is to evaluate the Hessian once more.

What factors affect error in the Hessian due to iteration lag? Clearly, the smaller the convergence tolerance is, the closer $\hat{\theta}_k$ and $\hat{\theta}_{k-1}$ will be, so that the closer the reported Hessian $H(\hat{\theta}_{k-1})$ will be to the desired Hessian $H(\hat{\theta}_k)$. If the convergence criterion is a function of the size of the score functions (or relative changes therein), then $\hat{\theta}_{k-1}$ and $\hat{\theta}_k$ may be quite different. This will typically occur when the likelihood surface is very flat in at least one direction, which in turn will occur when the Hessian is nearly singular. In this case computations based on the Hessian matrix will be unstable, reflecting instability in the statistical aspects of the problem. Finally, the third derivatives may be large at the solution, causing the Hessian there to differ markedly from the Hessian evaluated even at nearby points. This indicates a highly non-Gaussian shape to the likelihood at its maximum, which is itself of important statistical interest.

When $H(\theta)$ is not computed exactly but is approximated by finite differences as in the discrete Newton methods of section 4.3.4.1, the situation is compounded. All of the foregoing difficulties apply, together with the additional problems introduced by the approximations to the derivatives. In this case, the reported finite difference approximation $\hat{H}$ at the last iteration may be quite a bad approximation to H. This is most likely to occur when $H(\hat{\theta}_k)$ is nearly singular or if $H(\cdot)$ is changing rapidly for θ near $\hat{\theta}_k$. The discrete approximation is sensitive, of course, to the choice of the discretization parameters. If these parameters are too large, then the derivatives are poorly approximated; if they are too small, then the denominators of the finite differences are near zero, leading to numerical instability and round-off error. When using discrete-Newton methods, the discretization parameters are often chosen so that previously obtained function values can be re-used in the derivative approximation. This often leads to very large discretization steps in the early iterations and—particularly when the convergence tolerance is quite small—to numerically unstable derivative estimates in the final iterations.

The foregoing discussion assumes that the score functions can be computed exactly so that they can serve as the foundation for a discrete approximation to the Hessian. When the score functions must also be approximated, the Hessian estimates—now based on second differences—become correspondingly more sensitive. They also can become more expensive, as the number of function evaluations increases as the square of the dimension of the parameter. If the function F is too noisy (so that finite-difference approximations are unstable), or if second differences are too expensive to compute, then secant approximations to the Hessian are available which can be used. Dennis and Schnabel (1983) give a complete discussion, both of finite-difference approximations to the Hessian and of secant methods for optimization.

The best course of action when the Hessian is based on finite differences or when the solution method does not employ explicit approximations to the Hessian, is to perform a one-time calculation to obtain the Hessian at the final value of the iterate $\hat{\theta}_k$. In such a special calculation one can take extra care in ways that are too expensive to incorporate in the basic iteration. In the univariate secant method, for instance, the derivative at each step is approximated by the difference ratio

$$f'(x_i) \approx \frac{f(x_i) - f(x_{i-1})}{x_i - x_{i-1}}, \tag{4.4.1}$$

so that when $x_{i-1} > x_i$ the derivative is approximated from the right, otherwise from the left. More accurate derivative approximations can be achieved using *central differences*, in which the point at which the derivative is desired is in the center of the set of points whose function values are used in the approximation. Thus, a more accurate alternative to (4.4.1) is to define $h = x_i - x_{i-1}$ and then to use

$$f'(x_i) \approx \frac{f(x_i + h) - f(x_i - h)}{2h}. \tag{4.4.2}$$

Note that, while $f(x_i - h) = f(x_{i-1})$ has already been evaluated and can thus be re-used, an additional function evaluation (at $x_i + h = 2x_i - x_{i-1}$) is required to compute (4.4.2). This additional expense is not generally warranted during the course of the iteration, it is well worth doing once at the end of the iteration.

The multivariate Jacobian estimate based on central differences replaces equation (4.3.3) by

$$J_{ij}(x) \approx \frac{f_i(x + h_{ij}e_j) - f_i(x - h_{ij}e_j)}{2h_{ij}}. \tag{4.4.3}$$

Equation (4.3.4) can also be replaced by a central-difference approximation which is straightforward to write down, however, the features of (4.3.4) which make it an attractive alternative to (4.3.3) are lost in the transition to central differences, making that formulation less attractive in this setting.

Choosing appropriate values for the discretization parameters h_{ij} involves a trade-off between discretization error (which decreases as $h_{ij} \to 0$) and round-off error (which increases as $h_{ij} \to 0$). Rice (1983) argues that h_{ij} should be chosen to be a constant which depends on the floating-point precision ϵ, on the degree k of the derivative, and on whether central or backward differences are used in the approximation. Rice's rule of thumb is to take $\log h = (\log \epsilon)/(k + d + 1)$, where $d = 1$ if using central differences and $d = 0$ otherwise.

EXERCISES

4.31. [17] Show that (4.4.1) is exact whenever $f(x)$ is a linear function, but that (4.4.2) is exact whenever $f(x)$ is a quadratic function, thus demonstrating the *superaccuracy* of the central-difference approximation.

4.32. [12] Write out the central-difference approximation to the Jacobian corresponding to (4.3.4).

4.33. [22] Let $h = x_i - x_{i-1}$. Show that the error in (4.4.1) is $O(h)$, whereas the error in (4.4.2) is $O(h^2)$.

4.5 Optimization methods

Let $F(x)$ be a given real-valued function on $\mathcal{R}^p$, and let S be a given connected subset of $\mathcal{R}^p$. The general optimization problem is to find that value of $x \in S \subset \mathcal{R}^p$ at which $F(x)$ attains an optimum value, either a maximum or a minimum. The function F is called the *objective function*. When $S \neq \mathcal{R}^p$ the problem is said to be a *constrained optimization problem* and the set S is called the *constraint set*. When $S = \mathcal{R}^p$, the problem is said to be *unconstrained*. Since a maximum of $F(x)$ is a minimum of $-F(x)$, we may without loss of generality speak only of minimization.

Both maximum likelihood estimation and least-squares estimation are optimization problems, and both constrained and unconstrained examples arise naturally. If the function F is differentiable at its minimum value, which we shall denote by s, and if s is an interior point of S, then multivariate calculus tells us that the gradient function $F'(x) = \nabla F(x)$ has a zero at s. Hence, the solution to such a minimization problem necessarily is also a solution to the nonlinear system $F'(x) = 0$. The methods of section 4.3 automatically apply, so that those methods constitute an important special class of methods for solving optimization problems. For consistency of notation with section 4.3, we shall write $f(x)$, as a synonym for $F'(x)$.

Methods for nonlinear systems do not, however, completely solve the minimization problem, for several reasons. First, a solution to the system $F'(x) = 0$ may be a maximum, a minimum, or a saddle point of the objective function. A necessary condition for the solution s^* of the system to be a minimum is that $F''(s^*) = \nabla^2 F(s^*)$ be positive definite. Second, the solution to the system, even if $\nabla^2 F$ is positive definite at the solution, may represent only a local minimum and not the global minimum on S. Third, it is possible that the minimum of F on S occurs on the boundary of S, in which case the gradient will generally not be zero at the solution. Fourth, it is not unusual to seek minima of functions whose derivatives are either very difficult to obtain analytically or are very expensive to compute computationally. In this case, the nonlinear system whose solution we would seek is computationally difficult to work with. Fifth, it may be possible to specify an algorithm for evaluating $F(x)$ without having an explicit closed-form expression for F at all. In this case, it is impossible to compute the gradient $\nabla F(x)$ whose zero must be found.

We note here that Newton-like methods require both a gradient $\nabla F(x)$ and a Hessian matrix $\nabla^2 F(x)$. When neither of these is available analytically, discrete approximations must be used. For the Hessian matrix, either finite-difference methods or secant methods can be used. Moreover, the symmetry (and definiteness in statistical problems) of the Hessian should be incorporated in the approximation algorithm. Dennis and Schnabel (1983) devote a chapter to secant approximations in optimization problems, and they address these issues in detail. They also note, however, that there is no sufficiently accurate secant approximation to the gradient for use in minimization problems; finite-difference approximations *must* be used.

COMMENT. There is one more reason that methods for nonlinear systems of equations are not adequate for the general optimization problem, and that is that the objective function may not be differentiable! Because such problems arise only rarely in the course of data analysis, we shall discuss them only briefly here. Among problems of this nature arising in statistical applications, the two most likely to be encountered are nonlinear least absolute-value regression (NLAV) and nonlinear *minimax residual regression*. The latter problem minimizes the largest squared residual. Gill, Murray, and Wright (1981, Chapter 4) note that these problems can be transformed into smooth problems subject to nonlinear constraints for which special methods are available. Tishler and Zang (1982) propose an algorithm for the NLAV problem based on a smooth approximation to the actual objective function.

4.5.1 Grid search

When the objective function F is not well-understood, it is sensible to find out something about it, including such information as identifying a region of S in which the minimum of F is likely to lie, discovering how smooth F is, and determining whether F is likely to have multiple local minima. The *grid search* can sometimes provide such information.

The method is a classic example of the use of brute force. A lattice of points L is laid out on S (or on a plausible and manageable subset of S) and F is then evaluated at each point on the lattice. The point on L at which F attains its minimum is often a good initial estimate for the minimum on S. The values of F at nearby lattice points give some indication of the behavior of F, in particular whether it is relatively flat or peaked near the solution, and whether there are ridges near the minimizing value.

Every lattice $L \subset \mathcal{R}^p$ can be represented in the following form

$$L = \{x \mid x = x_0 + Ma, \quad a \in Z^p\}, \tag{4.5.1}$$

where x_0 is a given fixed point in $\mathcal{R}^p$ (called the *origin* of the lattice), and M is a $p \times p$ matrix with linearly independent columns. The columns of M are called the *Minkowski basis* for the lattice. The most common choice for M is a multiple of the identity matrix, but other choices are sometimes useful. Given the set of function values on $L \cap S$, it is useful to plot F in various cross-sections by holding all but one of the components of a fixed and varying the remaining component. Local non-monotonic behavior in these plots indicates that automatic root-finders must be used with great care.

In dimensions much higher than two, grid searches may not be practical, as the number of function evaluations grows exponentially in p. Even in low dimensions, one has the feeling that most of the function evaluations are being wasted, in that they occur far from the minimum. When little is known about the function, however, the information gleaned from the grid search may provide considerable insight into the stability and quality of any subsequent optimization solution, whose starting value may in turn have been obtained from searching on the grid.

In higher-dimensional problems the full grid search is impractical, but it is possible to search in one direction at a time by finding the minimum of F on a grid over which x_1 varies, with $x_2, \ldots, x_p$ held at fixed initial values. Then x_1 is fixed at this value, and x_2 is varied, and so forth. When no component of x changes after one full cycle, we terminate the iteration. We call this procedure *sequential search*. It is clear that this procedure will terminate under mild conditions on S or F (if S is bounded, for instance). It is equally clear that the resulting point need not be the global solution on the grid. The idea of the sequential grid search can be extended to use more efficient searching strategies on each, provided that F is sufficiently well-behaved.

The idea of sequential grid search is a simple one which can be refined considerably. The basic idea behind the sequential grid search is reminiscent of the Gauss-Seidel method for solving a system of equations, namely, we minimize $F(x_1, x_2, \ldots, x_p)$ with respect to one component at a time, holding all the others fixed at their most recently attained values. A single (major) iteration step consists of performing the (minor) computation just described p times, once for each parameter value.

In the grid-based search, three aspects of the problem are fixed from the outset: 1) the distance between possible evaluation points (*step size*), 2) the number of function evaluations at each minor step, and 3) the direction in which the search will take place at each step. Thus, for instance, we shall always be taking equally-spaced steps along a line (*transect*) parallel to the first coordinate axis at every p^{th} minor step. The improvements we can attain over sequential search arise from relaxing such restrictions, from incorporating additional information about F, or both.

A more general iteration can be obtained by specifying at each step 1) the line along which to search next (*search direction*), and 2) the strategy to employ in searching along the line (*step size*). Each component may depend upon the history of the iteration, as well as properties of F. The search strategy generally will involve taking steps of varying size along the selected line, the size of which will depend on progress made toward some interim goal. The next two subsections deal, respectively, with strategies for searching on a line and with minimization methods based on alternative ways for choosing the search direction.

4.5.2 Linear search strategies

We now turn to the problem of searching for minima along a given line. Suppose then that we are given a starting point x and a direction vector $d \in \mathcal{R}^p$, so that we are to explore the function $g(t) = F(x + td)$, for $t \in \mathcal{R}^1$, with a view toward finding a new point at which the objective function is smaller in magnitude. To achieve this goal in general, it is usually necessary for the function $g(t)$ to be decreasing at $t = 0$. Any direction d for which this is true is called an *admissible search direction;* we shall confine our attention to such choices of direction.

Linear searches can be undertaken with two rather different goals in mind. The first is to find the absolute minimum in the given search direction. The second is to find a step size which merely *reduces* the value of $g(t)$, but which also satisfies other conditions leading to desirable convergence properties. The latter problem is considered in section 4.5.3 in conjunction with our discussion of step sizes.

The first class of methods we consider for finding the minimum of g in a given direction is analogous to the class of bracketing methods for root-finding, discussed in section 4.2.4. To use any of these methods, we must start from an interval $[a, b]$ on which $g(\cdot)$ is known to have a minimum and on which $g(\cdot)$ is *strictly unimodal*. If such a bracketing interval is not known *a priori*, one can usually be found by evaluating g along a coarse grid of points, stopping as soon as $g(\cdot)$ begins to increase.

The strict unimodality of $g(\cdot)$ means that g is decreasing everywhere to the left of its minimum t^* and increasing everywhere to the right. Thus, if $a \leq t_1 < t_2 \leq b$, then $t^* \in [a, t_2]$ whenever $g(t_1) \leq g(t_2)$, and $t^* \in [t_1, b]$ whenever $g(t_1) \geq g(t_2)$. An alternative result is that, if $a \leq t_1 < t_2 < t_3 \leq b$, then $t^* \in [t_1, t_3]$ if and only if $g(t_1) > g(t_2)$ and $g(t_2) < g(t_3)$. To decrease the size of the interval, g must be evaluated at two interior points, t_1 and t_2. The same process can be applied repeatedly, narrowing the interval at each step, until the length of the resulting interval is acceptably small.

4.5.2.1 Golden-section search

How should the test points t_1 and t_2 be chosen? Note that the narrower interval that results will have either t_1 or t_2 as an interior point. Economy suggests that a good algorithm will require only *one* additional test point at each step, re-using the leftover test point from the previous iteration for the second interior test point. Absent additional information about the form of $g(\cdot)$, it seems sensible to place the points symmetrically in the interval, so that the two possible resulting subintervals and the interior test points they contain will be mirror images of one another. Finally, it is plausible to require that the relative sizes of the subintervals at each step be constant. These three conditions are sufficient to determine an algorithm called the *Golden-section search,* which is given in Algorithm 4.5.1. Each step reduces the interval length by the constant factor $\xi = (\sqrt{5} - 1)/2 \approx 0.618$.

Algorithm 4.5.1 (Golden-section search)

Choose a, b such that g (strictly unimodal) has minimum on $[a, b]$
$L := a$
$U := b$
$\alpha := (3 - \sqrt{5})/2$
$W := U - L$
$t_1 := L + \alpha W$
$t_2 := U - \alpha W$
for $i := 1$ **to** ∞ **until** convergence
    if $g(t_1) > g(t_2)$ **then**
        $L := t_1$
        $t_1 := t_2$
        $W := U - L$
        $t_2 := U - \alpha W$
    else $U := t_2$
        $t_2 := t_1$
        $W := U - L$
        $t_1 := L + \alpha W$

COMMENT. (OPTIMAL LINEAR SEARCH). We can define an *optimal linear search strategy* by associating with each possible strategy a "figure of merit" and declaring a procedure to be optimal if it achieves the minimum possible value (over all possible strategies) for this criterion. One such criterion is called the *minimax criterion,* for which the figure of merit is obtained as follows. Let s denote a strategy for linear search based on bracketing intervals, let g be the function to be searched, and let $L_M(g, s)$ denote the length of the final interval attained using strategy s with at most M function evaluations of g. The figure of merit associated with strategy s is, for fixed M, $m(s) \equiv \sup_g L_M(g, s)$, where the supremum is taken over some given class of functions, in our case, the strictly unimodal functions. A *minimax search strategy* s^* is one for which $m(s^*) = \inf_s m(s)$. If a strategy $s' = s'(\epsilon)$ depends upon a parameter ϵ in such a way that, for every positive ϵ, $m(s'(\epsilon)) \leq \inf_s m(s) + \epsilon$, then the strategy $s'(\epsilon)$ is said to be ϵ-*minimax*. Other figures of merit could be envisioned as well, such as the mean length of the final interval; such a criterion requires some probabilistic structure on the class of functions from which g is selected. The remainder of this section deals only with the minimax criterion.

If a fixed number M of function evaluations is to be carried out, the Golden-section search is not optimal, in the minimax sense, but it is not much worse than the optimal (the final interval is about 17% longer than that produced by a minimax procedure). A nearly optimal procedure, called *Fibonacci search* (Kiefer, 1953, 1957), differs from Golden-section searching in that the subinterval widths do not have constant ratio. Instead, the ratio of $(t_1 - L)$ to $(t_2 - L)$ at the kth step $(k < M - 1)$ is $\xi_k = \text{Fib}(M - 1 - k)/\text{Fib}(M - k)$, where $\text{Fib}(i)$ is the ith number in the *Fibonacci sequence*, defined by $\text{Fib}(0) = 1$, $\text{Fib}(1) = 1$, and $\text{Fib}(j) = \text{Fib}(j - 1) + \text{Fib}(j - 2)$ for $j > 1$. The formula above would have $\xi_{M-1} = 1$, which is impossible; the choice $\xi_{M-1} = 1/(1 + \epsilon)$ is ϵ-minimax; no minimax search procedure exists. The Golden-section search has a constant $\xi = (1 + \sqrt{5})/2$. Since $\lim_{k \to \infty} \xi_k = \xi$, and since this limit converges quite rapidly, the Fibonacci search and the Golden-section search are virtually identical except at the final few steps. Both Fibonacci search and Golden-section search are due to Kiefer (1953).

4.5.2.2 Local polynomial approximation

Analogies between root-finding and optimization are often helpful to consider. Golden-section search is similar in spirit to bisection, for instance. Another root-finding method with an analogous search strategy is the secant method (or *regula falsi*) on a bracketing interval.

When the zero of a smooth function on such an interval is sought, the secant method uses a straight line through the endpoints of the interval as a *local model* for the nonlinear function, and the zero of this approximating function is used as an estimate for the true zero. The approximate zero is then used as input to the next iteration, which refines the approximation. When seeking minima, particularly of convex functions on a bracketing interval, it is often helpful to use a quadratic function that matches the objective function at the endpoints and at an interior point as a local model for the objective function. The minimum of this quadratic can then serve as a second interior point, and a bracketing interval of reduced length can be determined as in the previous section. One can go further than this by using more interior points and using a cubic approximation. Alternatively, a cubic can be fitted to the objective function using three points if, in addition, information about the slope $g'(0)$ is used; Kennedy and Gentle (1980) discuss such a method and describe and algorithm due to Davidon (1959).

4.5.2.3 Successive approximation

A third method can be used when the point of minimization has been bracketed by an interval $[a, b]$, again assuming that $g(t)$ is strictly unimodal on this interval. This method, called *successive approximation*, is due to Berman (1966), a simple variant of which we describe first. The function $g(t)$ is evaluated on a grid which divides the current interval into $q > 2$ subintervals of equal length. Since g is strictly unimodal, it need not be evaluated at all of these points. Rather, we can stop evaluating g at new grid points as soon as we reach a grid point at which g increases. When this happens, we can then determine a narrower interval in which the minimum lies. The process is repeated until the resulting interval is sufficiently narrow. Because the function has already been evaluated at the endpoints, at most $M = q - 1$ function evaluations are required by this method, and the interval of uncertainty after a single full iteration is $2/q$ times the previous length.

It is difficult to provide a satisfactory analysis of the method of successive approximation, since the number of function evaluations at each step depends critically on the shape of g and the location of its minimum in the initial interval. (Berman's analysis, for instance, is faulty.) If one treats the minimum t^* as a random variable whose distribution is uniform on the interval $[a, b]$, and if one assumes that knowledge of g at points other than t^* does not affect the conditional distribution of t^* (so that the conditional distribution of t^* is uniform on every subinterval produced in the iteration), and if one assumes that the probability structure on g has suitable invariance properties, then it is possible to compute the expected behavior of this simple successive-approximation method (Thisted, 1986c). If the values of g at the endpoints of the current interval are inherited from the previous step, so that they do not need to be recomputed, then the average factor by which the interval length is reduced is 0.6961 for $q = 3$, and 0.6781 for $q = 6$; the average factor increases for $q > 6$. These average performance figures (obtained under rather unrealistic assumptions) are considerably worse than the constant performance of $\xi = 0.6180$ achieved by the Golden-section method.

The method just described is not the one proposed by Berman. While the algorithm above starts at the left-hand endpoint of the interval and marches in fixed steps to the right, Berman's method of successive approximation is slightly more refined. It starts at the *midpoint* of the interval and then takes a single step (in either direction). If the step results in an increase in g, we reverse field and take all subsequent steps on the other side of the midpoint, otherwise we continue taking fixed steps in the initial direction, continuing until g begins to increase (or an endpoint is reached). The refinement of starting at the middle makes it possible to use steps which are only half as large as the method described earlier, at the cost of some additional programming and the additional function evaluation resulting from taking a false step about half the time. The best choice of q for Berman's method is $q = 5$, so that at most 4 function evaluations occur at each step. The average interval reduction per function evaluation is 0.6216 in this case, only slightly worse than for Fibonacci search. Thisted (1986c) describes a modification to Berman's method that has average performance which is slightly *better* than that for Fibonacci.

4.5.3 Selecting the step direction

In section 4.5.2 we examined search methods for finding a minimum of the function $g(t) = F(x + td)$, $t \geq 0$, where d was a fixed search direction. We considered only directions d for which $g(t)$ was decreasing at $t = 0$; such directions are termed *admissible,* or *descent directions.* This section focuses on methods for determining a suitable direction in which to step.

4.5.3.1 Newton steps

The most natural direction to choose is that of the Newton-Raphson iteration applied to the Jacobian, that is, to use

$$d_N = -[F''(x)]^{-1}[F'(x)].$$

This will be a descent direction if $F''(x)$ is positive definite. Unfortunately, this need not be the case when x is not close to the minimum, so that the Newton direction d_N may not be admissible. A related phenomenon is that $F''(x)$ may be nearly singular, so that the size of d_N may be quite large. In this case, a full step in the Newton direction would move well away from the current value of the iterate x. Thus, even though a full Newton step is optimal when sufficiently near the solution, it may not be a strategic choice near the beginning of the iterative process, either in terms of direction or in terms of size. This observation has led to alternative methods which ameliorate these difficulties.

4.5.3.2 Steepest descent

The first criterion for a direction d is that it be a direction in which F is decreasing. Given this requirement, and the fact that Newton steps may not be admissible in this sense, an obvious approach is to choose that direction d_S in which F is decreasing most rapidly at x. This choice produces

$$d_S = -F'(x),$$

and a strategy based on this choice is called, naturally enough, the method of *steepest descent*. The method depends heavily on the local behavior of $F(x)$. It is often the case that the best local direction remains a good one only for a very short time, so that only very short steps in the direction "achieve the potential" of steep descent. Such a direction will be a good one only if a sufficiently large step can be taken in that direction before another iteration is required. Marquardt (1963) noted that in many practical problem the angle between d_N and d_S is between 80° and 90°, so that the best local direction can be quite different from the best global one!

4.5.3.3 Levenberg-Marquardt adjustment

It turns out to be fairly easy to adjust F'' so that it is always positive definite, hence, always producing an admissible search direction. Levenberg (1944) and Marquardt (1963) independently proposed a method which has several virtues: it is simple to compute, it always produces a descent direction, and near the solution, it behaves like Newton-Raphson. The idea is to add a small nonnegative constant to the diagonal elements of $F''(x)$ to obtain

$$d_{LM} = -[F''(x) + \lambda I]^{-1}[F'(x)],$$

where $\lambda > 0$. Note that this expression will be a descent direction whenever $\lambda > \lambda_{min}$, where λ_{min} is the smallest eigenvalue of $F''(x)$. As a consequence, the choice for λ depends upon x and on $F''(x)$, and may be zero when $F''(x)$ is safely positive definite. As the iteration progresses, the sequence of λs generally becomes smaller and smaller.

When $\lambda = 0$, of course, $d_{LM} = d_N$. On the other hand, when λ is large, the matrix $F''(x) + \lambda I$ is dominated by the multiple of the identity, so that the search direction itself is approximately a multiple of the steepest-descent direction d_S. Thus, the Levenberg-Marquardt direction may be viewed as a compromise between steepest descent and Newton-Raphson. This approach is sometimes referred to as the *Marquardt compromise*.

The Levenberg-Marquardt steps can be written in terms of the eigenstructure of $F''(x)$, and it is helpful to do so to interpret the iteration in statistical terms. Suppose that $F''(x)$ can be written in the form

$$F''(x) = \Gamma \Delta \Gamma',$$

where $\Gamma\Gamma' = I$ and Δ is a diagonal matrix of eigenvalues, say $\delta_1 \geq \ldots \geq \delta_p$. We can write the step direction as

$$
\begin{aligned}
d_{LM} &= -[F''(x) + \lambda I]^{-1}[F'(x)] \\
&= -[F''(x) + \lambda I]^{-1}[F''(x)][F''(x)]^{-1}[F'(x)] \\
&\equiv A(\lambda)d_N.
\end{aligned}
$$

Hence we can write

$$
d^*_{LM} \equiv \Gamma'd_{LM} = -\Gamma'A(\lambda)\Gamma\Gamma'd_N = \Delta(\lambda)d_N,
$$

where $\Delta(\lambda)$ is a diagonal matrix with elements $\delta_i/(\delta_i + \lambda)$. Whenever $F''(x)$ is positive semidefinite, all of the diagonal elements of $\Delta(\lambda)$ are less than unity, so that the Levenberg-Marquardt step can be seen as a "damped" version of the Newton step in which components associated with small eigenvalues of $F''(x)$ are discounted more heavily than those associated with the larger eigenvalues. Roughly speaking, the eigenvectors of $F''(x)$ corresponding to small eigenvalues represent directions in which there is (locally) very little curvature, hence conducive to large steps from the current location. If F were exactly quadratic, such large steps would be fine, but in every other case, the local behavior in such "flat" directions indicates essentially nothing about the local behavior in the vicinity of a point reached by taking such a large step. In statistical terms, we would say that there is little information in the data about the correct step-size to take in that direction, and that a priori we should discount heavily the apparent information. A Bayesian treatment of this problem is developed in the exercises for the special case of linear least squares.

One of the potential difficulties with the Newton step is that, when $F''(x)$ is nearly singular, $|d|$ can be large. Since the Newton method is based on a *local* quadratic model, we should only trust the model within a relatively small vicinity of the current value x, that is, for d which are not too large. This observation has led to algorithms called *model-trust region* methods which take a Newton step whenever $|d_N| < \mu$, and which take a Levenberg-Marquardt step otherwise, with λ chosen so that $|d_{LM}| = \mu$, and $\mu > 0$ a prespecified parameter. Dennis and Schnabel (1983) discuss this approach in depth.

In order to achieve quadratic convergence near the solution, it is necessary that the sequence $\lambda_i \to 0$ in an appropriate fashion. Dennis and Schnabel (1983) detail the considerable progress that has been made on this subject in the last decade, and also give algorithms for the most promising methods.

4.5.3.4 Quasi-Newton steps

In the context of solving nonlinear systems, we introduced the quasi-Newton methods of Davidon, Fletcher, and Powell, and of Broyden, Fletcher, Goldfarb, and Shanno. These methods can, of course, be applied in the optimization context as well, and each is designed to maintain positive definiteness of the update, hence, insuring a descent direction. The disadvantage of quasi-Newton methods, as mentioned above, is that the matrix which plays the role of the Hessian may not be an adequate approximation to the Hessian itself, even after convergence has been attained. Thus, such methods provide no reliable estimate which can be used for the asymptotic variance matrix of the estimated parameters in maximum-likelihood and nonlinear regression problems.

4.5.3.5 Conjugate-gradient methods

When the amount of available main memory is a constraint, it may not be possible to maintain a full $p \times p$ matrix corresponding to the Hessian. Particularly in the analysis of designed experiments with large numbers of parameters, it may be impractical to maintain the original data matrix, and the Jacobian, and the Hessian matrices simultaneously. The *method of conjugate gradients* is an optimization method which generates the next search direction without ever storing these matrices explicitly. While the need for such methods in statistical problems is rare—the problem must be too large to fit easily in main memory, but not so large that even the underlying data matrix must be stored in external storage—the method is an interesting and occasionally useful one. Because the Hessian matrix is not computed, there are no readily available standard errors for the parameters estimated using a conjugate-gradient algorithm. This fact, too, limits their general utility for statistical applications. For these reasons we do not elaborate further on the method here.

The conjugate-gradient method was originally devised by Hestenes and Stiefel (1952) for solving large linear systems; in this form the algorithm is quite simple and is referred to as the *linear conjugate-gradient method*. McIntosh (1982) discusses the statistical applications of conjugate gradients in considerable detail, with particular emphasis on linear and generalized linear models. Section 4.8.3 of Gill, Murray, and Wright (1981) is devoted to the numerical aspects of conjugate-gradient algorithms, their relationship to other methods such as quasi-Newton methods, and alternative algorithms for the limited-memory problem.

4.5.4 Some practical considerations

Scaling. When faced with an optimization problem and a software package to do the work, a statistician must make a few choices which can materially affect the progress—or even the success—of the iterative process. One of the choices often made by default is that of *scaling the variables* in the problem. While the Newton-Raphson iteration is unaffected by merely rescaling the variables, other methods such as steepest descent and Levenberg-Marquardt are affected considerably, as shown in the exercises. The best general advice that numerical analysts give is to scale the variables so that they are of comparable magnitude. Lying behind this advice is an implicit assumption that variables encountered in practice are often measured on different scales, but have comparable coefficients of variation, and in many scientific settings this is approximately true.

It is interesting to see what this advice entails in the statistical context of regression. In the linear-regression setting, we commonly include a constant term whose effect is first removed from all other independent variables in the model. This makes the independent variables all have "similar magnitude," in the sense that they all have zero mean. Now the solution to the linear least-squares problem is equivalent to taking a single Newton step from any starting estimate of the regression parameter β, so that after removing means, further adjustment to the magnitudes of the regressor variables would have no effect except, perhaps, in terms of an algorithm's numerical stability. In the nonlinear regression setting, however, further scaling can be all-important. Rescaling can change the direction of steepest descent, for instance, and spherical model-trust regions seem most appropriate when the different components of d are approximately on the scale. In statistical terms, this suggests making the variances of the regressor variables approximately equal in magnitude, that is, by reducing the $X'X$ matrix to correlation form. The problem can be carried out in this rescaled form, and then transformed back to the original scale at the end. In general optimization problems, the appropriate rescaling makes the diagonal elements of $F''(x)$ approximately equal. If the starting value for the iteration is fairly good, rescaling probably needs to be done only once, at the beginning of the optimization process. It is sometimes necessary, however, to do rescaling periodically during the iteration, as the shape of F'' changes.

The choice of scale also affects the stopping criterion. Suppose, for instance, one chooses to stop when the step sizes are sufficiently small, say when $|d| < \epsilon$. If the typical size of one of the components is small compared to the others its convergence (or lack of it) will play little role in the decision about stopping. The relative size of parameters in a regression depend entirely on the scaling of the independent variables. Whether the estimation process is said to have converged should not really depend upon whether time is measured in microseconds or decades. It is sensible, then, on pragmatic data-analytic grounds to give each of the parameters equal weight in the convergence test; for convergence criteria such as the one just mentioned, this entails rescaling the variables to have approximately correlation form.

Software. The choices to be made in implementing an optimization algorithm are many and varied, including such issues as choosing a step direction, choosing a step size, how to conduct a line search when that is necessary, how to compute or approximate derivative matrices, whether to use secant approximations, how to choose the damping parameter λ in Levenberg-Marquardt, and a host of others. Each combination of choices produces an essentially unique algorithm. It is generally possible to find subroutines or programs that implement one or more of these combinations, but no one of these choices dominates the others for the wide variety of problems encountered in practice. Since these subroutines must be obtained from a variety of sources and are seldom "plug-compatible," the strategy most often employed by statisticians needing an optimization routine was to pick one from among those most readily available at the local computation center, and then to hope that it would work adequately.

Schnabel, Koontz, and Weiss (1985) have recently described a modular system of algorithms for the unconstrained optimization problem, called UNCMIN. Their system really consists of an interface to a large collection of possible algorithms, each of which represents a combination of choices such as those outlined above. What is distinctive is that UNCMIN appears to the user to be a single optimization subroutine, the particular algorithm employed depending on a set of parameters in the subroutine call. Thus, for instance, to shift from an algorithm employing a finite-difference Hessian from analytic gradients to one using a BFGS update requires changing only a single parameter in the subroutine call. Altogether, UNCMIN allows eighteen different algorithmic combinations, all tightly coded and taking advantage of the best numerical practice.

4.5.5 Nonlinear least squares and the Gauss-Newton method

The *nonlinear least-squares problem*, or the *nonlinear regression problem*, is probably the context in which optimization most frequently arises in practical statistical work. We shall employ the familiar regression terminology in this section, in which the unknown parameter is denoted by θ and the regressor variables are denoted by x. Both θ and x may be multidimensional. Consider, then, the nonlinear regression model

$$\begin{aligned} y_i &= g(x_i, \theta) + \epsilon_i \\ &\equiv g_i(\theta) + \epsilon_i, \end{aligned} \tag{4.5.2}$$

where $i = 1, 2, \ldots, n$, θ is a $p \times 1$ vector of unknown parameters, and x_i is a $k \times 1$ vector of covariates corresponding to the ith observation. If we take $\epsilon_i \sim \mathcal{N}(0, \sigma^2)$, then the maximum likelihood estimator for θ is that which minimizes

$$F(\theta) = \sum_{i=1}^{n} (y_i - g(x_i, \theta))^2 = \sum (y_i - g_i(\theta))^2, \qquad (4.5.3)$$

the sum of squared residuals. Whether or not the errors have a normal distribution, this expression can be minimized to obtain an estimate for θ, which is naturally enough called the (nonlinear) least-squares estimator.

COMMENT. Note that in equations (4.5.2) and (4.5.3), we have written the model function in two different ways, as $g(x_i, \cdot)$ and as $g_i(\cdot)$. These different notations are used interchangeably; the first focuses on the function g, the common aspect of the model, whereas the latter notation focuses on the differences in the mean function from one observation to the next. In this section we shall also adopt the notational convention that g' denotes the vector of derivatives of g with respect to the components of θ, and $(g')^t$ will denote the transpose of this vector.

We can, of course, proceed as above to obtain the gradient vector and Hessian matrix of $F(\theta)$ and proceed as before. However, the full Newton approach can be quite costly. To see why this is so, let us examine the gradient and Hessian for $F(\theta)$, which are given by

$$F'(\theta) = -2 \sum (y_i - g_i) g_i'$$

$$\equiv -2 \sum r_i g_i' \qquad (4.5.4)$$

$$F''(\theta) = -2 \sum \left\{ (y_i - g_i) g_i'' - [g_i'][g_i']^t \right\}$$

$$= -2 \sum \left\{ r_i g_i'' - [g_i'][g_i']^t \right\} \equiv -2(A + G), \qquad (4.5.5)$$

where we have suppressed the dependence of g and its derivatives on θ. Note that g_i' is a $p \times 1$ vector, and g_i'' is a $p \times p$ matrix. Thus, in the nonlinear regression problem the function g must be evaluated n times just to obtain *one* evaluation of the objective function F. What is more, obtaining $F'(\theta)$ requires evaluation of n gradient vectors $g'(\theta)$, and $F''(\theta)$ requires that n Hessian matrices $g''(\theta)$ be computed. If n is at all large, these computations may become prohibitively expensive.

The Newton methods are based on approximating the objective function at a current estimate θ_0 by a quadratic, and then minimizing the approximating quadratic to obtain an update for θ. Another approach, similar in spirit, can be applied in the nonlinear regression problem. Instead of approximating (4.5.3) by the leading terms of its Taylor series, we could do the same thing for the nonlinear function g in the model (4.5.2). Stopping at the linear term in this expansion about θ_0 produces a *linear* regression problem as an approximant to the nonlinear regression problem.

We write

$$y_i = g_i(\theta) + \epsilon_i$$
$$\approx g_i(\theta_0) + (\theta - \theta_0)^t g_i'(\theta_0) + \epsilon_i. \qquad (4.5.6)$$

The latter expression can be rearranged to obtain

$$y_i - g_i(\theta_0) = [g_i'(\theta_0)]^t (\theta - \theta_0) + \epsilon_i$$

or, redefining terms on each side,

$$Y_i = X_i^t (\theta - \theta_0) + \epsilon_i.$$

From this expression, we can solve for the updated $\theta - \theta_0$ using ordinary least-squares methods, obtaining

$$\theta - \theta_0 = (X^t X)^{-1} X^t Y = \left[\sum (g_i')(g_i')^t \right]^{-1} \sum g_i'(y_i - g_i)$$
$$= - \left[-\sum (g_i')(g_i')^t \right]^{-1} \left(\sum g_i'(y_i - g_i) \right) \qquad (4.5.7)$$
$$= -[F'' - A]^{-1}[F'] = -G^{-1}[F'].$$

This iterative procedure is known as the *Gauss-Newton method* for nonlinear least squares.

How well does Gauss-Newton do relative to other methods? It requires no fewer function or gradient computations than the full Newton approach, but it omits the n Hessian computations. There is a price to be paid for these computational savings—Gauss-Newton may converge very slowly or fail to converge at all, even from starting values close to the solution. To see why this might be the case, it is instructive to compare (4.5.7) to the Newton update, $-[F'']^{-1}[F']$, to which it is very similar in form. The exact Hessian can be written as in (4.5.5) as a sum of two terms $F''(\theta_0) = A + G$. The Gauss-Newton iteration uses only the second of these two terms as its approximate Hessian. The omitted component is $A = \sum r_i g_i''$, where r_i is the ith residual. When all of the residuals are nearly zero, this term is negligible. In this case, Gauss-Newton converges rapidly, and, because of the omitted expense associated with the second-order calculations, can produce superior performance to that of Newton-Raphson. (If A is exactly zero, the convergence is quadratic; if it is small relative to G the convergence is linear; see Dennis and Schnabel ,1983) Unfortunately, when the residuals at the solution are large—and this is often the case in statistical data analysis—the matrix A is large relative to G and the Gauss-Newton iteration either will diverge or will converge with painful timidity.

A compromise approach is described by Dennis, Gay, and Welsch (1981), who provide an adaptive algorithm for the nonlinear least-squares problem. Their algorithm, called NL2SOL, computes the matrix G exactly, but approximates the second-order term A using a secant approximation.

The algorithm incorporates an heuristic for deciding whether to use the estimate for A or not. Starting initially with Gauss-Newton, the algorithm switches from one method to the other whenever the current method produces a bad step while the alternative method would produce a good one. This algorithm converges more reliably than Gauss-Newton and variants based on it (such as Levenberg-Marquardt) in large-residual situations, while still being far more efficient than Newton-Raphson, or even secant methods such as BFGS and Davidon-Fletcher-Powell applied to the full Hessian matrix $F''(\theta)$ throughout the iteration. NL2SOL is notable for the care given to convergence criteria and to providing a good estimate of the asymptotic covariance matrix of the estimated parameters. It is interesting to note that such a carefully coded package is not trivial to program; the FORTRAN program for NL2SOL consists of more than 13,000 lines of code!

COMMENT. The matrix $G = -\sum (g')(g')^t$ in Gauss-Newton is often nearly singular. The Levenberg-Marquardt idea can be applied in this case, using $G + \lambda I$ instead of G, for some positive damping factor λ. Indeed, the Levenberg-Marquardt method was devised in the context of nonlinear least-squares.

COMMENT. Because the matrix G is often ill-conditioned in practical nonlinear least-squares problems, it is doubly important not to form the matrix G explicitly in either the Gauss-Newton or the Levenberg-Marquardt approach. Rather, a factored form of G should be employed here, just as in the linear least-squares problem. Maintaining either a Cholesky factorization, or, if there is sufficient memory, using the Householder orthogonalization algorithm, is generally satisfactory.

COMMENT. Gauss-Newton methods for nonlinear regression cause difficulty for statistical inference on the parameter vector θ, in that they rarely produce an acceptable estimate for the asymptotic variance matrix of the estimates. If the likelihood function is locally quadratic, then the inverse Hessian matrix of (4.5.5) is proportional to the asymptotic variance matrix. But we have seen that the matrix produced by the Gauss-Newton procedure can be markedly different from the correct Hessian matrix, even when the procedure converges. Since much software for nonlinear regression is based upon Gauss-Newton or related methods, the reported standard errors for parameter estimates can be quite unreliable.

COMMENT. The Gauss-Newton algorithm is a method to minimize the sum of squared residuals $R'R$, where $R = Y - G(\theta)$ is the $n \times 1$ vector of residuals. If the responses y_i are q-dimensional rather than one-dimensional, one can construct a model similar to (4.5.2), except that the model function $g_i(\theta)$ is now q-dimensional. The parameter θ can be estimated by minimizing some generalization of the residual sum of squares. For example, one could minimize the determinant of the residual SSCP matrix $R'R$. Bates and Watts (1984) give an algorithm that does just this, generalizing the idea of the Gauss-Newton method to multi-response data. Their algorithm is in FORTRAN and makes extensive use of LINPACK.

Because of its similarity to Newton's method, the Gauss-Newton approach can be used as the starting point for most of the modifications applied to Newton-Raphson, including quasi-Newton updates, discrete approximations to g', and other methods which "compromise" between the Newton direction and the Gauss-Newton direction for search. The interested reader is directed to Gill, Murray, and Wright (1981), and to Dennis and Schnabel (1983) for further discussion and references. A detailed survey of various quasi-Newton methods, which includes discussion of their relationships to one another, motivation, derivation, and convergence properties can be found in Dennis and Moré (1977).

4.5.6 Iteratively reweighted least squares

We now turn to an important special case in which the nonlinear regression (or maximum-likelihood problem) exhibits linear structure in the parameter vector θ. Consider, for instance, the model in which

$$y_i = g(x_i^t \theta) + \epsilon_i,$$

where now $g(\cdot)$ is a scalar function, and the usual assumptions of zero mean and constant variance apply to ϵ. This model asserts that the mean of y depends on x only through some *linear* combination of the covariates. What does the Gauss-Newton method produce when applied to nonlinear least-squares estimation of θ in this special context? First, we approximate the systematic part of y by a linear function in θ, expanding about an initial point θ_0, to obtain

$$y_i = g(x_i^t \theta) + \epsilon_i$$

$$\approx g(x_i^t \theta_0) + \left. \frac{d}{d\theta} g(x_i^t \theta) \right|_{\theta_0} (\theta - \theta_0) + \epsilon_i$$

$$= g(x_i^t \theta_0) + g'(x_i^t \theta_0) \cdot x_i^t (\theta - \theta_0) + \epsilon_i.$$

Writing

$$r_i = r_i(\theta_0) = y_i - g(x_i^t \theta_0)$$

and

$$w_i = w_i(\theta_0) = g'(x_i^t \theta_0),$$

so that

$$r_i = w_i x_i^t (\theta - \theta_0) + \epsilon_i,$$

we have expressed the original problem as (approximately) a weighted least-squares problem which can be written in matrix form as

$$R = W X (\theta - \theta_0) + \epsilon.$$

The solution to this WLS problem gives the update to θ as

$$\hat{\theta} - \theta_0 = [X^t W^2 X]^{-1} [X^t W R]. \tag{4.5.8}$$

Equivalently, we may write (4.5.8) as

$$\hat{\theta} - \theta_0 = [X^t W^2 X]^{-1} [X^t W^2 R^*],$$

where $R^* = W^{-1} R$; this makes it somewhat easier to recognize the update to θ as the weighted regression of $W^{-1} R$ on X, with weights w_i^2. Since both the residual vector R and the diagonal matrix of weights W depend upon the initial estimate θ_0, the entire process is repeated with the updated value of θ until convergence is obtained; this process is aptly described as *iteratively reweighted least squares (IRLS)*. The iteration can also be derived from the condition that, at the solution, the orthogonality condition

$$R(\theta)^t W(\theta) X = 0 \tag{4.5.9}$$

must be satisfied.

Several comments are in order. First, the development above is based upon the least-squares criterion. Nonlinear least squares and maximum likelihood will coincide here whenever the error distribution of ϵ is normal. Second, as a solution to a nonlinear least-squares problem, IRLS *is* Gauss-Newton, so that it may be expected to perform poorly in large-residual problems or when the function g is highly nonlinear over the range of the covariates at the solution.

IRLS can also be applied in maximum likelihood problems when the distribution of y depends on the parameters only through a linear function in θ, provided that the form of the likelihood is sufficiently nice. This is the case for the generalized linear models introduced in Chapter 3. In particular, suppose that y_i is from a (scaled) exponential family, so that we may write the likelihood for y_i as $\ell(\xi_i) = (y_i \xi_i - b(\xi_i))/a(\phi) + c(y_i, \xi_i)$. For simplicity, we shall consider the dispersion function to be constant, say $a(\phi) \equiv 1$. Then the likelihood equation for ξ_i is $\dot{\ell}(\xi_i) = y_i - b'(\xi_i) = 0$. Taking expectations of both sides, we have that $E(y_i) = b'(\xi_i)$. Suppose now that the mean of y_i has the special structure discussed above, namely, that we can write

$$E(y_i) = b'(\xi_i) \equiv g(x_i^t \theta).$$

A standard computation shows that the likelihood equations for θ are given by

$$\sum [y_i - g(x_i^t\theta)]\frac{g'(x_i^t\theta)}{b''(\xi_i)}x_i = 0,$$

where ξ_i is implicitly defined by $b'(\xi_i) = g(x_i^t\theta)$. What is more, this has precisely the form of the orthogonality conditions (4.5.9), where here $r_i = y_i - g(x_i^t\theta)$, and $w_i = g'(x_i^t\theta)/b''(\xi_i)$. Thus, the same weighted least-squares iteration can be used to obtain maximum-likelihood estimates for the generalized linear model.

COMMENT. McCullagh and Nelder (1983) provide an alternative derivation of the iteratively reweighted least-squares algorithm for generalized linear model, treating as well the nonconstant dispersion parameter. Their derivation is based on the Fisher scoring method in the scaled exponential family. It is interesting to note that the form of the nonlinear mean function g can be chosen so that the omitted portion of the approximating Hessian matrix (A in our discussion of Gauss-Newton) is identically zero. This will occur when $\xi_i \equiv x_i^t\theta$, which in turn occurs when the *canonical link function* is used in the theory of the generalized linear model. (The canonical link function is the one which makes the natural parameter ξ of the exponential family a linear function in the covariates.) In this case, IRLS is equivalent to Newton-Raphson, and the observed and the expected information matrices for θ coincide. As one would expect from the behavior of the general Gauss-Newton procedure, convergence difficulties often arise when using IRLS to fit generalized linear models using non-canonical links unless the model fits the data quite well.

Jørgensen (1983) extended the class of generalized linear models to include correlated observations and nonlinear predictors. He showed generally that the likelihood equations can be interpreted as weighted least-squares problems, and that Fisher's method of scoring produces an IRLS procedure in which the dependent variable is a modified response variable.

Stirling (1984) has shown that an IRLS iteration can be constructed that coincides with Newton-Raphson-based maximum likelihood estimation whenever the distribution of y_i depends only on $\eta_i = x_i'\theta$, and the observations y_i are independent observations with density, say $f(y_i, \eta_i)$. Writing $z_i^* = f'(y_i, \eta_i)$ and $w_i = -f''(y_i, \eta_i))$, we have $\dot{\ell}(\theta) = X'Z^*$ and $\ddot{\ell}(\theta) = -X'WX$, where X and Z have rows x_i and z_i^*, respectively, and $W = diag(w_i)$. The Newton-Raphson approach to solving the likelihood equations sets

$$\hat{\theta}_{new} = \hat{\theta} + (X'\hat{W}X)^{-1}X'\hat{Z}^*$$
$$= (X'\hat{W}X)^{-1}X'\hat{W}(X\hat{\theta} + \hat{W}^{-1}\hat{Z}^*) \qquad (4.5.10)$$
$$\equiv (X'\hat{W}X)^{-1}X'\hat{W}\hat{Z},$$

where $\hat{W}$ and $\hat{Z}^*$ denote the corresponding matrices evaluated at the current value of $\hat{\theta}$. The "response variable" in this weighted regression of Z on X is an adjusted estimate for the parameter η, that is,

$$\hat{z}_i = \hat{\eta}_i - \frac{f'(y_i, \hat{\eta}_i)}{f''(y_i, \hat{\eta}_i)}.$$

Not only does this formulation cover GLIM models with constant dispersion function, it can also be applied to cases outside the exponential family, including examples of censored, grouped, and truncated data, and estimation of regression models involving the beta-binomial and negative binomial distributions (Stirling, 1984).

Wedderburn (1974) has introduced the notion of *quasi-likelihood functions*, which can be used for parameter estimation based on models defined only in terms of second-order properties. More precisely, we have independent observations y_i with means $E(y_i) = \mu_i$ and variances $V(\mu_i)$, where $V(\cdot)$ is a function either known completely, or known up to a scalar multiple. The quasi-likelihood $K(y_i, \mu_i)$ is defined by treating the standardized quantity $(y_i - \mu_i)/V(\mu_i)$ as if it were a score function, that is, by solving the differential equation

$$\frac{\partial K(y_i, \mu_i)}{\partial \mu_i} = \frac{y_i - \mu_i}{V(\mu_i)}.$$

(In view of this definition, $K(y_i, \mu_i)$ should be thought of as a "quasi-*log*-likelihood function," but the extra verbiage and punctuation involved in actually calling it that are not worth the trouble.) Suppose now that μ_i is a function g_i of parameters $\theta_1, \theta_2, \ldots, \theta_p$. Then parameter estimates can be obtained by maximizing the quasi-likelihood function $\sum K(y_i, \mu_i)$; call the estimates so obtained *maximum quasi-likelihood (MQL) estimates*. When $V(\mu_i) = 1$, the MQL equations reduce to a least-squares problem, which can be solved, say by the Gauss-Newton method. In a strong sense, MQL estimation bears the same relationship to maximum-likelihood estimation in exponential families as least-squares does to normal-theory maximum likelihood. Wedderburn (1974) and McCullagh (1983) examine the theoretical aspects of MQL estimation. Of computational interest is the fact that MQL problems with general variance functions can be solved by an IRLS algorithm that is a weighted version of the Gauss-Newton regression with weights $1/V(\mu)$. Wedderburn shows that this algorithm is equivalent to the method of scoring for solving the quasi-likelihood equations.

Green (1984) gives a comprehensive overview of IRLS and its relationship to GLIM models, robust regression, exponential-family estimation, and Gauss-Newton algorithms.

COMMENT. One should note that it is often the case that a single maximum likelihood problem can be solved using one of *several* IRLS formulations. Gauss-Newton and Newton-Raphson generally produce different iterations, although each can often be expressed as an IRLS iteration. Different parameterizations produce different IRLS when each is derived using the weighted Gauss-Newton approach. Green (1984) shows how three distinct IRLS procedures can be derived quite naturally for the linear regression problem. (The three methods coincide when the errors have a normal distribution.) The speed of convergence and the sensitivity to starting values can depend critically on such choices as parameterization.

4.5.7 Constrained optimization

The emphasis in this chapter up to this point has been on optimization without constraints on the solution vector θ. Occasionally, problems with constraints do arise in statistical problems, and dealing with them constructively is not always easy. The additional complexities introduced by the constraints generally require a combinatorial search through a tree of unconstrained problems, in each of which a particular subset of the constraints is enforced. Because of their combinatorial nature, we defer a detailed review of these methods to Chapter 9.

There are, however, some rough approaches which have the double virtues of illustrating the nature of the problem and of sometimes producing the solution. These quick and dirty approaches are worth mentioning, because they often work tolerably well in one-shot problems. The easiest approach falling in this category is to solve the unconstrained problem, if that is possible, and to hope that it satisfies the constraints. This "solution" is rarely more than a pious hope, and more systematic approaches to the problem are required.

The general problem can be expressed in the form

$$\text{minimize } F(\theta)$$

subject to

$$c_1(\theta) = 0$$

$$\vdots$$

$$c_E(\theta) = 0 \qquad (4.5.11)$$

$$c_{E+1}(\theta) \geq 0$$

$$\vdots$$

$$c_{E+I}(\theta) \geq 0;$$

we have here E *equality constraints* plus I *inequality constraints*. The current state of the art varies widely, depending upon the nature of the objective function $F(\theta)$ (linear, quadratic, other), the linearity of the constraint functions $c_j(\theta)$, and the sense of the constraints (equality or inequality). (It is easy to show that any equality constraint can be re-expressed in terms of inequality constraints; in practice somewhat different methods are used in the two situations.) Two situations, for which $F(\theta)$ is either linear or quadratic, are relatively simple and have quite satisfactory solutions.

4.5.7.1 Linear programming

First, the *linear programming problem* has a linear objective function $F(\theta)$ and a set of linear constraint functions $c_j(\theta)$, which satisfy either equalities or inequalities. This problem has been widely studied and is beyond the scope of this book. The reader is referred to Chapter 8 of Fletcher (1981), and to Chapter 5 of Gill, Murray, and Wright (1981) for detailed discussion and bibliographies. Standard books on operations research treat linear programming at length. The relationship of linear programming to least absolute value regression is discussed in Section 3.12.1. For all but the simplest linear programs, it is advisable to use one of the many commercial computer programs available rather than to write a new program from first principles.

4.5.7.2 Least squares with linear equality constraints

The second "easy" problem is that of linear least-squares regression with *linear* equality constraints. In formulation (4.5.11), $F(\theta)$ is quadratic in θ and $I = 0$. This problem can be restated in the following form: minimize $|Y - X\beta|^2$ subject to $C\beta = 0$, where X is $n \times p$, Y is $p \times 1$, β is $p \times p$, and C is $m \times p$ with $m \leq p$. Assume that C is of full (row) rank. Let A represent any $(p - m) \times p$ matrix of rank $p - m$ whose rows are mutually orthogonal and are also orthogonal to the rows of C. Thus $CA' = 0$ and $AA' = I_{p-m}$. Then the matrix of regressor variables can be written as

$$X = ZA + WC, \tag{4.5.12}$$

where Z is $n \times (p - m)$ and W is $n \times m$. For the constraint to be satisfied $X\beta = ZA\beta \equiv Z\gamma$, where γ is now a parameter of dimension $p - m$. Thus, a set of m linear equality constraints simply reduces the dimension of the problem by m. The least-squares estimator for γ is $\hat{\gamma} = (Z'Z)^{-1}Z'Y$.

Under the stated conditions, the constrained least-squares estimate of β is given by $\hat{\beta} = A'\hat{\gamma}$; the proof is left as an exercise.

The estimate so obtained can be thought of in the following way. Each of the m constraints allows us to *eliminate* one of the components of β by solving for it in terms of those components that remain. This process is closely related to methods for solving linear systems by Gaussian elimination, and this process is called *solution by elimination* of the equality-constrained least-squares problem.

4.5.7.3 Linear regression with linear inequality constraints

One type of constraint that often arise in regression problems is that some or all of the components of θ be non-negative. The linear least-squares model subject to such constraints is an excellent setting in which to illustrate the general problem of constrained optimization. This problem is referred to in the optimization literature as the *quadratic programming problem*, a lucid overview of which may be found in Fletcher (1981). Note that a consistent set of linear inequalities of full rank can be reduced by reparameterization to a set of nonnegativity constraints; the proof is an exercise. Some notation will be helpful in discussing the problem. Let $\mathcal{I}$ denote the set of m indices j for which θ_j is required to be nonnegative, and denote by $\hat{\theta}$ the solution to the constrained problem. Thus, by definition, $\hat{\theta}_j \geq 0$ for all $j \in \mathcal{I}$. It is possible that $\hat{\theta}$ coincides with the unconstrained solution. This will happen whenever $\hat{\theta}_j > 0$ for all $j \in \mathcal{I}$, so that none of the constraints are brought into play. In this case, all m of the constraints are said to be *inactive*. More generally, however, one or more of the inequalities must be enforced, so that $\hat{\theta}_j = 0$; such inequalities are said to be *active constraints*.

Let $\mathcal{C}$ (for Constraint) denote the indices corresponding to active constraints, and let $\mathcal{M}$ (for Model) denote the set of all other indices (corresponding both to inactive constraints and to unconstrained parameters). Clearly the constrained solution to the overall problem is equivalent to the unconstrained solution to the regression problem whose model contains only those X_j for which $j \in \mathcal{M}$. So the constrained regression problem can be reduced to that of determining the optimal subset $\mathcal{C}$ (or $\mathcal{M}$) of parameter indices. Doing so usually requires computation of many different trial solutions to the regression problem, building up $\mathcal{M}$ from a sequence of regression models.

Determining $\mathcal{M}$ is similar both conceptually and computationally to the problem of finding the best subset of regressor variables of a given size (all-subsets regression). In each case there are obvious stepwise algorithms which proceed by omitting one variable from $\mathcal{M}$ at a time. Such algorithms need not give the optimal solution, but often produce an adequate one. Consider, for instance, the following method.

All variables begin in the set $\mathcal{M}$, and the set $\mathcal{C}$ is empty. We begin by fitting the linear model using the model set $\mathcal{M}$. If any of the non-negative parameters are estimated to be less than zero, then one of them is chosen to be moved from $\mathcal{M}$ into $\mathcal{C}$. This variable is then deleted from the model (in effect, constrained to be zero). When there is more than one such variable, the one whose t-statistic is largest in magnitude is selected. At this point the reduced model using model set $\mathcal{M}$ is fit. If the constraints are not satisfied, another variable is moved from $\mathcal{M}$ into $\mathcal{C}$, and the process is repeated. If, on the other hand, the constraints *are* now satisfied, then one of two situations must hold. Either the partial correlations of the response variable with the variables in $\mathcal{C}$ are all negative, or at least one is positive. In the former case, the parameter estimates are given by the most recent least-squares fit for variables in $\mathcal{M}$ and by zero for the variables in $\mathcal{C}$, and this is a solution (not necessarily optimal) to the constrained linear least-squares problem. In the latter case, the variable with the largest positive partial correlation is moved from $\mathcal{C}$ into $\mathcal{M}$, and the fitting process is repeated one more time.

This cyclical process is easy enough to carry out by hand for single problems of moderate size and complexity; we refer to it as the "manual method." It is easy to show that a variable which moved from $\mathcal{M}$ to $\mathcal{C}$, or vice versa, cannot move in the opposite direction at the next iteration, thus eliminating the possibility of cycles of length one. Whether cycles of greater length can occur, and under what conditions, has not been studied systematically. It is also not known how badly this method performs in the worst case.

With m nonnegativity constraints, 2^m regressions are in principle required to determine the optimal solution satisfying the constraints. However these regressions are closely related to one another, and several algorithms exploiting their relationships exist which require far fewer regressions to be computed. No algorithm has been proven to produce the optimal solution using a number of regression evaluations bounded in a polynomial in m, although in practice most perform acceptably. Kennedy and Gentle (1980) outline three approaches. Lawson and Hanson (1974) describe an approach similar in spirit to the "manual" method described above, based on the *Kuhn-Tucker optimality conditions*. They also give an efficient FORTRAN implementation. Gill, Murray, and Wright (1981) discuss linear constraints in the more general setting of constrained optimization of general objective functions $F(\theta)$. A readable overview of the least-squares problem with linear inequality constraints is given in Chapter 10 of Fletcher, (1981).

Farebrother (1986) considers inferential aspects of these models. He has undertaken an extensive study of testing the constraint hypothesis $R\beta = c$ against what might be termed an *orthant alternative*, $R\beta \geq c$, in which at least one of the inequalities holds strictly. When R $(k \times p)$ has full row rank, the null distribution of the test statistic is a weighted sum of χ^2 random variables. The weights are each p-dimensional normal orthant probabilities.

4.5.7.4 Nonquadratic programming with linear constraints

When the objective function $F(\theta)$ is not quadratic in θ, additional difficulties come into play since, given a model set $\mathcal{M}$, it is no longer possible to find the unconstrained optimum for this (restricted) problem in a single step. Most methods for the general problem are based on *active set algorithms*, of which the "manual method" of the previous section is an example. A value for θ is said to be *feasible* if it satisfies all of the constraints of the problem. At each stage of an active-set method, one must first find a feasible point exactly satisfying the constraints in the current active set $\mathcal{C}$. Then one must iterate (through a sequence of feasible points) until convergence is obtained or until another, inactive, constraint becomes active. The nature of the iterative components of these methods are simply variants of those employed in the unconstrained problem which preserve feasibility of the iterates. These algorithms can be decomposed into algorithms for selecting a feasible direction in which to search, and algorithms for determining an appropriate step length. See Chapter 5 of Gill, Murray, and Wright (1981) or Chapter 11 of Fletcher (1981) for an extensive discussion of the alternatives. Wright and Holt (1985) discuss algorithms for the problem of nonlinear least-squares regression with linear inequality constraints.

4.5.7.5 Nonlinear constraints

We have saved the hardest problem for last. Fortunately, it rarely arises in statistical problems. When some or all of the $c_j(\theta)$ are nonlinear in θ, the optimization problem becomes at once enormously complicated and extremely delicate. In the case of linear constraints, once a feasible point has been determined it is possible to obtain a sequence of improved iterates which, by construction, are also feasible. This is generally *not* the case when even a single constraint function is nonlinear. Such problems require great care as a result. Although there are several general approaches to the problem, none can be applied as a "black box" with much hope of success, and few general-purpose computer programs for the problem exist. The reader is referred to Chapter 6 of Gill, Murray, and Wright (1981) and to Chapter 12 of Fletcher (1981), which are devoted to the problems engendered by nonlinearity in the constraint functions.

4.5.8 Estimation diagnostics

The topic of regression diagnostics for linear least-squares computations now has a large literature. What is not generally recognized is that there are analogs to most of the standard regression diagnostics such as those discussed in Section 3.6 for nonlinear problems as well. These analogs are most easily interpreted in the context of nonlinear least squares, as we may think of them as applying to the linearized approximation to the problem which is the basis for the Gauss-Newton optimization methods.

Some work has been done to extend diagnostics methods more formally to particular nonlinear settings. An important example is the work of Pregibon (1981), in which influence diagnostics measuring the effect of each observation on the components of the maximum-likelihood estimate are obtained for the logistic regression model. Pregibon's approach is to take a single step of the Newton-Raphson iteration, and to construct an analog of the "hat matrix" $(X'X)^{-1}X'Y$ from ordinary least squares. These methods can be extended to generalized linear models in scaled exponential families. Moolgavkar, Lustbader, and Venzon (1984) study the same problem from a slightly different perspective. Their diagnostics ("MLV diagnostics") also assess the sensitivity of estimates to deletion of single observations in exponential-family regression; their approach is based on a geometric construction which leads to diagnostics based on Fisher scoring rather than on Newton-Raphson. Thus, the MLV diagnostics are based upon expected rather than observed information. Although the former leads to computational simplification, it is purchased only by giving up the ability to condition on the observed configuration of the data—a particularly valuable feature for model diagnosis. The MLV diagnostics are precisely the ones obtained from the one-step Gauss-Newton approximation. They coincide with diagnostics based on Newton-Raphson in generalized linear models using the canonical link function.

Of special interest for nonlinear maximum-likelihood estimation are the conditioning diagnostics of Section 3.6.5. These diagnostics in the linear case are based on the eigenvalues and eigenvectors of $(X'X)^{-1}$, which is the variance matrix of the estimate $\hat{\beta}$. The negative inverse of the Hessian $H(\theta)$ of the log-likelihood function is the asymptotic variance matrix of the parameter vector θ, so that the interpretation of the conditioning diagnostics in terms of variances of linear combinations of the parameter estimates can be employed in the nonlinear case as well. These diagnostics can be of particular value in discovering why an iteration fails to converge, or converges only very slowly. Nonlinear estimation problems are often overspecified (or nearly so), so that some function of the parameter estimates has infinite (or very large) variance, which can lead to poor convergence due to near singularity in the estimates for $H(\theta_i)$. When this is the case, the function describing the singularity can be expanded in a Taylor's series about the current iterate, producing a *linear* combination of the current parameter estimates whose variance is very large. This corresponds to the eigenvector associated with the largest eigenvalue of $H(\theta_i)^{-1}$, which in turn corresponds to the smallest eigenvalue of $H(\theta_i)$. The conditioning diagnostics can identify when this is happening, and at the same time gives an estimate for the linearized version of the singularity relationship. This, in turn, can assist the astute data analyst to a more appropriate parameterization of the problem, perhaps in terms of fewer parameters.

EXERCISES

4.34. [11] Why does the number of points in a grid search increase exponentially in the dimension p?

4.35. [19] Show that if S is bounded, then sequential search on a lattice L ultimately terminates.

4.36. [02] Give an example to show that sequential search need not terminate at the global minimum.

4.37. [05] Let g be a strictly unimodal function on $[a, b]$, and let $a \leq t_1 < t_2 \leq b$. If $g(t_1) = g(t_2)$, what can be concluded about the location of the minimum of g?

4.38. [26] Show that the Golden-section search algorithm can be deduced from the three requirements that 1) at each step the two interior trial points be symmetrically placed with respect to the endpoints, 2) only one new trial point be added at each step, and 3) ratios of subinterval lengths remain constant.

4.39. [18] An initial interval of length one is to be used for a Golden-section search. How many iterations are required before the minimum can be determined within ± 0.01?

4.40. [31] Show that no minimax strategy exists for searching for the minimum of a strictly unimodal function.

4.41. [28] For the simple successive approximation scheme, let q denote the number of subintervals into which the uncertainty interval $[a, b]$ is divided at each stage, so that $q - 1$ is the maximum number of times that g will be evaluated in a single iteration. Number the evaluation points from 0 to q, and let J denote the point on this grid at which g is smallest. Assume that $Pr(J = i)$ is $1/q$ for $0 < i < q$ and is $1/2q$ otherwise. What is the expected number of function evaluations and the expected interval reduction per function evaluation?

4.42. [21] Show that d_{LM} is a descent direction whenever λ is larger than the smallest eigenvalue of $F''(x)$.

4.43. [20] (Ridge regression). Consider the linear regression model $Y = X\beta + \epsilon$. Show how the Levenberg-Marquardt procedure can be applied in the context of linear least-squares to obtain the *ridge regression estimator* $\hat{\beta}(k) = (X'X + kI)^{-1}X'Y$, for $k > 0$.

4.44. [25] (Ridge regression and Bayes estimation). Show that the estimator of the previous exercise is a compromise between $\beta = 0$ and the usual least-squares estimator $\hat{\beta} = (X'X)^{-1}X'Y$. How might the Levenberg-Marquardt ideas be used to obtain a more general form of the ridge estimator?

4.45. [24] (Ridge regression and Bayes estimation, continued). Show that the generalized estimator obtained in the previous exercise is the posterior mean for a suitably chosen prior distribution for β.

4.46. [50] The conjugate-gradient method does not provide a Hessian which can serve as the asymptotic variance matrix of the parameters. Sometimes, however, only a small number of parameters are of direct interest, the other parameters being fit only to adjust for important covariates or to introduce blocking factors. Devise suitable hybrid methods that achieve the storage gains of conjugate-gradient methods for those parameters not of interest, but which provide asymptotic conditional variances for the remaining parameters.

4.47. [21] Let $\Delta = diag(\delta_1, \ldots, \delta_p)$ be a diagonal matrix of positive scaling factors, and define $y = \Delta x$. To use a rescaled version of Newton-Raphson from a point x, we would first transform to the y scale, take a Newton step, and then transform back. Show that Newton-Raphson iteration is unaffected by changing the scale of the variables in this fashion.

4.48. [18] Using the notation of the previous exercise, show that the direction of steepest descent does depend upon the scale of the variables.

4.49. [12] Show that the maximum-likelihood estimator for θ in the non-linear regression problem with normal errors is given by the minimizer of equation (4.5.3).

4.50. [19] Show how the orthogonality condition (4.5.9) leads to the iteratively reweighted least-squares update (4.5.8).

4.51. [22] Equation (4.5.8) is the basis for an IRLS estimator for the update to θ_0, that is, for $\hat{\theta} - \theta_0$. Derive an equivalent expression for $\hat{\theta}$ from which an IRLS procedure can be constructed.

4.52. [05] Show that an equality constraint $c(\theta) = 0$ can be re-expressed in terms of inequality constraints.

4.53. [18] Prove that, in the linearly constrained least-squares problem, the X matrix can be re-expressed in the form of equation (4.5.12).

4.54. [27] For the equality-constrained linear regression problem $Y = X\beta + \epsilon$ subject to $C\beta = 0$, with C $m \times p$, show that the least-squares estimator is given uniquely by $\hat{\beta} = A'\hat{\gamma}$, where A and $\hat{\gamma}$ are as defined in Section 4.5.7.2.

4.55. [29] Suppose that in the constrained optimization problem (4.5.11), $E = 0$, $I = m$, θ is $p \times 1$, and all of the constraint functions $c_j(\theta)$ are linear in θ. Suppose further that the constraints are consistent, and that the matrix C whose rows contain the coefficients of θ in the constraint functions has full rank m. Show that (4.5.11) can be reparameterized so that the constraints have the form $\beta_j \geq 0$, for $j = 1, 2, \ldots, m$.

4.56. [48] Under what conditions can the "manual" method of linear inequality constrained regression described in Section 4.5.7.3 fail to converge? How much worse than the optimal solution can the result found by the "manual" method be?

4.57. [48] Determine the worst-case and average-case number of regression computations required by a standard algorithm for linear least-squares regression with m nonnegativity constraints, as a function of m. Determine whether any algorithm exists for which this number is bounded above by a polynomial in m.

4.6 Computer-intensive methods

During the last ten years, a number of methods for statistical data analysis have been developed which depend intrinsically on being able to do substantial amounts of floating-point computation in a short amount of elapsed time. Although at the time they were introduced these methods strained the limits of computing power generally available to data analysts, they have now—thanks to the rapid increases in computing speeds accompanied by equally rapid reductions in computing costs—become economically feasible for at least occasional use by almost all data analysts. Due to the large volume of computation associated with these methods, they have come to be know collectively as *computer-intensive methods* for data analysis.

COMMENT. The term "computer-intensive methods" was coined by Efron and Diaconis (1983) to refer to sample-reuse methods such as the bootstrap, the jackknife, and cross-validation. These topics will be discussed in Chapters 8 and 9. The term now encompasses a wide range of methods which, roughly speaking, relinquish standard assumptions of classical statistics such as normal error distributions or linearity of response at the cost of massively increased computational effort. Many of the statistical and computational ideas now associated with this term have been developed by Jerome Friedman and his colleagues and students at the Stanford Linear Accelerator Center. Three such ideas are discussed below.

The three examples of these methods which we shall discuss in this section each make use of several very different computational ideas, including linear iterative methods, nonlinear optimization, smoothing, and sorting. They all rely for their practicality on efficient implementations of some of the computational ideas developed in Chapters 3 and 4. Each is a generalization of a standard linear method. Each method also requires a smoothing algorithm. A full discussion of the computational aspects of these methods requires material from Chapters 6 and 9 as well as material already presented. For the moment we shall emphasize the optimization aspects of the problem. The data structures employed for organizing the input data and the intermediate calculations can be chosen so as to support both the smoothing algorithm and the optimization process.

4.6.1 Nonparametric and semiparametric regression

We return now to the standard linear regression model, which can be written in the form

$$E(y_i) = E(y_i \mid x_{i1}, \ldots, x_{ip}) = \sum_{j=1}^{p} \beta_j x_{ij}, \qquad (4.6.1)$$

for $1 \leq i \leq n$. When this linear model inadequately represents the observed data, we can turn to generalizations of the linear model. Expression (4.6.1) for the mean of y_i can be interpreted in several (equivalent) ways, which in turn can be extended or generalized in several (different) ways. We begin by considering two particular generalizations.

First, we can think of (4.6.1) as saying that the mean of y depends *additively* on the regressor variables x_j. On this interpretation, the mean of y depends upon p functions (one for each x_j) which are then added together to produce the overall mean of y. In the linear model (4.6.1), these component functions are all very simple, namely, they are undetermined multiples of the identity function. A generalization based on this interpretation replaces x_j by a smooth function of x_j.

A second interpretation of (4.6.1) is that the mean of y depends only upon a single linear combination of the x's. On this interpretation, the mean of y grows linearly along a single direction in $\mathcal{R}^p$, and must therefore be constant in any orthogonal direction. In a sense, $E(y)$ grows only along a "ridge" in $\mathcal{R}^p$ defined by the function $g(x) = x'\beta$, $x \in \mathcal{R}^p$. More generally, functions which depend on $x \in \mathcal{R}^p$ only through a weighted sum of the components of x have the form $g(x) = f(x'\beta)$. Such functions, called *ridge functions*, are constant throughout a $(p-1)$-dimensional subspace of $\mathcal{R}^p$. A generalization based on this interpretation replaces $x_i'\beta$ in (4.6.1) by a sum of smooth ridge functions; here x_i' denotes the ith *row* of the matrix X.

The methods that we discuss below are variations on these two themes. In each case, smooth-function estimates are obtained by selecting among a parametric family of functions satisfying one or more smoothness constraints. The parameters involved in the smoothing process reflect local behavior of the mean function, and are introduced as technical devices which permit flexibility in the functions to be estimated while retaining computational practicality. We refer to methods based on fitting smooth mean functions as *nonparametric regression methods*, because the form of the estimated mean function is not rigidly specified in advance, and because the parameters for which estimates are obtained are intermediate results, of interest only insofar as they determine the form of the regression function.

Some of the models discussed below involve both model parameters of primary interest and general smooth functions to be estimated. These models involve parametric estimation after adjustment for (nonparametric) smooth functions; we call such models *semiparametric regression models*.

4.6.1.1 Projection-selection regression

A generalization of (4.6.1) along the lines of the first interpretation is an *additive model* in smooth functions of the x's rather than in the identity functions. This model can be expressed as

$$E(y_i) = \sum_{j=1}^{p} f_j(x_{ij}). \qquad (4.6.2)$$

This model is called the *projection selection model*, sometimes abbreviated *PSR*, for projection-selection regression (Friedman and Stuetzle, 1981). Roughly, the idea is to fit this model by least squares, or rather a variant thereof that minimizes

$$\sum_{i=1}^{n} \left[y_i - \sum_{j=1}^{p} f_j(x_{ij}) \right]^2 + P(f_1, \ldots, f_p)$$

with respect to the functions $f_1, f_2, \ldots, f_p$. Least-squares corresponds to $P \equiv 0$, but unfortunately this is impossible to do without a bit of additional structure. If the form of the functions $f_j(\cdot)$ were, up to one or two parameters each, given in advance, then the problem would be a straightforward one in nonlinear optimization. However the idea in projection selection is to estimate these functions from the data under the assumption that the functions are all sufficiently smooth. Since a very rough function can interpolate the data exactly, the sum of squares above can be made to equal zero. Clearly there is a trade-off between smoothness and fit. The factor $P(f_1, \ldots, f_p)$ is a *roughness penalty* that specifies implicitly what that trade-off shall be (see Chapter 6).

Each f is estimated using a nonparametric smoothing algorithm. We should think of the functions $f_j(\cdot)$ in (4.6.2) as replacing the regression coefficients β_j in (4.6.1) as parameters to be estimated. We still have a p-parameter problem, except for the minor complications that the "parameters" are each elements of some function space rather than points in $\mathcal{R}^1$! By suitable restrictions on the permissible candidates for the estimate functions $\hat{f}_j$, we can effectively reduce the problem to a one of relatively small dimension. (For instance, if we require that f_j be a cubic spline with k_j fixed knots, then f_j can be specified by $k_j + 2$ coefficients; see the discussion of spline smoothing in Chapter 6.) Most commonly this reduction is accomplished by disallowing functions with some degree of nonsmoothness. Smoothness is enforced either through introducing a linear parametric structure (as is the case with spline smoothing), or through fitting locally simple models (local averaging). Either method may employ weights which discount unusual or aberrant points.

Assume that the method for fitting a single function f of the form $E(y) = f(x)$ is settled upon. Assume, too, that the fitting method never increases the residual sum of squares. How might this algorithm be adapted to fitting the *set* of functions f_j in (4.6.2)? If we think of each f_j as a parameter, then the Gauss-Seidel method, in which each parameter is repeatedly estimated given the current estimates for the other parameters, is a natural choice. This approach is adopted in PSR. The linear and nonlinear versions of Gauss-Seidel are discussed, respectively, in Sections 3.11 and 4.3.4. The iteration begins by estimating f_1 by fitting the model $E(y_i \mid x_i 1) = f_{10}(x_{i1})$. Residuals are formed as $r_{10,i} = y_i - \hat{f}_{10}(x_{i1})$. Then an initial guess for f_2 is estimated by fitting $r_{10,i}$ to $f_{20}(x_{i2})$. The residuals $r_{20,i}$ from this fit are then used to estimate the next function, and so forth. When all p functions have been estimated, the process of cycling through the p functions estimating one at a time is repeated, in each cycle c finding the update $\hat{f}_{jc}$ to the jth smooth function by fitting f_{jc} to $r_{j-1,c}$, where by notational convention we take $r_{0,c} \equiv r_{p,c-1}$. The smooth functions $\hat{f}_j$ are generally stored as a set of n function values, and at each stage these values are updated by the most recently obtained $\hat{f}_{jc}$'s. (There is no reason to retain all of the functions $\hat{f}_{jc}$, as these are merely updating functions.)

COMMENT. It should be noted that, as in the linear case of the Gauss-Seidel algorithm, convergence should be speeded dramatically by using a double sweep rather than a single sweep through the p components. That is, after estimating f_{1c} through f_{pc}, the next cycle of estimates is taken by cycling through the components in reverse order ($f_{p-1,c+1}$ through $f_{1,c+1}$). The idea of a double sweep has not, at this writing, been applied in any implementation of PSR. Whether doing so would produce any benefit in practical applications, and under what circumstances, is an issue that has not yet been explored.

COMMENT. A suitable data structure for PSR is of some interest in its own right. The smoothing algorithm will be applied at each step to approximate the current residuals r as a smooth function of one of the regressor variables, say x_j. The smoothing algorithms require that the x_j variable be sorted, and this can be done once for each of the regressors at the beginning of the computation using the data structure described below. Suppose that the regressors are stored in the rectangular array $X[i,j]$, $1 \leq i \leq n$ and $1 \leq j \leq p$, and the current residuals are stored in the vector $R[j]$. Instead of sorting column j as such, the sorting routine can be made to generate an index matrix $P[i,j]$ whose (i,j)th element contains the index of the ith smallest element in column j of X. Thus, $X[P[i,j],j]$ is the value of ith smallest value of x_j. Moreover, $R[P[i,j]]$ is the value of the current residual corresponding to this element. Note that each column of P is a permutation of the integers 1 through n, so that as i ranges from 1 to n, $P[i,j]$ ranges through the same values, although in a different order.

If the smoothing algorithm accesses $X[P[i,j],j]$ and $R[P[i,j]]$ frequently, considerable computational effort can be devoted solely to the computation of array indices. If this is the case, additional efficiencies can be purchased at the expense of some additional preprocessing. Again, once at the beginning of the computation, the sorted columns of X can be computed and stored in a matrix Z, so that $Z[i,j] \equiv X[P[i,j],j]$. With care, the Z matrix can replace the X matrix in memory; an auxiliary vector of length n is required to accomplish this. Then at each iteration, the current residual vector R is copied in permuted form into the work vector $W[i] = R[P[i,j]]$, and the smoothing algorithm does its work with $Z[\cdot,j]$ and $W[\cdot]$. At the end of the iteration, assuming that the residual from the current fit has been saved in the vector W, the new residual vector is restored using $R[P[i,j]] = W[i]$. The update to the current fit is obtained in similar fashion at the end of each iteration.

This data structure is also well-suited to plotting the estimated smooth functions $\hat{f}_j$, since the current values of $\hat{f}_j$ would be stored in a matrix $F[i,j]$ whose elements corresponded to those of $Z[i,j]$. Since each column of $Z[i, j]$ is in ascending order, scaling and plotting are simplified.

COMMENT. As a diagnostic tool, PSR is likely to be of greatest benefit when applied to residuals from a standard linear additive fit, that is, the usual multiple regression model. (After such a PSR fit, the linear terms would have to be estimated again.) An integrated algorithm for applying PSR would construct the data structure of the previous comment *after* the initial linear fit had been accomplished, so that the X matrix could be restructured to contain sorted columns of X.

4.6.1.2 Projection-pursuit regression

An alternative generalization of model (4.6.1) focuses on the fact that the linear regression model represents $E(y)$ as a particularly simple *ridge function*, namely $E(y_i) = x_i'\alpha$, where x_i' denotes the ith row of the X matrix and α can be thought of as denoting a particular direction in p-dimensional space. The *projection-pursuit regression (PPR) model* generalizes this simple model by allowing $E(y)$ to be represented as an additive function of k smooth ridge functions, that is,

$$E(y_i) = \sum_{j=1}^{k} f_j(x_i'\alpha_j). \tag{4.6.3}$$

In this expression, each $f_j(\cdot)$ is a scalar function whose argument is determined from x_i by the direction vector α_j. Thus, in addition to the k smooth functions to be estimated from the data, the model depends upon the kp parameters in the vectors α_1 through α_p. Note that these latter parameters enter linearly in the columns of X in much the same way that the parameters enter in a GLIM model.

The PSR model is a special case of the PPR model. It has been shown by Diaconis and Shahshahani (1984) that, except for some pathological cases, general smooth p-variate functions can be approximated arbitrarily well by functions whose form is that of the right-hand side of (4.6.3).

Computationally, PPR is similar to PSR, but it does present a few new wrinkles. As with PSR it is convenient to consider the problem of fitting the model (4.6.3) by least squares. In effect, there are two sets of parameters to fit, the f_j's and the α_j's. [In practice, k must be estimated, too, but we shall consider k fixed for the moment.] Denote the former set by F, and the latter set by A; A has dimensionality kp, while the dimensionality of F depends upon the smoothing methods that will be employed and may. strictly speaking, be indeterminate. The Gauss-Seidel algorithm once again can be used, applying it to the superparameters F and A. Starting with an initial estimate for A, hold A fixed and fit F. Then, given the current F, find a new A, and then repeat the process.

Given A, how does one fit the set of functions that comprise F? Since the direction vectors α_j are now fixed, one can consider the ridge functions $z_j = X'\alpha_j$ to be a new set of k regressors, to which the PSR algorithm can be applied.

Given F, how does one fit the set of directions α_j that comprise A? Once again, Gauss-Seidel will generally be the only feasible general method. Fixing all save one of the directions at their current values, the remaining direction α_j is estimated so as to reduce the residual sum of squares. One then rotates through the k directions to be updated. This generates yet another subproblem, namely that of estimating α_j given the other directions (and the current f_j's). Smoothers with special structure (such as splines or explicitly represented local linearity) may be amenable to optimization methods similar to those employed in GLIM models such as Gauss-Newton. More generally, optimization at this level may require yet another Gauss-Seidel computation.

Note that PPR computation requires a complicated Gauss-Seidel iteration *at each step* of a larger Gauss-Seidel iteration. Each of these minor iteration steps, in turn, may involve something as complex as fitting a smooth function or minimizing a multivariate function. These methods are indeed computing intensive!

COMMENT. Because of the great expense in carrying out each iteration fully, it may be advisable, especially during the early stages of the fitting process, *not* to iterate to convergence, but rather to halt a given portion of the iteration when the improvement in the objective function it produces (generally the residual sum of squares) is small in percentage terms. This is particularly true when PPR is being used as an exploratory method to discover potentially important nonlinearities in the structure of a multivariate problem. The added insight produced by reducing R^2 by another fraction of a percent is almost never worth the computational effort required to do so.

COMMENT. In our treatment above, we have treated k as being fixed. In practice, one would begin by fitting the usual model (4.6.1), and then, working from the residuals of this model, perform PPR with $k = 1$. If the further reduction in residual sum of squares is sufficiently large to be worthwhile, one then considers expanding the process to fit the PPR with $k = 2$. This process is continued until either there is no important reduction in the residual sum of squares, or until the residual sum of squares is already acceptably small.

COMMENT. In practice, we have suggested first fitting the linear least-squares model, and then applying PSR or PPR to the residuals therefrom. Formally, then, the PPR model would be

$$E(y_i) = \sum_{j=1}^{p} x_{ij}\beta_j + \sum_{p+1}^{p+k} f_j(x_i'\alpha_j).$$

As in any estimation problem, the estimates of β would change when estimates for the other parameters (the f_j's and α_j's) changed. It would seem, then, that in each major Gauss-Seidel loop, one step would be to re-estimate the linear regression coefficients given the current values for the other variates. This need not be the case however, if the functions f_j are constrained to be orthogonal to the columns of X. The ramifications of this observation have not been explored, as it would require that the overall structure of X be taken into account by the smoothing algorithm. The additional expense of doing one regression per major iteration is minor, but still is probably not worth the effort to include in the program, as the primary purpose for doing PPR in the first place is to discover substantial nonlinearities in the data; if such nonlinearities are found, then the small adjustments to the linear part of the estimated regression surface will likely be of little concern. Rather, the emphasis should then be on constructing a nonlinear model that parsimoniously captures the bulk of the newly found structure.

COMMENT. The process of estimating the f_j's and α_j's successively using the nonlinear Gauss-Seidel method is termed *backfitting* in Friedman and Stuetzle's 1981 paper, as well as in other papers which build on and extend the technology described there.

COMMENT. The term *"projection pursuit"* was coined by Friedman and Tukey (1974) to describe an iterative search for a direction vector in a p-dimensional data space, along which the data exhibit interesting structure (such as well-separated clusters, for instance). The general methodology defines a figure of merit which measures "interestingness" along a given projection, and then computes by numerical optimization the direction which maximizes the given figure of merit. This idea had earlier been proposed by Kruskal (1969) and by Switzer (1970) in the clustering context.

Since its introduction, the notion of projection pursuit has been elaborated and extended in many directions, subsuming many of the standard multivariate methods based upon normal theory. Huber (1985) gives a comprehensive overview of the topic of projection-pursuit methods in general, including applications to clustering, regression, density estimation, time series, and computed tomography.

4.6.1.3 Additive spline models

In discussing PSR we noted that the computation of the smooth functions f_j could be reduced to a problem of relatively low dimension with a substantial degree of linear parametric structure by requiring the f_j to be splines. The same can be said in the context of PPR. This approach is developed fully by Friedman, Grosse, and Stuetzle (1983) in a paper on multidimensional additive spline approximation. Given the current set of directions α_j, the knots in the splines are selected heuristically. Given the knots and the directions, the spline coefficients are then found by linear least-squares techniques. These spline coefficients are then held constant while a search is conducted for a new direction, and the process is repeated. An advantage of this approach is that only the directions, spline coefficients, and knot positions need to be saved in order to reconstruct the fitted regression surface, thus providing a highly parsimonious representation.

The additive regression approach has been studied from a theoretical perspective by Stone (1985), who shows that for $X \in [0, 1]^p$, the additive spline estimates converge to the best (minimum mean-squared error) additive approximating function to the regression function $E(Y \mid X)$. Stone and Koo (1985) give a practical outline of additive spline models in the more useful setting where X is not restricted to the unit p-cube; the theoretical convergence properties of such procedures have not been worked out, but the examples given in this paper suggest that practical application need not await a fully-developed theory. Wahba (1986) has generalized spline-regression ideas beyond strict additivity by introducing the notion of *interaction splines;* first-order interaction splines, for instance, are bivariate splines involving pairs of the regressor variables. Wahba (1986) also discusses a wide variety of semiparametric models, which she refers to as *partial spline models.* These models have the practical advantage that software for computing them is available in an integrated package of portable FORTRAN subroutines called GCVPACK (Bates, Lindstrom, Wahba, and Yandell, 1985).

4.6.2 Alternating conditional expectations

Another computer-intensive method similar in spirit to projection-pursuit regression is the estimation of optimal transformations in regression problems (Breiman and Friedman, 1985). The theme on which this variation is based is once again the usual linear regression model (4.6.1). It is often the case that either the response variable y or some of the predictors x_j are first transformed before an additive model is fit, so that equation (4.6.1) is replaced by

$$E(\theta(y_i)) = \sum_{j=1}^{p} \phi_j(x_{ij}) \qquad (4.6.4)$$

$$\equiv \Phi(x_{i1}, x_{i2}, \ldots, x_{ip}). \tag{4.6.5}$$

The idea behind Breiman and Friedman's work is that these functions θ and $\{\phi_j\}$ can be estimated from the data using smooth functions which maximize the squared correlation between $\hat{\theta}(y)$ and $\{\hat{\phi}_j(x_j)\}$. As in the projection-pursuit methods, the estimated smooth functions need not be monotone. If any of the predictors x_j are categorical, the method can be used to obtain real-valued scores for each of the categories which maximize the correlation between the scores and the response y (or some function of y). When applied to a single y and a single x, the method estimates the so-called *maximal correlation*—the largest possible correlation between a function of y and a function of x. The paper of Breiman and Friedman, together with the published discussion that follows, examines the theory behind the method as well as applications and computational matters.

The computational algorithm is of greatest interest to us here, and its basic features are clearest in the bivariate case. Our description closely follows that of Breiman and Friedman (1985). Without loss of generality, assume that all transformations θ and ϕ are centered and that θ is scaled to have mean zero and variance one. Suppose that (Y, X) has a known bivariate distribution. Suppose for a moment that ϕ is a particular known function of X, and consider choosing θ so as to minimize the mean squared error $E[\theta(Y) - \phi(X)]^2$. It is well known that the solution is given by the regression of $\phi(X)$ on Y, that is, $\theta_1(Y) \equiv E[\phi(X)|Y]/c_1$. (The constant c_1 is the standard deviation of the numerator random variable and is needed to maintain the condition that $E(\theta^2(Y)) = 1$.) The mean-squared error is symmetric in $\theta(Y)$ and $\phi(X)$, however, so that the same argument can be applied to finding the optimal choice for ϕ, given the function $\theta_1(Y)$, namely, $\phi_1(X) \equiv E[\theta_1(Y) \mid X]$. Note that no scaling condition is needed. These two conditional expectations are repeated in alternation, hence, the descriptive name *ACE*, for *alternating conditional expectations*. The objective function to be minimized here can be thought of as a generalization of a sum of squares. As a consequence, the convergence of bivariate ACE to the optimal solutions can be thought of as a generalization to function space of the nonlinear Gauss-Seidel theorem.

The multivariate case is only scarcely more complex, and proceeds in the now-familiar manner of Gauss-Seidel. An outer loop alternates conditional expectations between θ and Φ of equation (4.6.5). Obtaining Φ for given θ, in turn, requires an inner iteration in which at each stage θ and all but one component ϕ_j in (4.6.4) are held constant. This inner iteration cycles through the ϕ_j's until no further reduction in mean-squared error can be achieved, whereupon Φ has been obtained for a single stage of the outer iteration. Breiman and Friedman prove that this algorithm converges to the minimum mean-squared error transformations.

As stated in the previous section, ACE applies to random variables whose joint distribution is known. In practice, a data-based version is used to obtain estimates for the optimal transformations by replacing the mean-squared error, the variance of $\theta(y)$, and the conditional expectations by estimates. The first two of these quantities are obtained by using the conventional estimates. The conditional expectations are obtained using a smoothing algorithm. In actual practice, then, the estimated smooth functions would be displayed graphically, although they can also be approximated using cubic splines, which can be parsimoniously represented.

COMMENT. The *canonical correlation problem* is that of finding, for random vectors $Y \in \mathcal{R}^q$ and $X \in \mathcal{R}^p$ the linear combinations $\alpha \in \mathcal{R}^q$ and $\beta \in \mathcal{R}^p$ for which the squared correlation of $\alpha'Y$ and $\beta'X$ is maximized (subject to a normalizing condition). As Breiman and Friedman note, the ACE algorithm can be easily modified to generalize canonical correlation in precisely the same way that ACE itself generalizes multiple regression, by choosing transformations $\theta_1, \ldots, \theta_q$ and $\phi_1, \ldots, \phi_p$ to minimize

$$E \left[\sum_{k=1}^{q} \theta_k(y_k) - \sum_{j=1}^{p} \phi_j(x_j) \right]^2 ,$$

where again the means of all functions are taken to be zero, and the variance of $\sum \theta_k(y_k)$ is taken to be one.

COMMENT. As the previous comment indicates, ACE is closely related to the canonical correlation problem. In general the structure of solutions to canonical correlations problems depends upon the eigenstructure of Σ_y relative to Σ_x, the respective variance matrices of the Y and X variates. The eigenvalues and eigenvectors in these problems are solutions to a variational problem that is considerably more complex than the usual least-squares optimization on which multiple regression is based. In a stimulating paper, Buja (1985) shows that understanding the behavior of ACE must be based on spectral properties rather than on least squares. He shows how, as a consequence, ACE can produce nontrivial transformations even when X and Y are stochastically independent. A more bothersome feature is that small perturbations in the input data can cause abrupt changes in the ACE transform. This latter property is related to the fact that the eigenvector corresponding to the largest eigenvalue generally does not depend continuously on the input data. What Buja points out are properties of ACE that one would not immediately deduce from the definition of the method. Understanding these properties is essential to using ACE well in data analysis.

Tibshirani (1986) has proposed a method which he calls "RACE" which appears from empirical evidence not to suffer from some of the troublesome behavior described above for ACE, and which is more natural in regression settings. The method is based upon variance-stabilizing transformations for the response variable instead of conditional mean-fitting transformations.

4.6.3 Generalized additive models

The parametric and semiparametric regression methods generalize ordinary least squares by allowing much more general functional forms for the mean of y as a function of $X = ((x_1, \ldots, x_p))$. We have also discussed another generalization of OLS in our discussion of GLIM models; in these models a linear structure in X is retained, but the mean of y depends on this linear structure through a link function, and the errors need not be Gaussian. This generally means that maximum likelihood and least squares no longer coincide. These two ideas have been successfully merged. Models in which the mean of y depends on an *additive* model (rather than a linear one) through a link function are called *generalized additive models*. We shall use the acronym *GAIM* to refer to such models.

The basic structure of GLIM is retained, however the linear predictor $\sum \beta_j x_j$ is replaced by an additive predictor $\sum f_j(x_j)$, where the functions $f_j(\cdot)$ are smooth functions estimated from the data. This extends smooth nonlinear modeling to such common problems of data analysis as logistic regression and Poisson regression.

Computationally, the GAIM approach must abandon some of the special structure, due to linearity, that GLIM models can take advantage of. Hastie and Tibshirani (1986) propose an estimation method that they call *local scoring* for generalized additive models. Just as ACE empirically minimizes the residual sum of squares, Hastie and Tibshirani's method empirically maximizes the expected log likelihood; standard maximum-likelihood estimation in the GLIM model is a special case. Portable interactive software for GAIM similar in spirit to GLIM is available, in which the f_j's are estimated by a locally linear smoothing algorithm similar to that used in Friedman and Stuetzle's implementation of projection-pursuit regression. As in PPR, the estimated smooth functions must be adjusted for the additive contributions of subsequent components, hence Gauss-Seidel (backfitting) adjustments must be incorporated at each step.

An alternative approach to the general additive model is to restrict the f_j's to be spline functions. This introduces considerable computational simplification due to the linear structure of spline functions. O'Sullivan, Raynor, and Yandell (1986) develop spline-based methods for this problem. They show how the resulting estimates can be viewed as *maximum penalized likelihood* estimates; here a solution represents a trade-off between likelihood maximization and a penalty for lack of smoothness in the solution. They also develop the necessary methods for carrying out the computational details in practice. Stone (1985) has examined general models exhibiting additive structure and has described theoretical properties of certain spline estimators, including convergence rates, in this context.

EXERCISES

4.58. [12] Given the matrices X and P in the discussion of projection-selection regression, why is a work vector of length n needed to replace X by the matrix Z whose columns are the sorted columns of X?

4.59. [28] For the single-sweep version of PSR, suggest an alternative data structure to the index matrix $P[i,j]$ that eliminates the need to restore the residuals from the working vector using $R[P[i,j]] = W[i]$.

4.60. [45] Examine the convergence properties of the Gauss-Seidel algorithm as applied in projection-selection regression, and determine whether the "double sweep" modification produces noticeable improvements in performance.

4.61. [06] Why can the projection-selection regression model be thought of as a special case of the the projection-pursuit regression model?

4.62. [49] Investigate whether the use of Gauss-Seidel iterations as used in projection-pursuit regression or in projection-selection regression can be generalized usefully by using the method of successive over-relaxation instead.

4.63. [29] Show how to extend the ACE algorithm to deal with the canonical correlation problem.

4.64. [38] (Fowlkes and Kettenring, 1985) Discuss how ACE, or a modification of it, can be used to search for transformations ϕ_j for which $\sum \phi_j(x_j)$ has nearly zero variance. This generalizes the notion of principal components.

4.65. [02] What does the "I" stand for in the acronym *GAIM?*

4.7 Missing data: The EM algorithm

The problem of incomplete data or missing information is as important as it is common. Every seasoned data analyst has had to face it; younger ones can look forward to doing so. In a seminal paper on incomplete observations, Orchard and Woodbury (1972) note that, "obviously the best way to treat missing information problems is not to have them." Those of us who can't arrange to avoid such problems are fortunate that a large body of work has been developed to make missing-data problems considerably more tractable than they once were. Many of the advances in the area have resulted in effective methods for computing maximum-likelihood estimators. Indeed, there are now problems for which the easiest computational approach involves reformulating a (complete-data) problem in terms of a more easily solved incomplete-data problem!

In this section we shall focus on parameter estimation by maximum likelihood when some relevant random variables are unobserved. One particularly attractive general method for obtaining MLEs is the *EM algorithm* of Dempster, Laird, and Rubin (1977); the computational aspects of this method are quite interesting.

Formally, suppose that we have a random variable Z whose density function $f(z \mid \theta)$ depends on a parameter of interest, θ. If Z were observed we could estimate θ by finding that value $\hat{\theta}$ for which $L(\theta \mid Z) \equiv f(z \mid \theta)$ is maximized, with the bonus that the observed information matrix evaluated at this estimate provides information about the precision of $\hat{\theta}$. Suppose, however, that $Z = (Z_p, Z_m)$, where Z_p is observed (*Present*) and z_m is not (*Missing*).

An intuitively appealing approach—often advocated in the literature—is to treat Z_m as if it were an unknown parameter. This "parameter" is estimated (predicted, if you prefer) and the estimates $\tilde{Z}_m$ are treated as if they had been observed. Thus, θ is then estimated by maximizing $L(\theta \mid \tilde{Z})$, where $\tilde{Z} \equiv (Z_p, \tilde{Z}_m)$. In effect, we "complete" the data vector and use maximum likelihood with the *imputed* data; this is equivalent to maximizing $f(z_p, z_m \mid \theta)$ with respect to (z_m, θ). Unfortunately this approach produces estimates which in general are severely biased (Little and Rubin, 1983). What is doubly unfortunate is that the major statistical analysis packages provide no missing-data options more sophisticated than this one, and even this method is available only in those circumstances in which the best predictor of Z_m is the mean of the components in Z_p.

The actual likelihood function that should be maximized is given by

$$L(\theta \mid z_p) = f(z_p \mid \theta) = \int f(z_p, z_m \mid \theta) f(z_m) dz_m. \qquad (4.7.1)$$

By the early 1970's a number of algorithms for estimating θ based on z_p in (4.7.1) had been developed for particular problems having special structure such as, for example, regression problems with nested patterns of missing data (Anderson, 1957). All of these methods, developed in piecemeal fashion, were based on essentially the same ideas. Orchard and Woodbury (1972) enunciated a "missing information principle" which subsumed most of the work done on special cases of maximum-likelihood estimation for missing-data problems. Dempster, Laird, and Rubin (1977) recognized that all of these approaches could be described in terms of a general iterative algorithm containing two steps, which they called the *EM algorithm*. This algorithm has the twin virtues of being conceptually simple and (often) relatively easy to implement. It has the disadvantages that it often converges very slowly, and it does not automatically produce an asymptotic covariance matrix of the resulting parameter estimates. They also examined some of the theoretical properties of the algorithm, and they provided a large catalog of useful statistical applications. The EM algorithm is similar in form to other optimization methods we have discussed, such as the ACE algorithm and Gauss-Seidel iteration.

The EM algorithm is most practically useful when the maximum likelihood estimate for θ based on the *complete* data $Z = (Z_p, Z_m)$ is easy to compute. The algorithm is an iterative one, each iteration consisting of two steps. The first part of each iteration is called the *E-step* (for *E*xpectation), in which the expected log-likelihood function of the complete data given the observed data is computed, evaluated at the current estimate for θ, say $\hat{\theta}^{(i)}$. That is, we compute

$$\ell^{(i)}(\theta) \equiv E\left[\log f(Z_p, Z_m \mid \theta) \,\Big|\, Z_p, \hat{\theta}^{(i)}\right]. \tag{4.7.2}$$

Although in general this involves computing the expected value of a random function, in many practical situations it does not. In particular, if the distribution of Z is from an exponential family, then the right-hand side of equation (4.7.2) can be reduced to computing the expected value of the complete-data sufficient statistic for θ, which is often a one- or two-dimensional random variable. The second part of each iteration is called the *M-step* (for *M*aximization), in which the expected log-likelihood just computed is maximized. That is, we compute the updated estimate for θ as that value $\hat{\theta}^{(i+1)}$ which maximizes $\ell^{(i)}(\theta)$:

$$\hat{\theta}^{(i+1)} \equiv \max_{\theta}^{-1} \ell^{(i)}(\theta). \tag{4.7.3}$$

When the complete data are from an exponential family this step, too, is nearly trivial. In this case, one must find that value of θ for which the conditional expectation of the complete-data sufficient statistic agrees with the unconditional expectation.

Louis (1982) has shown how to obtain the conditional (as opposed to expected) information matrix for maximum likelihood estimates obtained using the EM algorithm. His algorithm depends on the first and second derivatives of the complete-data log-likelihood function, which are generally much easier to obtain than the corresponding values for the incomplete-data log-likelihood function.

An example will help to illustrate the EM approach. We return to the Swedish pension data of section 4.3.5.1, in which we estimated a binomial mixture of two populations. In population A each widow had exactly zero children. In population B, the distribution of children is Poisson with rate parameter λ. The probability of being in population A is ξ. The data (given in Table 4.3.1) are the number of widows n_i having exactly i children, $0 \leq i \leq 6$. We can think of this as an incomplete-data problem by thinking of n_0 as the sum of two unobserved random variables, n_A and n_B, the number of widows with no children from the two populations. If we could observe these random variables directly, the estimation problem would be trivial: we would use $\hat{\xi} = n_A/N$ to estimate the binomial fraction, and we would take $\hat{\lambda} = \sum_0^6 i \cdot n_i/(n_B + \sum_1^6 n_i)$ to be the estimate of the Poisson parameter. Indeed, these are just the maximum likelihood estimators for the two parameters based on the complete-data sufficient statistic $(n_A, n_B, n_1, \ldots, n_6)$. We can now describe the M-step: replace n_A and n_B in the previous sentence with their current estimates, $\hat{n}_A$ and $\hat{n}_B = n_0 - \hat{n}_A$.

For the E-step, we note that the complete data are from a multinomial distribution, an exponential family whose natural parameters in this case depend on ξ and λ in a simple way. Thus, we need only obtain the conditional expectation of the complete-data sufficient statistic given the observed data and the current estimates of the parameters. This reduces to finding $E(n_A \mid n_0, \hat{\xi}, \hat{\lambda})$, given by

$$\hat{n}_A \equiv E(n_A \mid n_0, \hat{\xi}, \hat{\lambda}) = \frac{\hat{\xi} n_0}{\hat{\xi} + (1 - \hat{\xi}) \exp(-\hat{\lambda})}. \qquad (4.7.4)$$

Starting with the estimates 0.75 and 0.40 for ξ and λ, respectively, a single iteration produces $\hat{\xi} = 0.614179$ and $\hat{\lambda} = 1.035478$. These estimates are each within half a percent of the MLE, and are comparable to the result after about four iterations of Newton-Raphson! The convergence behavior in subsequent iterations, however, is much more typical of the EM algorithm. Even though the first step produces a remarkably good answer, the algorithm requires more than thirty more iterations to converge to the same accuracy that Newton-Raphson achieves in a *total* of six iterations.

Once one gets in the spirit, it is quite easy to identify a "missing" random variable lurking in the shadows, observation of which would greatly simplify the problem at hand. In robust regression problems, for instance, matters would be greatly simplified if we could observe W_i, a random variable that has the value zero for normal data and the value one for "contaminated" data (Aitkin and Tunnicliffe Wilson, 1980). Factor analysis in some respects is just like regression, except that the predictor variables are unobserved (Dempster, Laird, and Rubin, 1977)! The EM algorithm can often be applied to mixture problems, of which the Swedish pension data is one example. For a review of the EM algorithm with special reference to estimating mixtures of exponential-family densities, see Redner and Walker (1984). Vardi, Shepp, and Kaufman (1985) use the EM algorithm to reconstruct the emission density in positron emission tomography (PET).

Even though convergence is often slow—Dempster, Laird, and Rubin show that it is generally only linear—the EM algorithm is widely known and used. The M-step frequently can be expressed in closed form and can be computed using a standard statistical package. The macro capabilities of interactive packages such as S, Minitab, and GLIM can often be used to implement the E-step in such cases, so that getting quick answers to relatively complex problems without expending much effort is possible.

COMMENT. Dempster, Laird, and Rubin's article on the EM algorithm contains a demonstration that the algorithm always converges to a maximum of the incomplete-data likelihood. As Wu (1983) notes, their convergence proof is in error. Wu examines the issue of convergence of the EM algorithm through its relationship to other optimization methods. He shows that when the complete data are from a curved exponential family with compact parameter space, and when the expected log-likelihood function satisfies a mild differentiability condition, then any EM sequence converges to a stationary point (not necessarily a maximum) of the likelihood function. (When there are multiple stationary points, the behavior of EM depends upon the starting point.) If the likelihood function is unimodal and satisfies the same differentiability condition, then the EM sequence will converge to the unique maximum likelihood estimator.

EXERCISES

4.66. [21] Show that the E-step of the EM algorithm can be reduced to finding the expected value of the complete-data sufficient statistic for θ when the complete data are from an exponential family.

4.67. [20] Show that the M-step of the EM algorithm when the complete-data likelihood is from an exponential family reduces to finding the value of θ for which the conditional and unconditional expectations of the complete-data sufficient statistic coincide.

4.68. [14] Explain why the E-step for the Swedish pension data reduces to finding $E(n_A \mid n_0, \hat{\xi}, \hat{\lambda})$.

4.69. [22] Show that the E-step for the Swedish pension data is given by equation (4.7.4).

4.70. [49] Explore the relationships between the EM algorithm, the ACE algorithm, and the nonlinear Gauss-Seidel algorithm. Under what circumstances does any one reduce to either of the others?

4.8 Time-series analysis

Data which are collected sequentially in time or space are called *time series*. Methods for analyzing such data differ from most of those discussed so far in that much of the structure in such data sets resides in the correlation between neighboring points of the series and in behavior that appears to be periodic in nature. Methods which focus on periodicities are called *frequency domain methods;* they describe the variations in a set of data in terms of a sum of sinusoidal functions at various frequencies. Bloomfield (1976) gives an introduction to the topic that is heavily grounded in computation. The computational building block of frequency methods is the *Fast Fourier transform*, which is the topic of Chapter 11. On the other hand, methods which focus on the correlational structure between observed points in the series are called *time domain methods*. These methods, pioneered by Box and Jenkins (1970), have been widely adopted in business and economics, partly because it is relatively easy to construct *forecasts* for future observations based on the past using time-domain models. New methods, based on the projection-pursuit paradigm have recently been proposed and are promising (McDonald, 1986).

A thorough treatment of computational aspects of time-series analysis could easily occupy an entire volume. Unfortunately, no such work has yet appeared, and a complete treatment here is not possible. Our focus in this section will be on time-domain analysis, with special emphasis on the general computing techniques that we have discussed earlier in Chapters 3 and 4.

The past ten years have seen major advances in computationally efficient methods for analysis of stationary Gaussian time series. The two biggest steps forward have been in obtaining maximum likelihood estimates of parameters, and in computing estimates and forecasts which adjust for prior information and missing observations. These advances are based, in turn, on advances in computing algorithms for obtaining values of the likelihood function of an *autoregressive moving-average (ARMA) process*, and for *Kalman filtering*.

Our study begins in Section 4.8.1 with an overview of ARMA models in general and an examination of the conditional likelihood methods which were used almost exclusively for ARMA estimation until the late 1970's. Sections 4.8.2 through 4.8.4 examine three approaches to computing the unconditional likelihood function (called the "exact likelihood" in the time-series literature).

4.8.1 Conditional likelihood estimation for ARMA models

The problem we consider is that of estimating the parameters α_1, α_2, ..., α_p, β_1 β_2, ..., β_q, and σ^2 from observations X_1, X_2, ..., X_n of a stationary Gaussian time series defined by

$$X_t = \sum_{j=1}^{p} \alpha_j X_{t-j} + a_t \qquad (4.8.1a)$$

$$a_t = \sum_{j=1}^{q} \beta_j \epsilon_{t-j} + \epsilon_t \qquad (4.8.1b)$$

$$\epsilon_t \sim \text{i.i.d. } \mathcal{N}(0, \sigma^2). \qquad (4.8.1c)$$

Equation (4.8.1a) expresses the current observation X_t in a regression-like model in terms of its p predecessors. The parameters in this equation (the α's) determine what is called the *autoregressive* structure in the series. The error structure (the a's) is described by equation (4.8.1b) as a weighted sum of independent underlying random variables (the ϵ's). These weights (β's) determine the *moving-average* structure of the errors. The model described by equations (4.8.1) is referred to as the Gaussian ARMA(p,q) model, where p and q refer to the number of autoregressive and moving-average terms, respectively. (If $q = 0$ we refer to the AR(p) model. Similarly, the MA(q) model is shorthand for the ARMA(0,q) model.)

A time series is said to be *stationary* if, for any set of subscripts $(t_1, t_2, \ldots, t_k)$, the joint distribution of $(X_{t_1}, X_{t_2}, \ldots, X_{t_k})$ is the same as that of $(X_{t_1+c}, X_{t_2+c}, \ldots, X_{t_k+c})$. That is, the distribution of a stationary series is unaffected by constant shifts of the time scale. These conditions imply that a Gaussian process (one for which each ϵ_t has the normal distribution) is stationary if and only if $E(X_t)$ and $\text{Var}(X_t)$ do not depend on t, and $\text{Cov}(X_t, X_s)$ is a function only of $|t - s|$.

COMMENT. For an ARMA process to be stationary, the autoregressive parameters must obey certain constraints, namely that the roots of the *characteristic equation* $1 - \sum_{j=1}^{p} \alpha_j z^j = 0$ must all lie outside the unit circle. Stationarity imposes no such constraints on the moving-average parameters, however it can be shown that a parameterization of the form (4.8.1b) can always be chosen so that the roots of the characteristic equation $1 - \sum_{j=1}^{q} \beta_j z^j = 0$ lie outside the unit circle. Such a representation is said to be *invertible;* invertibility of an MA process plays a similar role to that of stationarity for an AR process. Throughout this section we shall assume without further comment that both the stationarity and invertibility constraints are satisfied.

These models exhibit considerable linear structure, but only of a conditional sort—given p past X's, the mean of X_t can be expressed as a linear function of known quantities. It is this conditional linearity of means that makes ARMA models attractive for forecasting. What makes estimation in (4.8.1) more difficult than in ordinary regression is that the conditioning event changes from one observation to the next.

It is helpful to introduce the computing difficulties through examples. We first examine the autoregressive and moving average structures separately, and discuss the estimation problems which arise in each case, then we discuss how these structures combine in the general problem. Fuller (1976) describes computational methods for the conditional least-squares methods discussed below. The unconditional likelihood is generally highly nonlinear in the parameters, and methods of nonlinear optimization must be employed; to use these methods we require efficient means of calculating values of the log-likelihood function at fixed values of the parameters. The conditional computations provide a starting point for these computations. We now examine conditional methods in two special cases—the pure AR model and the pure MA model.

The AR(2) model. The stationary AR(2) model is the simplest nontrivial autoregressive model. For each k, let Y_k denote the $k \times 1$ vector containing the first k observations from the X series, that is, let $Y_k = (X_1, X_2, \ldots, X_k)$. The joint density can be factored as

$$
\begin{aligned}
f(y_n \mid \alpha_1, \alpha_2, \sigma^2) \\
&= f(x_n \mid y_{n-1}, \alpha_1, \alpha_2, \sigma^2) \times f(y_{n-1} \mid \alpha_1, \alpha_2, \sigma^2) \\
&= f(x_n \mid x_{n-1}, x_{n-2}, \alpha_1, \alpha_2, \sigma^2) \times f(y_{n-1} \mid \alpha_1, \alpha_2, \sigma^2) \\
&= \left[\prod_{j=3}^{n} f(x_j \mid x_{j-1}, x_{j-2}, \alpha_1, \alpha_2, \sigma^2) \right] \times f(x_1, x_2 \mid \alpha_1, \alpha_2, \sigma^2).
\end{aligned}
$$

(4.8.2)

Writing this in likelihood terms, we have

$$
\begin{aligned}
L(\alpha_1, \alpha_2, \sigma^2 \mid x_1, \ldots, x_n) =& L(\alpha_1, \alpha_2, \sigma^2 \mid x_1, x_2) \\
&\times \prod_{j=3}^{n} L_C(\alpha_1, \alpha_2, \sigma^2 \mid x_j, x_{j-1}, x_{j-2}).
\end{aligned}
$$

(4.8.3)

The factors L_C can be thought of as conditional likelihoods. If we focus for a moment on the product of the conditional likelihood terms, we notice something rather startling. If we write $Z_{j1} = X_{j-1}$ and $Z_{j2} = X_{j-2}$, for $j = 3, \ldots, n$, then the product of conditional likelihoods is exactly equal to the likelihood function for the regression problem

$$
X_j = \alpha_1 Z_{j1} + \alpha_2 Z_{j2} + a_j, \qquad j \geq 3. \tag{4.8.4}
$$

Here we are thinking of the values of the Z's as being fixed quantities whose values happen to coincide with the observed values of the X's. Thus, if we ignore the unconditional piece $L(\alpha_1, \alpha_2, \sigma^2 \mid x_1, x_2)$ of (4.8.3), then the maximum likelihood estimates for α_1, α_2 and σ^2 are very easy to obtain, namely $\hat{\alpha}_c = (Z'Z)^{-1} Z'Y$ and $\hat{\sigma}_c^2 = |X - Z\hat{\alpha}_c|^2/(n-5)$. These estimates are in fact often used, and are called the *conditional least-squares (CLS) estimators* (hence the subscript c).

Unfortunately, unless n is very large, that apparently insignificant unconditional part of the likelihood *cannot* safely be ignored. Let's look at this factor more closely. Since $Y_2 \equiv (X_1, X_2)$ has a bivariate normal distribution, we can deduce that

$$
\begin{aligned}
L(\alpha_1, \alpha_2, \sigma^2 \mid x_1, x_2) &= L(\alpha_1, \alpha_2, \sigma^2 \mid y_2) \\
&= |\Sigma|^{-1/2} \exp(-\frac{1}{2} y_2' \Sigma^{-1} y_2),
\end{aligned}
\tag{4.8.5}
$$

where

$$
\Sigma = \Sigma(\alpha_1, \alpha_2, \sigma^2) = \frac{\sigma^2}{1 - \delta_2^2} \begin{pmatrix} 1 & \delta_1 \\ \delta_1 & 1 \end{pmatrix},
\tag{4.8.6}
$$

and where $\delta_1 = \alpha_1/(1 - \alpha_2)$ is the *first-order autocorrelation*, and $\delta_2^2 = \alpha_1^2 + \alpha_2^2 + 2\alpha_1 \alpha_2 \delta_1$ must be less than one in order for the process to be stationary. When (4.8.5) is introduced into equation (4.8.3), the result is strong nonlinearity in the parameters. The problem is particularly acute when there is strong autoregressive structure, since large values for $|\alpha_1|$ and $|\alpha_2|$ imply that the denominator $(1 - \delta_2^2)$ on the right-hand side of (4.8.6) is very small.

For the general AR(p) model, we would have to condition on the first p observations, and the expression corresponding to (4.8.4) would have p terms containing the *lagged values* of the X's. Expression (4.8.5) would again hold, with y_p replacing y_2, although (4.8.6) would of course be a different, and more complicated, expression.

To obtain maximum likelihood estimates for the parameters based on (4.8.3) and (4.8.5), it is generally necessary to use nonlinear optimization. A straightforward approach is simply to supply a subroutine for computing the likelihood function (or its logarithm) to a general nonlinear optimizer such as UNCMIN, discussed in Section 4.5.4. For the pure autoregressive process, it is relatively easy to compute the log likelihood, which separates into a regression-like sum of squares, plus a quadratic form, plus a log determinant. The latter two components involve a covariance matrix (Σ) whose elements are easily computed, given the current values of the parameters.

The MA(1) process. The simplest moving-average process has $q = 1$, so that we can write $X_t = \epsilon_t + \beta \epsilon_{t-1}$. Since there is only a single moving-average parameter, we have omitted the subscript on β. We shall again use Y_n to denote the vector of the first n observations $(X_1, \ldots, X_n)$. We can obtain a conditional likelihood in the MA case in a fashion similar to that of the AR case, but now we must condition on an unobserved random variable, ϵ_0.

We begin by writing $X_t = \epsilon_t + \beta\epsilon_{t-1} \equiv \epsilon_t + \hat{X}_t$, where

$$\hat{X}_t = - \left[\sum_{j=1}^{t-1} (-\beta)^j X_{t-j} + (-\beta)^t \epsilon_0 \right] \tag{4.8.7}$$

$$\equiv \beta\hat{\epsilon}_{t-1}(\beta, \epsilon_0).$$

Conditional on Y_{t-1} and ϵ_0, X_t has a normal distribution with mean $\hat{X}_t$ and variance σ^2. The hats in our notation emphasize that, given β, ϵ_0, and the previous observations in the series, we can obtain an unbiased forecast $(\hat{X}_t)$ of the next observation. Applying a factorization similar to that of equation (4.8.2), we may write the joint density as

$$f(y_n \mid \beta, \sigma^2) = \int f(y_n, \epsilon_0 \mid \beta, \sigma^2) \, d\epsilon_0, \tag{4.8.8}$$

where

$$
\begin{aligned}
f(y_n, &\epsilon_0 \mid \beta, \sigma^2) \\
&= f(y_n \mid \epsilon_0, \beta, \sigma^2) \times f(\epsilon_0 \mid \sigma^2) \\
&= f(x_n \mid y_{n-1}, \epsilon_0, \beta, \sigma^2) \times f(y_{n-1} \mid \epsilon_0, \beta, \sigma^2) \times f(\epsilon_0 \mid \sigma^2) \\
&= \left[\prod_{j=1}^{n} f(x_j \mid y_{j-1}, \epsilon_0, \beta, \sigma^2) \right] \times f(\epsilon_0 \mid \sigma^2) \\
&= \left[\prod_{j=1}^{n} f(x_j \mid \hat{x}_j, \sigma^2) \right] \times f(\epsilon_0 \mid \sigma^2). \tag{4.8.9}
\end{aligned}
$$

The likelihood interpretation of this expression is quite interesting. The product term is a product of n conditional likelihoods, and the final term involves none of the observed data! Rather, the final term involves an unobserved random variable (ϵ_0). Thus, we have expressed the MA(1) estimation problem as a missing-data problem, a situation in which the EM algorithm may prove useful.

Note that if β is close to zero, then the effect of ϵ_0 is negligible in all but a few of the $\hat{X}_t$'s; this suggests fixing ϵ_0 at some value and then ignoring the unconditional last term of (4.8.9). In this situation ϵ_0 is generally set equal to its expectation (zero) and then the estimate for β is obtained by nonlinear least squares, minimizing the log-likelihood $\sum_{j=1}^{n}(X_j - \hat{X}_j)^2$. (This maximizes the product term and ignores the last term of (4.8.9).) It is important to remember that $\hat{X}_j$ is a $(j+1)$st order polynomial in β, so that estimating even a single MA parameter leads to nonlinear methods. The solution so obtained is called the *CLS (conditional least-squares)* estimator for the moving-average parameter, and is a function of the selected *starting value* chosen for ϵ_0.

COMMENT. The CLS method for moving-average processes treats the missing datum (ϵ_0) as if it were a parameter, whose value is heuristically imputed to be zero. One could proceed a step further and "estimate" ϵ_0 by simultaneously maximizing (4.8.9) with respect to β and ϵ_0. Even with this choice of imputed value the procedure does *not* produce the maximum-likelihood estimate for β, since the quantity being maximized in (4.8.9) is not a likelihood when ϵ_0 is considered to be a parameter, and the estimates so obtained need not maximize the actual likelihood (4.8.8). As we discussed on page 240 in our discussion of the EM algorithm, estimates based on such an approach can be severely biased, particularly in small samples.

For the general MA(q) model, we would have to condition on q unobserved ϵ's preceding the observed data. While the expressions for $\hat{X}_j$ would change, as before, each would be computable in terms of the preceding X's and the conditional values of the q unobservables. If the model fits adequately, Gauss-Newton methods are generally acceptable for estimating MA parameters using conditional least-squares. Fuller (1976) outlines the details of such computations, and discusses an approximate method for imputation of starting values. The unconditional problem is more difficult, requiring a method for computing values of the likelihood at given parameter values to be used by a general optimization program.

The general ARMA(p,q) model. When fitting parameters in a model with both moving-average and autoregressive components, the methods described above can be combined. Generally, we can reduce the likelihood function to a product of two terms, one with simple structure conditional on the first p observations X_j and on q errors ϵ_j preceding ϵ_1. A conceptually simple way of thinking about the problem is in terms of a Gauss-Seidel iteration. First, we rewrite equation (4.8.1a) to define

$$Z_t \equiv X_t - \sum_{j=1}^{p} \alpha_j X_{t-j} = a_t. \qquad (4.8.10)$$

If we knew the values of the α's, the conditional distribution of $(Z_t \mid Y_{t-1})$ would be MA(q), so that we could apply our computational methods for the pure MA process to obtain estimates of the β's. In practice, we can do just that, using our current estimates for the autoregressive parameters. Once we have estimates for the moving-average parameters, we can then compute the variance matrix Σ_β of the errors a_t in the autoregressive formulation (4.8.1a). The conditional likelihood given the first p observations (and the variance matrix Σ_β) is simply that of a GLS problem, with weighting matrix Σ_β, so we can estimate the α's by straightforward GLS. The process then repeats, alternating between the autoregressive and moving-average parameters.

The procedure just described produces CLS estimators; whenever $q \geq 1$ the CLS approach requires imputation of q starting values, with all of the attendant difficulties outlined in the MA(1) model. Maximum-likelihood parameter estimates for the mixed model can be obtained as in the two pure models by using a nonlinear optimization package, provided that an efficient method for computing values of the likelihood function or its logarithm is available. The remaining subsections examine three approaches to computing the exact likelihood.

4.8.2 The method of backward forecasts

The first method for obtaining the exact likelihood was also the first proposed, and although computationally inefficient, is still in wide use. The method is based on two properties of stationary Gaussian processes. First, every stationary ARMA process has a pure AR representation (possibly of infinite order) and a pure MA representation (also possibly of infinite order). Second, since the joint distribution of (X_t, X_s) depends only on $|t - s|$, it follows that the distribution of $(X_1, \ldots, X_n)$ is the same as that of $(X_n, X_{n-1}, \ldots, X_1)$. Thus, the distribution of the series is invariant under *time reversal*.

Box and Jenkins (1970) show that the exact log likelihood for an ARMA(p,q) series can be written in the form

$$\ell(\alpha, \beta, \sigma^2) = -\frac{n}{2} \log \sigma^2 - \frac{1}{2} \log \det \Sigma_{\alpha\beta} - \frac{1}{2\sigma^2} S(\alpha, \beta), \qquad (4.8.11)$$

where

$$S(\alpha, \beta) = \sum_{t=-\infty}^{n} [E(\epsilon_t \mid y_n, \alpha, \beta)]^2$$

$$\equiv \sum_{t=-\infty}^{n} [\epsilon_t]^2. \qquad (4.8.12)$$

In these expressions we have written $\alpha = (\alpha_1, \ldots, \alpha_p)$, and $\beta = (\beta_1, \ldots, \beta_q)$, and we have adopted a succinct notation ($[\epsilon_t]$) to denote conditional expectation given the parameters and the observations $Y_n = (X_1, \ldots, X_n)$. Note that the summation has an infinite lower limit; in computational practice the lower limit must be replaced by $-Q$, where Q is a suitably large integer.

The variance matrix $\Sigma_{\alpha\beta}$ is a highly nonlinear function of the parameters. Box and Jenkins assert that the second term of (4.8.11) which involves this matrix is generally negligible in size compared to $S(\alpha, \beta)$, except possibly for small n. They suggest approximating the exact likelihood by omitting the covariance-matrix term in (4.8.11), leaving a nonlinear least-squares problem. This problem is *not* equivalent to the corresponding CLS problem, which has a summation over different terms extending over a finite range. Because of its origin as a least-squares approximation to an exact maximum likelihood estimator, estimates obtained by minimizing (4.8.12) have come to be called *exact least squares (ELS) estimators*. This terminology is inappropriate, since the ELS estimates rely on two levels of approximation—dropping a term from the likelihood and replacing an infinite sum by a finite one.

Whether or not one ignores the log-determinant term, the sum of squares $S(\alpha, \beta)$ must still be computed. The terms in the sum on the right-hand side of (4.8.12) are not easy to obtain exactly for general ARMA(p,q) processes. Using (4.8.1a) and (4.8.1b), we can write for all t

$$\epsilon_t = X_t - \sum_{j=1}^{p} \alpha_j X_{t-j} - \sum_{j=1}^{q} \beta_j \epsilon_{t-j}. \qquad (4.8.13)$$

Taking conditional expectations we may then write

$$[\epsilon_t] = [X_t] - \sum_{j=1}^{p} \alpha_j [X_{t-j}] - \sum_{j=1}^{q} \beta_j [\epsilon_{t-j}]. \qquad (4.8.14)$$

Note that, given p preceding $[X]$'s and q preceding $[\epsilon]$'s, we may compute either $[X_t]$ or $[\epsilon_t]$, given the other. Note, too, that $[X_t] = X_t$ for $1 \le t \le n$, and that $[\epsilon_t] = 0$ for $t > n$. In particular, we can obtain *forecasts* of X_t arbitrarily far into the future. What we need, though, are the values of $[\epsilon_t]$ for $t \le 0$, stretching arbitrarily far back into the past, but we don't have ready access to them.

Now for the sleight of hand—watch closely! We invoke the time-reversibility of stationary processes to note that the stochastic structure of the X's remains unchanged if we view X_1 as the *end* of the series and X_n as the *beginning*. Thus, there must be a set of uncorrelated errors, let's call them $\{\delta_t\}$, for which we can write

$$\delta_t = X_t - \sum_{j=1}^{p} \alpha_j X_{t+j} - \sum_{j=1}^{q} \beta_j \delta_{t+j}. \qquad (4.8.15)$$

In terms of conditional expectations, we have $[\delta_t] = 0$ for $t \le 0$, so that using the recurrence analogous to (4.8.14), we can obtain values for $[X_t]$ for $t \le 0$, given the $[X]$'s and $[\delta]$'s to the *right* of the current value of t. From this perspective, the roles of past and future have been interchanged, and in this setting we can "forecast" X_t arbitrarily far back into the past. Estimates for $[X_t]$ obtained in this way are called *backward forecasts*, or *backforecasts*. But once we have these values, we can then use them in (4.8.14) to obtain estimates for the corresponding $[\epsilon_t]$'s which we needed in the first place.

We use the last p observations as starting values for the backforecasts. Any unknown values for $[\delta_t]$ that are needed to start the iteration are taken to be zero as an initial estimate. As we move t from right to left we can, at each stage, solve for $[\delta_t]$, until $t = 1$. From $t = 0$ onward we know that $[\delta_t] = 0$, so that we can then begin to solve for the unknown $[X_t]$'s. As $t \to -\infty$, the values for $[X_t] \to 0$. When these values are sufficiently small, say for $t = -Q$, we turn around and start moving t in the forward direction, from left to right. Again, we start with the p most recently obtained values of $[X_t]$ to begin the forward iteration, that is, from $[X_{-Q}]$ through $[X_{p-Q-1}]$. The corresponding values for $[\epsilon_t]$ are taken to be zero. On this pass through, we obtain values for $[\epsilon_t]$ (since the values of $[X_t]$ are known) until $t = n$. At this point, we have estimates for the terms in $S(\alpha, \beta)$, and we could stop. However, these estimates were based on the assumption that $[\delta_t] = 0$ for $n - p + 1 \leq t \leq \infty$. To get a better estimate, we could proceed to forecast $[X_t]$ into the future, say through $t = Q$, taking advantage of the fact that $[\epsilon_t] = 0$ for $t > n$. With these forecasts in hand, we could turn around once more and obtain $[\delta_t]$'s, this time obtaining nonzero values for $n - p + 1 \leq t \leq Q$. These new values could then be used as starting values for a second iteration of the backforecasting process. The backforecasting procedure is one involving alternating conditional expectations, and can be viewed as an instance of the ACE algorithm.

If the roots of either the moving-average or the autoregressive characteristic equations are close to the unit circle, convergence of the backforecasting procedure is very slow, and Q must generally be quite large before the backforecasts are negligible in size. Most computer programs using backforecasting employ a fixed Q (often $Q = 100$), and perform only a single iteration of the process described above.

To obtain the exact likelihood one must evaluate both $S(\alpha, \beta)$ and the log determinant of $\Sigma_{\alpha\beta}$. The latter calculation requires a method for computing the variance matrix of a general ARMA(p,q) process, and then taking its determinant. Efficient methods for doing so can be combined with algorithms discussed in Section 4.8.3 for obtaining $S(\alpha, \beta)$ *exactly*, so that backfitting is used primarily in conjunction with ELS, which omits this computation altogether. Omitting the second term in (4.8.11) can lead to markedly inferior estimators, particularly when the roots of the characteristic equations are close to the unit circle. Still, ELS estimators continue to be used because they are relatively easy to compute, and are somewhat more appealing than CLS estimators.

4.8.3 Factorization methods

The most straightforward way of writing (4.8.11) is in terms of the raw observations themselves, which we have denoted by Y_n:

$$\ell(\alpha, \beta, \sigma^2) = -\frac{n}{2} \log \sigma^2 - \frac{1}{2} \log \det \Sigma_{\alpha\beta} - \frac{y_n' \Sigma_{\alpha\beta}^{-1} y_n}{2\sigma^2}. \qquad (4.8.16)$$

For given α and β one can compute $\Sigma_{\alpha\beta}$ using algorithms described by McLeod (1975) and by Tunnicliffe Wilson (1979). The latter method is more efficient when p or q is large, as may be the case when estimating models with seasonal components. Once $\Sigma_{\alpha\beta}$ has been obtained, it can be used to evaluate (4.8.16). Generally this would involve working with an $n \times n$ matrix which, in many cases, would be too large to deal with effectively.

A more satisfactory computing scheme is proposed by Ansley (1979), based on a simple idea. Letting $m = \max(p, q)$, and defining Z_t as in equation (4.8.10), consider the transformation

$$W_t = \begin{cases} X_t, & \text{for } 1 \leq t \leq m; \\ Z_t, & \text{for } m < t \leq n. \end{cases} \qquad (4.8.17)$$

The vectors W and X are related by a simple linear transformation $W = KX$, where K is an $n \times n$ lower triangular with coefficients depending only on the autoregressive parameters. Fewer than np of these coefficients are nonzero. Note also that this transformation has unit Jacobian. The covariance matrix $\Sigma_w \equiv \text{Cov}(W)$ is now a band matrix of bandwidth m in the first m rows, and of bandwidth q thereafter. The upper $m \times m$ matrix can be computed using a general algorithm for the theoretical autocovariance function of an ARMA(p,q) process, while the remainder can be obtained from the MA(q) part alone. The Cholesky decomposition can be used to write $\Sigma_w = LL'$, where L is a lower-triangular band matrix. The final step is based on the transformation $E = L^{-1}W$. The log likelihood written in terms of E is particularly simple, namely,

$$\ell(\alpha, \beta, \sigma^2 \mid x) = -\frac{n}{2} \log \sigma^2 - \log \det L - \frac{1}{2\sigma^2} \sum_{j=1}^{n} e_t^2. \qquad (4.8.18)$$

The determinant in this expression is easy to compute. Since the matrix involved is lower triangular, its determinant is simply the product of its diagonal elements. (For numerical stability it is preferable to compute the sum of the logarithms of the diagonal elements rather than taking the logarithm of the product.)

COMMENT. The method of McLeod (1975) for obtaining the theoretical covariance matrix of an ARMA(p,q) process can be extended to the vector case (Kohn and Ansley, 1982). Special algorithms for the

Cholesky decomposition of a band matrix take advantage of the special structure introduced by the transformation (4.8.17); see Martin and Wilkinson (1965).

4.8.4 Kalman filter methods

The factorization method is based on the idea of transforming the data so as to replace the original covariance matrix $\Sigma_{\alpha\beta}$ by a factored version of the covariance matrix that is much easier to compute with. In Ansley's work, $\Sigma_{\alpha\beta}$ is replaced by a band matrix whose Cholesky factor is easily obtained. The bandwidth of this matrix is $m = \max(p,q)$. If n is much larger than m, we can think of the process as making $\Sigma_{\alpha\beta}$ more nearly diagonal, corresponding to a stochastic process whose increments are "more nearly" orthogonal. The idea behind the *Kalman filter* approach is to transform the original series into a derived series whose increments are precisely orthogonal. Although the derivation is more complex, the result is a computationally attractive scheme for recursive computation of the likelihood function.

To do this, we begin by factoring the joint density in the following way. Once again, we shall write $Y_t = (X_1,\ldots,X_t)$ to denote the data observed through time t. Then we may write

$$f(y_n \mid \alpha,\beta,\sigma^2) = f(x_n \mid y_{n-1},\alpha,\beta,\sigma^2) \times f(y_{n-1} \mid \alpha,\beta,\sigma^2)$$
$$= \prod_{j=1}^{n} f(x_j \mid y_{j-1},\alpha,\beta,\sigma^2), \qquad (4.8.19)$$

where we define $f(x_1 \mid y_0,\alpha,\beta,\sigma^2) \equiv f(x_1 \mid \alpha,\beta,\sigma^2)$. Let us consider the terms $f(x_j \mid y_{j-1},\alpha,\beta,\sigma^2)$ in the product. Since the process is Gaussian, we know that this conditional distribution of X_j must also be normal, say $X_j \sim \mathcal{N}(\mu_j,V_j)$. Clearly, $\mu_j = E(X_j \mid y_{j-1},\alpha,\beta,\sigma^2)$, and $\sigma^2 V_j = \mathrm{Var}(X_j \mid y_{j-1},\alpha,\beta,\sigma^2)$. What is more, the conditional distribution of X_j given μ_j and V_j is independent of that of Y_{j-1}. Thus we can write the log likelihood as

$$\ell(\alpha,\beta,\sigma^2 \mid y_n) = -\sum_{j=1}^{n}\left[\log\sigma^2 V_j + \frac{(x_j-\mu_j)^2}{\sigma^2 V_j}\right]. \qquad (4.8.20)$$

At the maximum, σ^2 must satisfy the usual relationship $\sigma^2 = n^{-1}\sum(x_j-\mu_j)^2/V_j$. When this expression is substituted into equation (4.8.20) and constant terms are omitted, the expression reduces to

$$\ell(\alpha,\beta \mid y_n,\hat\sigma^2) = -\sum\log V_j - n\log\sum(x_j-\mu_j)^2/V_j. \qquad (4.8.21)$$

Now if only μ_j and V_j could be computed, the likelihood would be easy to evaluate. We can think of μ_j as the one-step-ahead forecast of X, based on data through time $j - 1$, and of V_j as the one-step-ahead prediction variance. Let's write $x(j \mid t)$ to denote the conditional mean of X_j given data through time t. If $j \leq t$, then $x(j \mid t)$ is simply x_j itself. Similarly, write $v(j \mid t)$ to denote the conditional variance of X_j given data through time t. The quantities of interest to us can be expressed easily in these terms, namely, $\mu_j = x(j \mid j - 1)$ and $V_j = v(j \mid j - 1)$.

Consider the AR(p) model for a moment. In this model $x(j \mid j - 1)$ is a function of the p previously observed values of X. Shifting perspective slightly, we see that we first begin to accumulate information about μ_j in this model p steps earlier, when we observe X_{j-p}. As we observe each of the next $p - 1$ observations, we obtain more of the information needed to compute the one-step forecast. At each of these p stages, we can summarize what we know about X_j in terms of its current conditional expectation. At time $j - p$, for instance, we can construct the p-step predictor $x(j \mid j - p)$. To obtain μ_j we must keep track of accumulating information about X_j for the preceding p steps. Thus, at any given time (say at time $j - 1$) we are "working on" p different forecasts, namely $x(j \mid j - 1)$ through $x(j + p - 1 \mid j - 1)$, and their associated variances. Note that the first of these quantities (with its variance) is precisely what we need to compute the likelihood contribution for X_j. After observing X_j, we can then proceed to update our forecasts and proceed recursively.

More formally, let $m = \max(p, q + 1)$, and let $\alpha_j = 0$ for $j > p$ and $\beta_j = 0$ for $j > q$. Let $Z(t)$ denote the $m \times 1$ vector of observations starting at time t, that is, $Z(t) = (X_t, \ldots, X_{t+m-1})'$. Note that the components of $Z(t)$ extend into the *future* from time t. We define the *state vector* $Z(t \mid t)$ of the process at time t by

$$Z(t \mid t) = \begin{bmatrix} x(t \mid t) \\ x(t + 1 \mid t) \\ \vdots \\ x(t + m - 1 \mid t) \end{bmatrix} = E[Z(t) \mid y_t].$$

We also define the conditional state vector $Z(t \mid j)$ as the conditional expectation $E[Z(t) \mid y_j]$, and we denote the conditional covariance matrix of $Z(t)$ given data through time j by $P(t \mid j)$.

We now show how to obtain the likelihood contribution associated with X_{t+1}, given $Z(t \mid t)$ and $P(t \mid t)$, and how then to update the latter two quantities to obtain $Z(t+1 \mid t+1)$ and $P(t+1 \mid t+1)$ so that the recursion can proceed. We follow closely the derivation given in Jones (1980), wherein omitted details can be found.

Although it is not immediately obvious, the state vector at time $t + 1$ can be written as

$$Z(t + 1 \,|\, t + 1) = FZ(t \,|\, t) + G\epsilon_{t+1}, \qquad (4.8.22)$$

where F is an $m \times m$ matrix with ones on the superdiagonal, with coefficients $\alpha_m, \ldots, \alpha_1$ in the last row, and with zeroes elsewhere, and where G is an $m \times 1$ vector with coefficients g_j recursively defined by $g_j = \beta_{j-1} + \sum_{k=1}^{j-1} \alpha_k g_{j-l}$. Since ϵ_{t+1} is independent of the observations at or before time t, we can use (4.8.22) to write the one-step forecast as

$$Z(t + 1 \,|\, t) = FZ(t \,|\, t).$$

This forecast has conditional variance $P(t+1|t) = FP(t|t)F' + \sigma^2 GG'$, and the conditional mean of the next observation—the one-step prediction—is $\mu_{t+1} = x(t + 1 \,|\, t) = HZ(t + 1 \,|\, t)$, where $H = (1, 0, \ldots, 0)$.

The next observation (X_{t+1}) can now be used to update all of the quantities in the computation. Now let $B_{t+1} = (HP(t + 1 \,|\, t)H')^{-1}PH'$. Then it can be shown that

$$Z(t + 1 \,|\, t + 1) = Z(t + 1 \,|\, t) + B_{t+1}[x_{t+1} - x(t + 1 \,|\, t)], \qquad (4.8.23)$$

and that

$$P(t + 1 \,|\, t + 1) = P(t + 1 \,|\, t) - B_{t+1}HP(t + 1 \,|\, t). \qquad (4.8.24)$$

Finally, $V_{t+1} = HP(t + 1 \,|\, t)H' = P_{11}(t + 1 \,|\, t)$.

The equations above make it possible, starting from an initial state vector $Z(0 \,|\, 0) = 0$ and an initial covariance matrix $P(0 \,|\, 0)$ to obtain all of the components of the log likelihood expression (4.8.21), and to do so in a single pass through the data. The details of obtaining the initial-state covariance matrix $P(0 \,|\, 0)$ are given in Jones (1980).

The Kalman filtering approach to obtaining the likelihood has several advantages over the factorization method. The filtering algorithm can be readily adapted to series observed with noise (Jones, 1980), to multivariate series (Ansley and Kohn, 1983), to series with missing or aggregated data (Ansley and Kohn, 1983), and to nonstationary series (Ansley and Kohn, 1984). The method also makes it possible to treat series observed at points unequally spaced in time, to allow for time-varying parameters, and to incorporate a Bayesian prior by suitable adjustment to $Z(0 \,|\, 0)$ and $P(0 \,|\, 0)$. A unified theory is presented in Ansley and Kohn (1985).

The fastest known algorithms for computing an ARMA likelihood are based on Kalman filtering ideas. Mélard (1984a) gives such an algorithm for univariate series coded in FORTRAN. It uses a somewhat more efficient variant of the Kalman filter than that described above. It does not allow for missing data. In a subsequent paper, Mélard (1984b) shows how this algorithm can be extended to the regression problem with ARMA errors, and he shows in principle how to obtain the score function of the ARMA(p,q) likelihood. If the latter can be implemented efficiently, it holds promise for improved numerical optimization of the likelihood.

At this writing, no publicly available programs exist for computing the exact likelihood for multivariate time series.

4.8.5 A note on standard errors

Even if a general nonlinear optimization routine is used to obtain the maximum likelihood parameter estimates based on an exact likelihood algorithm, one need *not* rely on the optimizing software to provide an asymptotic variance matrix for the estimates. The asymptotic variance structure of the MLE is known theoretically as a function of the parameters, and efficient algorithms exist for computing both the information matrix and its inverse (Godolphin and Unwin, 1983). Thus, one can select an optimization algorithm without regard to the quality of the Hessian matrix since a better estimate, more directly obtained, is available at little cost.

EXERCISES

4.71. [24] Consider the AR(1) model. What conditions on α_1 and σ^2 will guarantee that the series is stationary?

4.72. [16] Derive equation (4.8.2).

4.73. [22] Show that the product of conditional likelihoods in (4.8.3) is the same as the (unconditional) likelihood for the regression problem of (4.8.4).

4.74. [01] Why is the factor in the denominator of $\hat{\sigma}_c^2$ equal to $n - 5$?

4.75. [27] Derive the variance matrix for the AR(2) process given by equation (4.8.6).

4.76. [03] In the first line of equation (4.8.9), why does the term $f(\epsilon_0 \mid \sigma^2)$ appear instead of $f(\epsilon_0 \mid \beta, \sigma^2)$?

4.77. [10] Justify the last step in equation (4.8.9).

4.78. [12] Write out the jth term of the product in the last line of equation (4.8.9).

4.79. [14] The transformation $W = KX$ of equation (4.8.17) has unit Jacobian. Why?

4.80. [17] Show that the conditional distribution of X_j given μ_j and V_j is independent of that of Y_{j-1}.

4.81. [12] Show that σ^2 must satisfy the relationship $\sigma^2 = n^{-1} \sum (x_j - \mu_j)^2 / V_j$ when equation (4.8.20) is maximized.

4.82. [14] Verify equation (4.8.21).

4.83. [29] Prove equation (4.8.23).

5 NUMERICAL INTEGRATION AND APPROXIMATION

In the course of everyday statistical work, whether theoretical or applied, one often encounters an integral whose value must be obtained. After all, most of the familiar quantities with which statisticians deal are averages of one sort or another, and the theoretical values or the normalizing constants are averages obtained by integrating functions over some domain. Probabilities, means, variances, mean squared errors, can all be thought of as appropriately defined integrals.

In like fashion, one often needs to be able to obtain values for a tabled function such as the normal cumulative distribution function (cdf) inside a computer program—a place where books of tables are hard to use! So we need algorithms for approximating these functions, and methods for deriving new approximations for functions which are of interest to us.

These days, the training received by most statisticians in the numerical evaluation of integrals and computationally useful approximation of functions is meager. Most have been exposed to little more than the passing mention accorded to the trapezoidal rule and to Simpson's rule in a college calculus course; the topics covered in this chapter are largely not on the menus of most undergraduate programs in mathematics and graduate programs in statistics. Engineers, physicists, and numerical analysts, by contrast, have generally been more fortunate, because they have been exposed to some of the elegant, enjoyable, and intriguing aspects of a most interesting subject.

Our treatment of *quadrature* (an older name for numerical integration) is necessarily incomplete. As in the rest of this book, we shall limit our discussion to those methods which are of greatest use in statistics. The most comprehensive single reference to the topic is Davis and Rabinowitz (1984). A lovely and practical book by Press, Flannery, Teukolsky and Vetterling (1986) has not only a clear presentation of the ideas behind quadrature and approximation methods, but also FORTRAN subroutines for many of the methods discussed here. For serious use, one should know about *QUAD-PACK*, a package of well-designed, carefully-coded FORTRAN routines for univariate quadrature (Piessens, de Doncker-Kapenga, Überhuber, and

Kahaner, 1983) A gold mine of formulæ, numerical procedures, approximations, and tables can be found in Abramowitz and Stegun (1964).

The plan of this chapter is to begin with the integration rules which are most familiar, namely, weighted sums of function values evaluated on an equally-spaced grid of points. These methods have the generic name of *Newton-Cotes rules*. We then show how to adapt these and other methods to cope with infinite or semi-infinite ranges of integration. Relaxing the requirement that the evaluation points be equally spaced leads to the *Gaussian quadrature rules*, and relaxing the requirement that the evaluation intervals be fixed in advance leads to *adaptive quadrature rules*, which devote more effort to those areas in which the integrand is more difficult. If the integrand is very expensive to evaluate, or perhaps is known only through tabulation, it may be convenient to approximate the integrand by *spline functions*. In statistical problems the integrand may be a probability density function not easily expressible in a form suitable for computation, but at the same time it may be easy to sample from the distribution; in such cases *Monte Carlo integration* is a plausible candidate. Bayesian statisticians often must investigate features of posterior and predictive distributions, computation of which involves multiple integration of moderate to high order. We discuss some approaches both to multiple integration in general and to Bayesian analysis in particular. We then discuss methods for function evaluation and approximation, with particular emphasis on computing cumulative distribution functions, their complements, and their inverses, and we close with a section containing some useful approximations and pointers to available software.

Throughout this chapter, we shall assume that we are interested in evaluating the integral of a function $f(x)$, possibly multiplied by a weight function $w(x)$, over a range (a, b). Ultimately, we shall allow either a or b to be infinite, but we shall begin by considering the case in which both a and b are finite.

The quality of an integral approximation depends upon the degree of smoothness of the integrand. It is generally true that the smoother the integrand is, the smaller the error will be, up to a limiting factor which depends upon the particular algorithm. It is also generally the case that algorithms converge more rapidly for smoother functions. "Smoothness" of a function can be quantified in terms of the number of continuous derivatives it has, the magnitude of these derivatives, and the analyticity of the function. Table 5.1 contains a few classes of functions to which we shall refer in this chapter, in decreasing order of smoothness.

Error estimates are associated with most of the integration rules we shall discuss. They are generally of the form $O(n^{-k} f^{(k)}(x^*))$, where x^* is a (generally unknowable) point in the range of integration, f is the integrand, and n is the number of subintervals into which the range of integration has

Symbol	Defining property
P_n	Polynomials of degree at most n
$S_n(x_1, \ldots, x_m)$	Splines of degree n with knots $x_1, \ldots, x_m$
$C^n[a, b]$	Functions with n continuous derivatives on $[a, b]$
$PC^n[a, b]$	Functions with piecewise continuous nth derivative on $[a, b]$
$C^1[a, b]$	Functions with continuous first derivative on $[a, b]$
$C^0[a, b]$ or $C[a, b]$	Functions continuous on $[a, b]$
$R[a, b]$	Bounded, Riemann-integrable functions on $[a, b]$

TABLE 5.1.1 *Some classes of real-valued functions, in decreasing order of smoothness. Classes* S_n *and* C^n *are not comparable; however* $P_n \subset S_n, C^n \subset PC_n$. *When the interval on which the functions are defined is omitted, it is assumed to be the entire real line.*

been divided. In every case the validity of such an error bound requires that the integrand f be k times continuously differentiable, that is, $f \in C^k[a, b]$. We shall make this differentiability assumption implicitly throughout this chapter, although we do discuss what happens on those occasions when such a smoothness condition is not satisfied.

5.1 Newton-Cotes methods

The easiest approach to evaluating $\int_a^b f(x)dx$ approximates the integrand by a well-behaved function that is easy to integrate exactly (such as a polynomial) and that agrees with $f(x)$ on a grid of points. Any integration rule for which this grid consists of *equally spaced* points and which integrates polynomials of a certain degree exactly is called a *Newton-Cotes integration rule*. (Algorithms for numerical integration are rarely called algorithms; they are called "rules." The terms are interchangeable, although reference to "Simpson's algorithm" generally elicits only puzzled stares.) These rules are generally easy to program because their structure is so simple. Despite their simplicity, they are the building blocks which underly even some of the most modern and effective of automatic numerical integration programs.

5.1.1 Riemann integrals

The simplest methods for numerical approximation of integrals begin with the definition of the Riemann integral itself, which we shall now describe. For each n, subdivide the interval $[a, b]$ into n disjoint subintervals Δx_{in} whose union equals $[a, b]$, and from each subinterval select a point $x_{in} \in \Delta x_{in}$. Choose these subdivisions so that the maximum length of the subintervals $\max_i |\Delta x_{in}| \to 0$ as $n \to \infty$. Define $S_n = \sum_{i=1}^n f(x_{in})|\Delta x_{in}|$. Then the *Riemann integral of f on* $[a, b]$ is defined to be $I(a, b) = \lim_{n \to \infty} S_n$. If this limit exists, $f(x)$ is said to be Riemann integrable. The value of the integral is independent of the particular sequence of subintervals chosen and of the method of choosing the points x_{in} at which the integrand is evaluated.

Since the integral is independent of the particular sequence of grids, the most natural approximation to the integral is to divide $[a, b]$ into, say, n subintervals of equal length. We shall denote the endpoints of these intervals by x_i, where $x_0 = a$, $x_n = b$, and in general, $x_i = a + ih$, where $h \equiv (b - a)/n$. For each of the subintervals $[x_i, x_{i+1}]$ the function is evaluated at a representative point, say (for convenience) at the left endpoint. This leads to the *(left) Riemann approximation:*

$$I_R^{(n)}(a, b) = h \sum_{i=0}^{n-1} f(a + ih), \qquad (5.1.1)$$

where $h \equiv (b - a)/n$, as above. Except when it is absolutely necessary, we shall suppress the n in our notation above and write instead $I_R(a, b)$. By definition of the Riemann integral,

$$\int_a^b f(x)dx = \lim_{n \to \infty} I_R^{(n)}(a, b).$$

There are several things to notice about equation (5.1.1) which carry over to more accurate rules. These properties are illustrated in Figure 5.1.1. First, it is composed of the same basic rule applied repeatedly to subintervals. The left Riemann estimate for $\int_{x_i}^{x_{i+1}} f(x)\,dx$ is simply $hf(x_i) \equiv (x_{i+1} - x_i)f(x_i)$. Thus, to obtain the estimate of the integral over $[a, b]$, we have "pasted together" Riemann estimates for $[a, x_1]$, $[x_1, x_2]$, and so forth. The rules of this section are all *extended* or *composite rules*, in the sense that the integral over a collection of subintervals is obtained by using a simple rule on each subinterval and then adding the results. In the remainder of the section we shall define integration formulæ in terms of "basic rules" which are then combined to obtain the standard composite rules. For example, the basic left Riemann rule is given by $I_R(x_0, x_1) = hf(x_0)$.

Second, as Figure 5.1.1 makes clear, the Riemann approximation treats the integrand as if it were constant on each subinterval. As a consequence, the Riemann approximation is *exact* whenever $f(x)$ is a constant. In this case, we could obtain the integral over all of $[a, b]$ exactly by choosing $n = 1$! Indeed, each rule we examine will produce the exact integral when the integrand is a polynomial, typically of low degree. To the extent that the integrand is well approximated by such a low-order polynomial, these rules will do quite well. We can think, then, of the Riemann estimate as

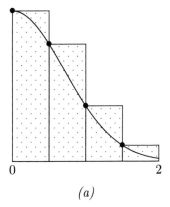 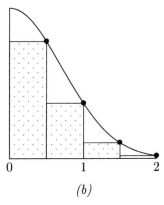

(a)   (b)

FIGURE 5.1.1 *Left and right Riemann rules* (I_R *and* I'_R, *respectively) for integrating a portion of the normal density function. The large points indicate the points at which the integrand is evaluated.*

first approximating the integrand by a piecewise constant function, and then performing the integration exactly for the approximating function.

A third observation is that we could just as easily have evaluated the function at the right-hand endpoints of the subintervals, giving as an alternative the *(right) Riemann approximation:*

$$I'_R(a,b) = h \sum_{i=1}^{n} f(a + ih). \qquad (5.1.2)$$

Clearly if $f(x)$ is, say, a decreasing function on $[a,b]$, then I'_R underestimates the integral and I_R overestimates it; see Figure 5.1.1. This suggests that a compromise between the two estimates might lead to an improved approximation, and indeed this is the case. More generally, increased accuracy can often be obtained by combining lower-order rules in a suitable way.

The Riemann approximation is so easily programmed that it is often used to obtain a rough estimate of the value of an integral, and for some problems that will suffice. To get very accurate estimates of an integral using this method, however, the integrand must be evaluated at a very large number of points, and it is usually the cost of evaluating the integrand which is the major factor in quadrature problems. One can almost always do very much better than applying brute force and relying on the Riemann approximation.

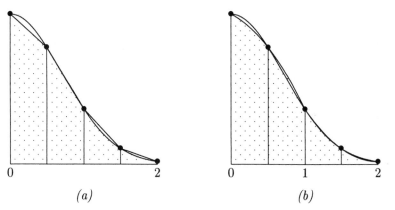

FIGURE 5.1.2 *Trapezoidal and Simpson's estimates based on the same integration points for a portion of the normal density. The large plotted points indicate the five points at which the integrand is evaluated in each of the rules.*

5.1.2 The trapezoidal rule

The left- and right-Riemann estimates on $[a, b]$ are both zero*th*-order methods, in the sense that they are exact for zero-order polynomials (constant functions). How might we obtain a better estimate? For monotone functions, the two approximations bracket the answer, so a plausible improvement would be to take the average of the two rules, giving the basic *trapezoidal rule:*

$$I_T(x_0, x_1) = \frac{x_1 - x_0}{2}(f(x_0) + f(x_1)) = \frac{h}{2}(f(x_0) + f(x_1))$$
$$= I(x_0, x_1) + O(h^3 f''(x^*)), \tag{5.1.3}$$

where x^* is an unknown point in $[x_0, x_1]$. The second half of (5.1.3) gives the order of the error in the approximation. The composite form gives the *extended trapezoidal rule*, illustrated in Figure 5.1.2(a):

$$I_T(a, b) = \frac{h}{2}f(a) + h \sum_{i=1}^{n-1} f(x_i) + \frac{h}{2}f(b). \tag{5.1.4}$$

The basic trapezoidal rule is exact for linear functions, and the composite version replaces the integrand by a piecewise linear approximation.

How good is the trapezoidal rule when $f(x)$ is *not* linear? Equation (5.1.3) implies that, when the integrand is twice continuously differentiable, the extended trapezoidal rule has an error expression given by

$$I(a, b) = I_T(a, b) + O\left(\frac{f''(x^*)}{n^2}\right), \tag{5.1.5}$$

where x^* is an (unknown) point in the interval $[a, b]$. Thus, not only does the error depend upon the curvature of the integrand (through f''), but also through the square of the number of points. Thus doubling the number of points in the extended rule reduces the error by a factor of four (asymptotically).

In practice, a sequence of trapezoidal estimates is computed, based on increasing numbers of points. The most practical method doubles the number of points at each step, so as not to lose the benefit of previously computed function values; this is discussed in more detail in Section 5.1.5. Let I_0, I_1, I_2, ..., denote a sequence of such estimates. For the trapezoidal rules, I_j will be based upon $2^j + 1$ function evaluations. The error in estimate j can be approximated by $E_j = |I_j - I_{j+1}|$. Successive terms in the sequence are computed until $|E_j|$ is sufficiently small, in either an absolute or a relative sense, at which point I_{j+1} is used as the estimate for the integral. The resulting error estimate E_j will only give an indication of the adequacy of the result. Without additional conditions on the integrand, there is no guarantee that a small value of E_j corresponds to a small value of $|I - I_j|$.

COMMENT. Relative error tolerances are often specified when testing for convergence of a sequence of integral estimates. Since the integral estimates can be close to or equal to zero, a suitable test in a computer program will be of the form "**if** $E_j \leq I_{j+1} \cdot tolerance$ **then stop**," rather than "**if** $E_j / I_{j+1} \leq tolerance$ **then stop**," since the latter can be unstable.

5.1.3 Simpson's rule

The basic Riemann approximation evaluates the integrand at just one point and produces the exact result for zero-order polynomials. The basic trapezoidal rule evaluates the integrand at two points and produces an exact result for first-order polynomials. As we might expect, one can construct a k-point basic integration rule that exactly integrates polynomials through degree $k - 1$. The extended forms of these rules in effect approximate the integrand by a piecewise polynomial function. Generally, the higher the polynomial degree, the better the approximation for smooth integrands. We can think of an integration rule as being based on a *model* for the shape of the integrand.

The standard basic three-point method is called *Simpson's rule:*

$$I_S(x_0, x_2) = \frac{h}{3}[f(x_0) + 4f(x_1) + f(x_2)]$$
$$= I(x_0, x_2) + O(h^5 f^{(4)}(x^*)),$$

(5.1.6)

where $h = x_i - x_{i-1} = (b-a)/2$, and the second expression gives the order of the error. As in expression (5.1.3), x^* is an unknown point in $[x_0, x_2]$. The composite form, illustrated in Figure 5.1.2(b), gives the extended version of Simpson's rule which is frequently mentioned in calculus textbooks:

$$I_S^{(2n)}(a,b) = \frac{h}{3}\left[f(a) + 4\sum_{i=1}^{n}f(x_{2i-1}) + 2\sum_{i=1}^{n-1}f(x_{2i}) + f(b)\right]. \quad (5.1.7)$$

Although the basic rule (5.1.6) evaluates the integrand at only three points, Simpson's rule has the remarkable property, evident from (5.1.6), that it is exact for polynomials of degree 3 or less. (We would expect only to be able to evaluate quadratics exactly based on only three points.) Provided that the integrand has four continuous derivatives, the extended rule has an error which can be expressed as

$$I(a,b) = I_S(a,b) + O\left(\frac{f^{(4)}(x^*)}{n^4}\right), \quad (5.1.8)$$

where, as before, x^* is an unknown point in $[a, b]$. This unusual property is due to cancellation of the error terms associated with cubic behavior of the integrand and, curiously, is a consequence of the error structure of the trapezoidal rule, to which it is related. We discuss this property more fully in Section 5.1.6. For the moment it is sufficient to note that Simpson's rule is one of the most "cost-effective" of integration rules. Indeed, doubling the number of points reduces the error by a factor of approximately sixteen.

5.1.4 General Newton-Cotes rules

The three rules discussed so far share two properties. They evaluate the integrand at equally spaced points, and they integrate certain polynomials exactly. In particular, the k-point rule integrates polynomials of degree (at least) $k - 1$. More generally, k-point rules satisfying these properties exist; such rules are called *Newton-Cotes integration formulæ.*

The trapezoidal rule and Simpson's rule also share another property. The endpoints of each interval are among the points at which the integrand is evaluated. Rules with this additional property are said to be "closed." Closed basic rules can easily be strung together to obtain extended rules, in which the integrand is evaluated on an equally-spaced grid of points. We have illustrated the extension process for the trapezoidal rule and for Simpson's rule.

Sometimes the integrand exhibits a singularity at one or both end-points, so that a closed rule cannot be used. The χ_1^2 density is a common statistical example with an integrable singularity at the origin. *Open Newton-Cotes rules* are based on equally spaced points placed on the *interior* of the integration interval. Although they are occasionally useful, open rules have the serious drawback that they cannot be combined to produce extended rules. What is more, integrable singularities at endpoints of the range of integration can often be handled more satisfactorily by transformation, since such functions are rarely well-approximated by low-order polynomials near the singularity. For virtually all practical problems, the Gaussian quadrature rules discussed below are superior to rules based on open Newton-Cotes.

Rules can also be constructed that use points outside the range of integration; these rules are called *Newton-Cotes extrapolation rules.* The k-point basic rule works by taking the right-hand endpoint as the first step, and then taking $k - 1$ additional steps ahead. Thus, for instance, the basic two-point rule with error bound is given by

$$\int_{x_0}^{x_1} f(x)\, dx = \frac{h}{2}\,[3f(x_1) - f(x_2)] + O(h^3 f''(x^*)), \qquad (5.1.9)$$

where x^* is a point in the interval $[x_0, x_2]$. This rule extrapolates to the right and is open on the left (since x_0 is not used), so we might call (5.1.9) the "right-extrapolative two-point rule." There is a corresponding left-extrapolative rule which is open on the right. Although extrapolative rules may seem quite unnatural since the integrand isn't evaluated anywhere on the interior of the interval, we have already encountered one such rule that seems quite natural: the one-point right-extrapolation rule is simply the right-Riemann formula from which (5.1.2) is constructed! Indeed, the main use for extrapolative rules is as building blocks for special-purpose extended rules.

5.1.5 Extended rules

Generally, no low-order polynomial is even a moderately satisfactory approximation to a function over the entire interval of integration. For this reason, basic integration rules are almost never used on the whole interval. Rather, the integration interval is subdivided into smaller intervals, on each of which a basic rule is employed. When these smaller intervals are of equal length and the same basic closed Newton-Cotes rule is used on each, the points at which the function must be evaluated lie on an equally-spaced grid of points $\{x_i\}$ on $[a, b]$. What is more, the weights associated with x_i are generally very simple functions of i. These facts combine to make extended Newton-Cotes rules particularly easy to implement. (This is not generally true if open rules are used, since then the endpoints of the subintervals are not points at which $f(x)$ is evaluated, so that the grid is no longer equally spaced.)

In evaluating such rules, the amount of computation is proportional to the number of times the integrand $f(x)$ must be evaluated. Extended rules are usually formed by using exactly the same basic rule on each subinterval. However it is possible and occasionally desirable to begin at the left using one basic rule and then to switch to another basic rule for the remaining subintervals, possibly using yet another rule at the right-hand end. For example, one could start with the trapezoidal rule (on the interval $[x_0, x_1]$), and then proceed using Simpson's rule on intervals $[x_1, x_3]$ through $[x_{n-3}, x_{n-1}]$, and then finish with trapezoidal again on $[x_{n-1}, x_n]$. Note that the two subintervals on the end are half as long as the intervals in the middle; had we made all of the intervals of equal size it would have been impossible to maintain equal spacing of the evaluation points. For this reason it is desirable to take the number of points as fundamental rather than the number of subintervals.

The equal point spacing in extended Newton-Cotes rules can be exploited computationally. Suppose that we have estimated an integral using such a rule based on $n + 1$ points, and that we wish to increase our accuracy by increasing the number of points used. If we increase the number to $2n + 1$, then two things happen. First, the old evaluations points are a subset of the new ones, so that the integrand need only be evaluated at n additional points. Second, these additional points are equally spaced themselves. What is more, it is possible to arrange the computation so that all of the old function values need not be stored; a few summary values are stored instead.

Let I_n denote the trapezoidal estimate based on $2^n + 1$ points. If the $f(x)$ is Riemann integrable, then the sequence $I_0, I_1, \ldots, I_j, \ldots$ will converge to the actual integral. This suggests that the approximation stop when I_{j-1} and I_j are sufficiently close (in either absolute or relative terms). Early stopping in this fashion can save a considerable number of function evaluations for well-behaved integrands.

The extended trapezoidal rule is an example of fundamental importance. Let $n = 2^j$, and let h here denote $(b-a)/(2n)$, the distance between evaluation points used to obtain I_{j+1}. Written in these terms,

$$2I_{j+1} - I_j = h \sum_{i=1}^{n} f(x_{2i-1}) \equiv hP_{j+1}. \tag{5.1.10}$$

Consequently, we have $I_{j+1} = (I_j + hP_{j+1})/2$. In this notation, $I_0 = (b-a) \cdot [f(a)+f(b)]/2$. A simple iterative procedure results, and is expressed in Algorithm 5.1.1.

Algorithm 5.1.1 (Trapezoidal rule)

Set K such that $2^K + 1 = $ maximum number of function calls
$h := b - a$
$n := 1$
$I_0 := h[f(a) + f(b)]/2$
for $j := 1$ **to** K **until** $\{I_j$ converges$\}$
$\quad t := a + h/2$
$\quad p := 0$
$\quad$ **for** $i := 0$ **to** $n - 1$ **do** $p := p + f(t + ih)$
$\quad I_j := (I_{j-1} + hp)/2$
$\quad h := h/2$
$\quad n := 2n$

COMMENT. The extended trapezoidal rule is often quite adequate for integrating relatively nonsmooth functions. The benefits of higher-order methods only accrue for functions which are adequately smooth. Simpson's rule, for instance, requires the integrand to be four times continuously differentiable in order to guarantee the $O(n^{-4})$ convergence behavior. The trapezoidal rule, by contrast, requires only continuous second-order derivatives, and its performance often degrades only a little when even this requirement is not met.

5.1.6 Romberg integration

In Section 5.1.3 we noted that Simpson's rule is actually a third-order method, even though the basic algorithm is only a three-point rule. The reason for this behavior is clear from a more detailed understanding of the trapezoidal rule's error structure.

The trapezoidal rule is actually the first two terms in the *Euler summation formula* for an integral. If all of the indicated derivatives exist, this formula gives

$$\int_{x_0}^{x_n} f(x)\,dx = h \sum_{i=0}^{n-1} f(x_i) + \frac{h}{2}\left(f(x_n) - f(x_0)\right)$$
$$- \sum_{j=1}^{k} \frac{B_{2j} h^{2j}}{(2j)!}\left(f^{(2j-1)}(x_n) - f^{(2j-1)}(x_0)\right) + R_{2k}.$$

$$(5.1.11)$$

The coefficients B_{2j} are the *Bernoulli numbers*, which are defined to be the coefficients in the Maclaurin series expansion of $t/(e^t - 1)$. Table 5.1.2

$n =$	0	1	2	4	6	8	10
$B_n =$	1	-1/2	1/6	-1/30	1/42	-1/30	5/66

TABLE 5.1.2 *A short table of Bernoulli numbers. Note that, except for $n = 1$, $B_n = 0$ whenever n is odd.*

gives the first few Bernoulli numbers. The absolute value of the error term R_{2k} is no larger than twice that of the next term in the series, provided that the appropriate derivatives exist. (The error terms may, and generally do, increase after some value of k, so that equation (5.1.11) is not a convergent series as $k \to \infty$.) A detailed (and instructive!) development of the summation formula can be found in Section 1.2.11.2 of Knuth (1968).

What is noteworthy about (5.1.11) is the fact that with the exception of the first two terms (which constitute the trapezoidal formula) there are no odd powers of h, and hence of n^{-1}, in the expansion (since $h = n^{-1}(b-a)$). As in the previous section, let $n = 2^j$, $h = (b - a)/(2n)$, and let I_j denote the trapezoidal estimate based on $2^j + 1$ points. (Thus, h is the step size for evaluating I_{j+1}.) Using (5.1.11) we can write the integral in two ways, either as

$$I(a,b) = I_j + \frac{B_2}{2!}\left(\frac{2}{n}\right)^2 (f'(b) - f'(a)) + O(n^{-4}f^{(4)}(x_1^*)) \qquad (5.1.12)$$

or as

$$I(a,b) = I_{j+1} + \frac{B_2}{2!}\left(\frac{1}{n}\right)^2 (f'(b) - f'(a)) + O(n^{-4}f^{(4)}(x_2^*)). \qquad (5.1.13)$$

By taking an appropriately weighted sum of these expressions, we can make the error term in n^{-2} disappear. In particular, we can write

$$\begin{aligned} I(a,b) &= \frac{4}{3}I_{j+1} - \frac{1}{3}I_j + O(n^{-4}f^{(4)}(x_3^*)) \\ &\equiv S_{j+1} + O(n^{-4}f^{(4)}(x_3^*)). \end{aligned} \qquad (5.1.14)$$

Thus, with only two extra multiplications and a single extra addition we can transform Algorithm 5.1.1 into an algorithm with fourth-order convergence, just like Simpson's rule. In fact, S_j turns out just to *be* Simpson's rule based on $n = 2^j + 1$ points, and Algorithm 5.1.2 gives an efficient method for calculating it.

Algorithm 5.1.2 (Simpson's rule)

Set K such that $2^K + 1 = $ maximum number of function calls
$h := b - a$
$n := 1$
$I_0 := h[f(a) + f(b)]/2$
for $j := 1$ **to** K **until** $\{S_j$ converges$\}$
$\quad t := a + h/2$
$\quad p := 0$
$\quad$**for** $i := 0$ **to** $n - 1$ **do** $p := p + f(t + ih)$
$\quad I_j := (I_{j-1} + hp)/2$
$\quad S_j := (4I_j - I_{j-1})/3$
$\quad h := h/2$
$\quad n := 2n$

By taking a weighted sum of two consecutive trapezoidal estimates, we saw that it was possible to eliminate the error term of order n^{-2}. This is a special case of a more general result which makes it possible to eliminate higher order error terms by taking weighted sums of k successive trapezoidal estimates. This method is called *Romberg integration*. It is a special case of a general method termed *Richardson extrapolation*, which takes a sequence of estimates which depend upon a parameter h and then using the value obtained by extrapolating h to its limit (in our case, $h \to 0$). Romberg integration is discussed in Press, Flannery, Teukolsky and Vetterling (1986); a general treatment of Richardson extrapolation can be found in Henrici (1964).

EXERCISES

5.1. [12] Prove that, if $f(x)$ is a nondecreasing function on $[a, b]$ then $I_R \leq I'_R$.

5.2. [17] Prove that, if $f(x)$ is an increasing function on $[a, b]$, then I_R underestimates the integral and I'_R overestimates it.

5.3. [10] Prove that expression (5.1.4) follows by extending the first part of (5.1.3) as $I_T(a, b) = \sum_{i=1}^{n} I_T(x_{i-1}, x_i)$.

5.4. [20] Prove that the first part of expression (5.1.3) is exact when $f(x)$ is linear, that is, prove that $I_T = I$.

5.5. [20] Prove that $I(x_0, x_1) = I_T(x_0, x_1) + E_1$, where the error $E_1 = -\int_{x_0}^{x_1}(x - m)f'(x)\,dx$, and where $m = (X_0 + x_1)/2$ is the midpoint of the interval.

5.6. [26] Using the previous result, prove the second half of (5.1.3), assuming that $f''(x)$ exists and is continuous on $[x_0, x_1]$.

5.7. [02] Where is the trapezoid in the trapezoidal rule?

5.8. [14] Let $I = \int_0^1 f(x)\,dx$, let I_j denote the trapezoidal estimate based upon $2^j + 1$ points, and let E_j denote the error estimate $I_j - I_{j+1}$. Give an example of a function for which $E_0 = E_1 = \ldots = E_n = 0$, for some fixed $n > 1$, but for which $I_j \neq I$, for $j \leq n$. (Thus, the sequence of estimates would appear to converge, but without producing an accurate estimate of the integral.)

5.9. [23] Let $x_0 \leq x_1 \leq \ldots \leq x_n$, $n \geq 1$, and let $f(x)$ be a continuous function on the interval $[x_0, x_n]$. Let the points y_1 through y_n be arbitrary points satisfying the conditions $y_i \in [x_{i-1}, x_i]$. Prove that $\sum_{i=1}^n f(y_i) = nf(y^*)$, for some $y^* \in [x_0, x_n]$.

5.10. [26] In the second expression in (5.1.6), the coefficient of the leading term in $O(h^5 f^{(4)})$ does not depend upon the interval (x_0, x_2), so that one could actually write $h^5 c f^{(4)}(x^*) + o(h^5)$ in place of the final term. Using this fact, prove that the error for the extended form of Simpson's rule given in equation (5.1.8) follows from (5.1.6).

5.11. [19] Prove (5.1.10).

5.12. [09] If $n = 2^j$, show that $(b - a)/(2n)$ is the distance between evaluation points used to obtain the trapezoidal estimate I_{j+1}.

5.13. [08] Show that $S_{j+1} = (4I_{j+1} - I_j)/3$ is Simpson's rule based on $2^{j+1} + 1$ points.

5.14. [10] For sufficiently smooth functions, Newton-Cotes integration rules have error estimates of the form $O(n^{-k})$ for some k. However, for inadequately smooth integrands this error estimate is optimistic. Let I_n denote a Newton-Cotes estimate based on $2^n + 1$ points. How would one estimate the effective order k for a particular integrand?

5.15. [30] (Continuation) Using the order estimates of the previous exercise, estimate the order of both the trapezoidal rule and Simpson's rule for the integral $\int_1^2 \sqrt{t}\,dt$.

5.16. [18] (Continuation) Repeat the previous exercise for $\int_0^2 \sqrt{t}\,dt$. How do the answers differ?

5.17. [31] (Continuation) In the previous exercise, *why* do the answers differ as they do?

5.18. [28] Evaluate $\int_1^2 (\ln t + e^t)\, dt$ using Simpson's rule and the trapezoidal rule. How many function evaluations are required to obtain a relative error of less than 10^{-5}? To obtain an *estimated* relative error of the same magnitude?

5.19. [25] Show that the error term of order n^{-4} can be eliminated from Simpson's rule by applying Romberg integration to consecutive Simpson's estimates.

5.2 Improper integrals

Consider two integration problems typical of those arising in statistics. The first problem is to evaluate the prior expectation of $g(\theta)$, where θ is a binomial parameter and the expectation is with respect to the noninformative prior $p(\theta) \propto 1/\sqrt{\theta(1-\theta)}$. This expectation, if it exists, is given by

$$E(g(\theta)) = \frac{\int_0^1 g(\theta)p(\theta)\, d\theta}{\int_0^1 p(\theta)\, d\theta} = \frac{1}{\pi} \int_0^1 \frac{g(\theta)}{\sqrt{\theta(1-\theta)}}\, d\theta. \qquad (5.2.1)$$

The constant of integration from the denominator of the middle expression is easily evaluated by calculus (in this case), but the remaining integral, given on the right, may be impossible to obtain analytically.

As a second example, consider a *truncated gamma density*, given by

$$p(x) = \begin{cases} c^{-1}e^{-x}x^{\gamma-1} & x > 1, \\ 0 & \text{otherwise,} \end{cases} \qquad (5.2.2)$$

for some $\gamma > 0$, the problem being to evaluate the constant c. The answer is clearly $c = \int_1^\infty e^{-x}x^{\gamma-1}\, dx$.

In neither of these two problems can we apply the methods of the previous section directly. In the first case, the integrand (or at least a factor in the integrand) is infinite at both end points of the range of integration, so that closed Newton-Cotes rules cannot be used. In the second case, the integrand is well-behaved, but the range of integration is infinite, so that equal spacing of integration points is not possible. In this section we indicate how to deal with such problems.

5.2.1 Integrands with singularities at the end points

Integral (5.2.1) is an example in which the integrand cannot be evaluated at the end points, in this case because the integrand approaches infinity there. In other problems the integrand may approach a finite limit, but may still not be directly evaluable at the end point; $x \log(x)$ as $x \to 0$ is such an example. Although closed rules cannot be used in these situations, it turns out that there is an open rule analogous to Algorithm 5.1.2. Recall that that algorithm obtains Simpson's rule as an extrapolation from successive trapezoidal estimates. Algorithm 5.1.2 works because the error expansion of the trapezoidal rule depends only on even powers of h, the width of the subintervals. There is an open rule which also has this property; it is called the *midpoint rule*, the basic form of which is given by the formula

$$I_M(x_0, x_1) = hf\left(\frac{x_0 + x_1}{2}\right)$$

$$= I(x_0, x_1) + O(h^3 f''(x^*)).$$

$$(5.2.3)$$

The extended midpoint rule evaluates the integrand at the midpoint of each subinterval and, although it is awkward to write it is easy to compute:

$$I_M(a, b) \equiv h \sum_{i=1}^{n} f\left(a + h\left(i - \frac{1}{2}\right)\right)$$

$$= I(a, b) + O\left(\frac{f''(x^*)}{n^2}\right).$$

$$(5.2.4)$$

To use the trick of Richardson extrapolation as we did to obtain Simpson's rule, it is necessary to construct a sequence of extended midpoint rules $\{M_j\}$ in such a way that the evaluation points used in M_j are a subset of those used in M_{j+1}. Since the evaluation points are *midpoints* of intervals (and not endpoints), it is not possible to double the number of subintervals and retain this property. It is possible to treble the number of subintervals, however, as Press, Flannery, Teukolsky and Vetterling (1986) note. (See Exercise 5.21.)

Denote by M_j the extended midpoint rule based on $n = 3^j$ intervals (hence, 3^j function evaluations). When Richardson extrapolation is applied to this sequence, the result is the set of rules of *Maclaurin type*. Because of its similarity to Simpson's rule, we shall refer to the first of these Maclaurin-type rules as the *midpoint-Simpson rule*. The resulting computation is a minor variant of Algorithm 5.1.2, but is worth presenting in full.

Algorithm 5.2.1 (Midpoint-Simpson rule)

Set K such that $3^K + 1 = $ maximum number of function calls

$h := b - a$

$n := 1$

$M_0 := h \cdot f((a+b)/2)$

for $j := 1$ to K until $\{MS_j$ converges$\}$

$\quad \delta := h/3$

$\quad t_1 := a + \delta/2$

$\quad t_2 := t_1 + 2\delta$

$\quad p := 0$

$\quad$ for $i := 0$ to $n - 1$ do $p := p + f(t_1 + ih) + f(t_2 + ih)$

$\quad M_j := (M_{j-1} + hp)/3$

$\quad MS_j := (9M_j - M_{j-1})/8$

$\quad h := h/3$

$\quad n := 3n$

It should be emphasized that, although the midpoint rule and the midpoint-Simpson rule can be used when the integrand is singular at one endpoint or the other, the rate at which the sequences converge can be abysmal in such situations. The quoted orders of these methods apply only if the integrand has sufficiently many continuous derivatives on the *closed* interval $[a, b]$. In general, kth-order methods require integrands to be k times continuously differentiable.

COMMENT. The integral of equation (5.2.1) provides a case in point. The constant of integration is given by $\int_0^1 \theta^{-1/2}(1 - \theta)^{-1/2} \, d\theta = \pi$. It is not easy to evaluate this integral using Algorithm 5.2.1 because of the singularity at each endpoint. After $2187 = 3^7$ function evaluations, the midpoint estimate is only good to two decimal places. The midpoint rule is only of order $1/2$ in this case, and because of the singularity, the midpoint-Simpson rule can be no better. This means that tripling the step size increase the accuracy by only 70%! In evaluating the expectation $E(\theta(1 - \theta)) = 1/8$ the singularities at both ends are no longer a problem, although there remain singularities in the first derivatives at each end. The midpoint rule is of order $3/2$, as is midpoint-Simpson. Here, after only 9 evaluations the midpoint estimate is good to almost three places, and after 2187 evaluations is good to six places. In order for the midpoint-Simpson rule to outperform the simple midpoint algorithm at all, it would be necessary for the integrand to be more than twice continuously differentiable. To achieve Simpson's full potential, the integrand would need to be four times continuously differentiable.

5.2.2 Integration over infinite intervals

When an integral is hard to evaluate or an integrand is ill-behaved, a general rule is to convert it into an easier problem which has the same answer, and then solve the easier problem. By introducing a change of variables into improper integration problems it is possible to convert infinite endpoints into finite ones. For example, we can rewrite $\int_a^\infty f(x)\,dx$ as $\int_0^{1/a} f(1/t)/t^2\,dt$. More generally, let $u(x)$ be a continuously differentiable strictly monotone function on $[a, b]$. Then,

$$\int_a^b f(x)\,dx = \int_{u(a)}^{u(b)} \left[\frac{f(u^{-1}(t))}{u'(u^{-1}(t))} \right] dt. \qquad (5.2.5)$$

The idea is to select a transformation function $u(\cdot)$ which is easily invertible and which makes the bracketed expression in (5.2.5) as near to a low-order polynomial as possible. This is simply the old change-of-variable trick from calculus, but with a new twist. The inverse transformation must be easily computed for the computation to be possible, and the resulting integrand must be amenable to numerical integration using simple rules. When an end point is infinite, $u(\infty)$ denotes the limiting value. This generally means that the new integrand cannot be evaluated explicitly at such an end point, so that the midpoint sequence rather than the trapezoidal sequence must be used.

In statistical problems, the most frequently useful transformations are $u(x) = e^{-x}$ and $u(x) = 1/x$, both for $x > 0$. The choice depends upon the tail behavior of the integrand. The former transformation works better with integrands exhibiting exponential decay, while the latter works better with integrands whose tails fall off geometrically. When possible, it is advisable to plot $f(u^{-1}(t))/u'(u^{-1}(t))$ to determine by inspection the suitability of using the selected transformation.

COMMENT. As a general rule, one should break up integrals to infinity into two or more pieces, so that the integral involving the tail has begun to exhibit its asymptotic behavior. The range of the remaining integral is then often one on which the integrand is well approximated by polynomials *without* further transformation, so that this "middle piece" can be evaluated using Simpson's rule.

As an example, consider evaluating $P(X > 1)$, where X is a standard normal variate. The answer is proportional to $\int_1^\infty e^{-x^2/2}\,dx$. Figure 5.2.1 shows the effects of two alternative changes of variable on $g(t) = f(u^{-1}(t))/u'(u^{-1}(t))$ for this problem. The negative exponential transformation better matches the tail behavior of the normal density than the reciprocal transformation does, with the result that the former transformation results in an integrand which is smoother and which uses the full range

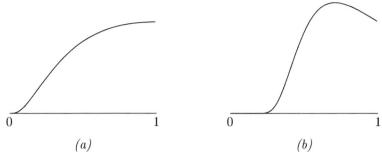

(a)   (b)

FIGURE 5.2.1 *Computing the integral of the normal density on $(1, \infty)$ using the transformations (a) $t = \exp(-x)$ and (b) $t = 1/x$. The horizontal and vertical scales in the two figures are the same. Transformation (a) uses more of the range and is more easily approximated piecewise by polynomials than is (b), but in each case Romberg integration using the midpoint rule should behave well.*

of integration. The reciprocal transformation still results in a very smooth function, however, so that the midpoint and midpoint-Simpson rules will still perform well.

A more difficult example is the problem of evaluating the integration constant c for the truncated gamma density of equation 5.2.2 when the parameter γ is less than one. Figure 5.2.2 compares the modified integrand $g(t) = f(u^{-1}(t))/u'(u^{-1}(t))$ for evaluating $\int_1^\infty e^{-x} x^{-1/2} \, dx$ for the negative-exponential and the reciprocal transformations; this is the case with $\gamma = 1/2$. Although Figure 5.2.2(a) looks appealing, deep trouble lurks behind the sharp downward bend on the left. Consider trapezoidal estimates of the area under this curve. (Midpoint estimates will have identical behavior.) Virtually all of the error in this estimate comes from the left-most of the subintervals, the integrand being nearly linear on most of the unit interval. Cutting the step size in half will cut the width of the left-most interval in half and (approximately) cutting the corresponding error in half. This qualitative argument suggests that the trapezoidal rule may be no better than first order. In fact, the problem is almost this bad. Near zero the behavior of the integrand is that of $h(t) = (-\log(t))^{-1/2}$. As t goes to zero, $h(t)$ approaches zero at a faster rate than any positive power of t. The extended midpoint rule is approximately of order 1.16 for this problem, and the trapezoidal rule of order 1.08, the difference due to the fact that the midpoint rule avoids an evaluation at the left end, where the integrand has its singularities in the derivatives.

By contrast, the reciprocal transformation (Figure 5.2.2(b)) has far superior behavior. Indeed, the midpoint sequence does quite well using

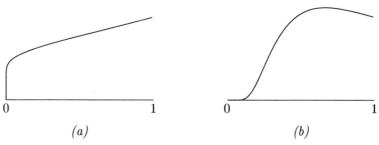

FIGURE 5.2.2 *Computing the integral of the truncated gamma density $f(x) \propto \exp(-x)x^{\gamma-1}\, dx$ on $(1, \infty)$, with $\gamma = 1/2$, using the transformations (a) $t = \exp(-x)$ and (b) $t = 1/x$. The horizontal and vertical scales in the two figures are the same. Transformation (a) exhibits $(-\log(t))^{-1/2}$ behavior on the left, so that the midpoint rule and all Romberg extrapolations of it will be approximately first order. The reciprocal transformation (b), on the other hand converges more rapidly, the midpoint rule being approximately second order and the Simpson-midpoint rule fourth order.*

this variable substitution. The midpoint and midpoint-Simpson rules have order two and four, respectively.

COMMENT. We have discussed only two transformations in any detail. Of particular note is $u(x) = x/(1 + x)$, which transforms $[0, \infty]$ into $[0, 1]$. Other changes of variable are discussed at somewhat greater length in Chapter 3 of Davis and Rabinowitz (1984).

COMMENT. The problem of choosing $g(t) = f(u^{-1}(t))/u'(u^{-1}(t))$ to be nearly polynomial is similar to the problem of selecting an importance function in the Monte Carlo method known as importance sampling, in which one samples from an easily generated probability density which is "nearly proportional" to the desired integrand. In the present case, if $g(t)$ is constant, the numerical integration is trivial. But to achieve $g(t) = c$ requires that we already know the answer we seek! Similarly, exact proportionality in importance sampling also requires knowing the answer.

COMMENT. Note that Algorithm 5.2.1 needs virtually no modification to be used with the change-of-variable methods discussed here. Except for adjusting the limits of integration, one need only redefine the subroutine which computes $f(x)$ by one that computes the modified integrand $g(t) = f(u^{-1}(t))/u'(u^{-1}(t))$. Since the latter involves the former, the net effect is to introduce a single additional layer of subroutine calls between the integrator and the integrand. This makes it very easy to investigate different transformations, since only a single subroutine is involved.

EXERCISES

5.20. [05] In Algorithm 5.2.1, MS_j is defined by $(9M_j - M_{j-1})/8$. Where did the coefficients in this expression come from?

5.21. [21] Let M_j denote the extended midpoint rule based on 3^j points. Show that the points on which M_j is based are a subset of the points which comprise M_{j+1}.

5.22. [26] Let $f(x)$ be a function on $[a, b]$ with $f''(x) \geq 0$ (f is convex). Show that $I_M(a, b) \leq I(a, b) \leq I_T(a, b)$. Under what conditions do equalities hold?

5.23. [13] When equation (5.2.5) is applied to the integral $c \int_1^\infty e^{-x^2/2} \, dx$ with $u(x) = e^{-x}$, the new range of integration should be $[0, e^{-1}]$. The range of integration plotted in Figure 5.2.1 is $[0, 1]$ for this problem. What transformation did the author *really* use to obtain the figure, and why?

5.24. [30] Would the transformation $u(x) = e^{-x^2}$ be a good transformation to use for the evaluating $\int_1^\infty e^{-x^2/2} \, dx$?

5.25. [22] Show that the derivative of $h(t) = (-\log(t))^{-1/2}$ becomes infinite at a faster rate than does t^α for any $\alpha > 0$.

5.3 Gaussian quadrature

The Newton-Cotes integration rules are defined on a grid of equally spaced points. Given these points, how are the coefficients obtained? In general, the basic formula using $n + 1$ points can integrate exactly polynomials of degree at most n. Such a rule is said to be an *nth-order integration formula*. It is easy to compute the coefficients in a Newton-Cotes formula by solving a set of linear equations that express this fact. If we write $NC(n) = h \sum_{i=0}^{n} a_i f(a + ih)$ then the monomials 1, x, ..., x^n can be integrated exactly as follows:

$$\frac{1}{j+1} = \int_0^1 x^j \, dx = \frac{1}{n} \sum_{i=0}^{n} a_i \left(\frac{i}{n}\right)^j, \qquad (5.3.1)$$

for $0 \leq j \leq n$. In effect, by giving ourselves the freedom to select $k + 1$ parameters (the coefficients $\{a_i\}$) we can construct a kth-order integration rule.

In Newton-Cotes integration, the abscissæ at which the integrand is evaluated are given. What happens if we allow ourselves the freedom not only to select the weights in the integration rule, but also to choose the abscissæ? It turns out that by appropriately choosing k evaluation points $\{x_j\}$ and k corresponding weights, we can obtain a $(2k - 1)$st-order integration rule. Such rules are called *Gaussian quadrature rules*.

The flexibility afforded by choosing both abscissæ and weights makes it possible to obtain high-order rules for more general integrands of the type

$$I(a,b) = \int_a^b f(x)w(x)\,dx, \qquad (5.3.2)$$

where a and b are each allowed to be finite or infinite and $w(x)$ is a nonnegative *weighting function*. In the parlance of numerical analysis, a weighting function is said to be *admissible* if $\int_a^b w(x)\,dx > 0$ and if, for all $k \geq 0$, $\int_a^b x^k w(x)\,dx < \infty$. If the weight function is normalized so that $\int_a^b w(x)\,dx = 1$, then admissible weight functions are merely those which are probability densities for which all moments are finite. If X is a random variable with proability density $w(x)$, we then recognize that the integral (5.3.2) is simply the expectation $E[f(X)]$. From the statistical standpoint, it means that high-order integration formulæ have already been derived for evaluating the moments of functions of random variables with several common distributions.

The theory of Gaussian quadrature is intimately intertwined with that of *orthogonal polynomials*. For a given interval (a,b) and admissible weight function $w(\cdot)$, there is a family of orthogonal polynomials the zeroes of which give the abscissæ of Gaussian quadrature formulæ.

We shall begin this section by examining the most commonly used family Gaussian quadrature rules, that for $w(x) = 1$ on a finite interval. These rules are called the *Gauss-Legendre rules*, for reasons that shall become clear momentarily.

COMMENT. In general, Gaussian quadrature formulæ can be obtained by solving systems of equations similar to (5.3.1). In general these systems are nonlinear; the details of a simple example are worked through in the exercises. Unfortunately, these systems rapidly become highly ill-conditioned, so that very high precision arithmetic must be used to obtain some of the higher-order rules by this method. More accurate methods rely on stable forms of recurrences relating orthogonal polynomials. The most commonly used Gaussian quadrature rules need not be obtained from scratch, since extensive tables are available (Abramowitz and Stegun, 1964; Stroud and Secrest, 1966). For routine use it is advisable to incorporate the constants directly into the integration subroutines by copying them (carefully!) from tables rather than by computing them afresh. It is better still to use one of the subroutines from a standard software library such as the NAG library or QUADPACK (1983). There are, however, several algorithms which are fairly stable for computing the weights and abscissæ which can be used when a number of different Gaussian integration rules are all to be used in the context of a single problem. Davis and Rabinowitz (1984) give a routine for Gauss-Legendre rules. Subroutines for obtaining other Gaussian quadrature formulæ can be found in Stroud and Secrest.

5.3.1 Gauss-Legendre rules

By a change of variable, any integral such as (5.3.2) defined on a finite interval (a, b) can be transformed to an integral on $(-1, 1)$. If the weight function $w(\cdot)$ is constant before transformation, it is constant afterward, so to examine Gaussian integration on finite intervals with respect to $w(x) \equiv 1$ it is sufficient to consider only the interval $(-1, 1)$. Rules for this interval with respect to the constant weight function are called *Gauss-Legendre* rules. The term *Gaussian quadrature* without any additional qualification generally refers to Gauss-Legendre quadrature.

A k-point Gaussian quadrature rule for a particular interval (a, b) with respect to a given weight function $w(x)$ is given by a set of k points x_i at which the integrand $f(x)$ is evaluated, and a set of weights a_i corresponding to these function values. Unlike Newton-Cotes rules, the set of evaluation points for the k-point rule will generally *not* be a subset of those for any m-point rule with $m > k$. This means that rules (that is, their sets of coefficients) must be tabulated for each k. It also means that the sort of convergence tests discussed in Section 5.1 cannot be made at no cost, since the benefit of previous function evaluations is ordinarily lost. A clever modification of Gaussian quadrature that nearly overcomes this deficiency is discussed in Section 5.3.5.

Legendre's name appears in conjunction with these quadrature rules because the abscissæ of the k-point rule are the zeros of $P_k(x)$, the kth *Legendre polynomial*. $P_k(x)$ is a polynomial of degree k, and is an even function of x when k is even, and an odd function when k is odd. This implies that the abscissæ of the integration rule are symmetric about zero. Given this fact, it can be deduced that the weights for $+x_i$ and $-x_i$ are equal. These facts simplify tabulation, as only abscissæ and weights for $x_i \geq 0$ need be listed. Tables of weights and abscissæ can be found in Abramowitz and Stegun (1964). Tables 5.3.1 and 5.3.2 give Gauss-Legendre rules for $k = 5$ and $k = 10$.

$\pm x_i$	a_i
0.00000 00000 00000	0.56888 88888 88889
0.53846 93101 05683	0.47862 86704 99366
0.90617 98459 38664	0.23692 68850 56189

TABLE 5.3.1 *Abscissæ x_i and weights a_i for the 5-point Gauss-Legendre integration rule, used for obtaining $\int_{-1}^{1} f(x)\,dx$. From Abramowitz and Stegun (1964).*

$\pm x_i$	a_i
0.14887 43389 81631	0.29552 42247 14753
0.43339 53941 29247	0.26926 67193 09996
0.67940 95682 99024	0.21908 63625 15982
0.86506 33666 88985	0.14945 13491 50581
0.97390 65285 17172	0.06667 13443 08688

TABLE 5.3.2 *Abscissæ x_i and weights a_i for the 10-point Gauss-Legendre integration rule, used for obtaining $\int_{-1}^{1} f(x)\,dx$. From Abramowitz and Stegun (1964).*

It should be noted that Gaussian quadrature does well for functions that are well-approximated by a polynomial *over the entire range* $(-1, 1)$. In practical terms, understanding the nature of the integrand can pay big dividends. If the range of integration can be split into two subregions on each of which the integrand behaves like a low-order polynomial, Gaussian quadrature applied to these subintervals separately may well produce better results than doubling the number of quadrature points on the full interval.

5.3.2 Orthogonal polynomials

To understand Gaussian quadrature in general, it is necessary to know something of the theory of orthogonal polynomials. In this discussion, we shall use the notation $p_k(x)$ or $q_k(x)$ to denote a (generic) polynomial of degree exactly k. It is convenient to take the leading coefficient to be positive. (Be sure to distinguish the generic jth-degree polynomial from P_j, the jth Legendre polynomial.) We shall also be concerned only with admissible weight functions, that is, weight functions which are probability densities (up to a positive constant multiple) with all moments finite.

Given an interval (a, b) and an admissible weight function $w(x)$, a function f for which $\int_a^b [f(x)]^2 w(x)\,dx < \infty$ is said to be *square integrable* with respect to w on the interval. Denote the class of all such functions by L_w^2. Note that $p_k \in L_w^2$ for any $k \geq 0$. We can define an inner product for $f, g \in L_w^2$ by

$$\langle f, g \rangle \equiv \int_a^b f(x)g(x)w(x)\,dx. \tag{5.3.3}$$

Two functions $f, g \in L_w^2$ are said to be *orthogonal* with respect to this inner product if $\langle f, g \rangle = 0$. As a matter of convenience, we shall also speak of functions being orthogonal with respect to w, the range of integration and the inner product determined by it being understood.

For a given combination of interval and weight function, there is a sequence of polynomials $p_j(x)$ which have the property that $\langle p_j, p_k \rangle = 0$ whenever $j \neq k$. Of course, if p_j is orthogonal to p_k, so is $c \cdot p_j$, for any constant c. By imposing a *standardization* on the sequence we can remove this ambiguity and, having done so, the sequence so defined is the unique set of *orthogonal polynomials* with respect to w on (a, b).

COMMENT. Choosing a standardization is largely a matter of convention, but it is important to know in any particular instance just which convention is being used. Common conventions include specifying $p_j(0)$, specifying $p_j(a)$ or $p_j(b)$, or requiring the leading coefficient in $p_j(x)$ to be one (in which case, the polynomials are said to be *monic*). It is customary, for example, to normalize the Legendre polynomials by requiring that $P_k(1) = 1$. This is equivalent to requiring that the coefficients sum to one. Another way of standardizing a set of orthogonal polynomials is to require that $|p_j|^2 \equiv \langle p_j, p_j \rangle = 1$. Such functions are said to be *orthonormal*.

The orthogonal polynomials with respect to $w(x)$ form a basis for L_w^2, so that for any function $f(x) \in L_w^2$ we may write $f(x) = \sum_{i=0}^{\infty} a_i p_i(x)$, and $a_i = \langle f, p_i \rangle / \langle p_i, p_i \rangle$.

Orthogonal polynomials have some beautiful properties from which many properties of Gaussian quadrature rules can be deduced. A long list of formulæ and interrelationships involving orthogonal polynomials can be found in Abramowitz and Stegun (1964). Among the most important properties are the following, which we present without proof.

The zeroes of an orthogonal polynomial with real coefficients are real, simple, and are located on the interior of (a, b). Thus, for instance, all of the integration points of a Gauss-Legendre rule are on the interior of the interval of integration $(-1, 1)$. If p_j and p_{j+1} are elements of the same set of orthogonal polynomials, then each root of p_j lies between two roots of p_{j+1}. This interleaving property makes it possible to find the roots of p_{j+1} using Newton-Raphson iterations, using a starting point for each iteration based upon the bracketing roots of p_j.

When the weight function $w(x)$ is symmetric about the midpoint of (a, b), or about zero for the interval $(-\infty, \infty)$, then the abscissæ are symmetrically placed about the midpoint, and the weights attached to symmetric points are equal.

One of the nicest properties of all is that orthogonal polynomials satisfy a recurrence relation of the form

$$p_j = (A_j + xB_j)p_{j-1} - C_j p_{j-2}, \qquad (5.3.4)$$

where the constants A_j, B_j, and C_j are often relatively simple functions of j. For the Legendre polynomials, for instance, $A_j = 0$, $B_j = (2j - 1)/j$, and $C_j = (j - 1)/j$. This makes them easy to compute in a computer program, as the coefficients in $p_j(x)$ can easily be deduced from those in $p_{j-1}(x)$ and $p_{j-2}(x)$ using (5.3.4).

COMMENT. The coefficients in the recursion (5.3.4) have an explicit representation. Let $\{p_j(x)\}$ be any system of polynomials orthogonal with respect to the inner product $\langle \cdot, \cdot \rangle$. Denote by $k_j > 0$ the leading coefficient of $p_j(x)$, that is, write $p_j(x) = k_j x^j + \ldots$, and then set $\gamma_j = k_{j+1}/k_j$. Then for $j \geq 1$ in (5.3.4) we have

$$A_j = -\gamma_{j-1} \frac{\langle x p_{j-1}, p_{j-1} \rangle}{\langle p_{j-1}, p_{j-1} \rangle}$$

$$B_j = \gamma_{j-1}$$

$$C_j = \begin{cases} 0 & j = 1 \\ \dfrac{\gamma_{j-1}}{\gamma_{j-2}} \dfrac{\langle p_{j-1}, p_{j-1} \rangle}{\langle p_{j-2}, p_{j-2} \rangle} & j \geq 2 \end{cases}.$$

$$(5.3.5)$$

This result depends only on the inner product satisfying the condition that $\langle xf, g \rangle = \langle f, xg \rangle$. Note that the constants k_j (and hence, γ_j) merely reflect the chosen standardization. For computational purposes, any convenient choice may be made.

Table 5.3.3 gives a short table of the most commonly encountered orthogonal polynomials in statistics, together with the intervals on which they are defined, their weight functions, and the usual standardization. Recurrence weights for equation (5.3.4) are given in Table 5.3.4.

The weight functions for some of the polynomials in Table 5.3.3 are among some of the most important of probability densities. These include the beta distributions (Jacobi polynomials), the exponential distribution (Laguerre polynomials), the gamma distributions (Generalized Laguerre polynomials), the uniform distribution (Legendre polynomials), and of course, the normal distribution (Hermite polynomials).

COMMENT. From a statistical standpoint, some of the most important systems of orthogonal polynomials satisfy a differential equation called *Rodrigues's Formula*, which has

$$p_j(x)w(x) = D_j \frac{d^j}{dx^j} \{[g(x)]^j w(x)\}. \qquad (5.3.6)$$

Here, $g(x)$ is a polynomial which is independent of j, and D_j is a constant. This fact is useful because it makes it possible to write probability densities in terms of an asymptotic expansion whose terms are the orthogonal polynomials. In effect, equation (5.3.6) asserts that the derivatives of w (times a fixed polynomial function) can be expressed as a polynomial multiple of w itself. It also turns out that the series of polynomials $p_j'(x) = dp_j(x)/dx$ is also a set of orthogonal polynomials. Only the Jacobi, Legendre, Generalized Laguerre, and Hermite polynomials satisfy Rodrigues's Formula, so that only the beta, gamma,

Name	$p_j(x)$	(a, b)	$w(x)$	Customary Standardization
Jacobi	$P_j^{(\alpha,\beta)}(x)$	$(-1, 1)$	$(1-x)^\alpha(1+x)^\beta$	$P_j^{(\alpha,\beta)}(1) = \binom{j+\alpha}{j}$
Jacobi	$G_j(p, q, x)$	$(0, 1)$	$(1-x)^{p-q}x^{q-1}$	$LC = 1$
Chebyshev I	$T_j(x)$	$(-1, 1)$	$(1-x^2)^{-1/2}$	$T_j(1) = 1$
Chebyshev II	$U_j(x)$	$(-1, 1)$	$(1-x^2)^{1/2}$	$U_j(1) = j+1$
Chebyshev I (shifted)	$T_j^*(x)$	$(0, 1)$	$x^{-1/2}(1-x)^{-1/2}$	$T_j^*(1) = 1$
Chebyshev II (shifted)	$U_j^*(x)$	$(0, 1)$	$\sqrt{x(1-x)}$	$U_j^*(1) = j+1$
Legendre	$P_j(x)$	$(-1, 1)$	1	$P_j(1) = 1$
Legendre (shifted)	$P_j^*(x)$	$(0, 1)$	1	$P_j^*(1) = 1$
Laguerre	$L_j(x)$	$(0, \infty)$	$\exp(-x)$	$LC = (-1)^j/j!$
Generalized Laguerre	$L_j^{(\alpha)}(x)$	$(0, \infty)$	$x^\alpha\exp(-x)$	$LC = (-1)^j/j!$
Hermite	$H_j(x)$	$(-\infty, \infty)$	$\exp(-x^2)$	$LC = 2^j$
Hermite	$He_j(x)$	$(-\infty, \infty)$	$\exp(-x^2/2)$	$LC = 1$

TABLE 5.3.3 *Common orthogonal polynomials, their symbols, integration intervals, and weight functions. In the table above, $LC = 1$ indicates that the leading coefficient should be taken to be one. Chebyshev I and II refer respectively to Chebyshev polynomials of the first and second kind.*

uniform, and normal distributions can be used in this fashion for asymptotic expansions.

The most familiar example is the *Gram-Charlier expansion* of a mean-zero probability density on $(-\infty, \infty)$. If $f(x)$ is such a density with central moments μ_j, and if we denote the standard normal density by $\phi(x) = (2\pi)^{-1/2}\exp(-x^2/2)$, then under suitable regularity conditions we may write

$$f(x) = \phi(x)\left(1 + \sum_{j=1}^{\infty} c_j He_j(x)\right), \qquad (5.3.7)$$

where $c_1 = 0$, $c_2 = (\mu_2 - 1)/2$, $c_3 = \mu_3/6$, and so forth. The jth coefficient is a linear combination of central moments of order j and below; when j is odd (even) only the odd (even) moments are involved.

Name	$p_j(x)$	A_j	B_j	C_j
Jacobi	$P_j^{(\alpha,\beta)}(x)$	see Abramowitz and Stegun(1964)		
Jacobi	$G_j(p,q,x)$	see Abramowitz and Stegun(1964)		
Chebyshev I	$T_j(x)$	0	2	1
Chebyshev II	$U_j(x)$	0	2	1
Chebyshev I (shifted)	$T_j^*(x)$	-2	4	1
Chebyshev II (shifted)	$U_j^*(x)$	-2	4	1
Legendre	$P_j(x)$	0	$(2j-1)/j$	$(j-1)/j$
Legendre (shifted)	$P_j^*(x)$	$(1-2j)/j$	$(4j-2)/j$	$(j-1)/j$
Laguerre	$L_j(x)$	$(2j-1)/j$	$-1/j$	$(j-1)/j$
Generalized Laguerre	$L_j^{(\alpha)}(x)$	$(2j-1+\alpha)/j$	$-1/j$	$(j-1+\alpha)/j$
Hermite	$H_j(x)$	0	2	$2j-2$
Hermite	$He_j(x)$	0	1	$j-1$

TABLE 5.3.4 *Recurrence coefficients for common orthogonal polynomials. The coefficients are those for the recurrence* $p_j(x) = (A_j + B_j x)p_{j-1}(x) - C_j p_{j-2}(x)$.

The simplicity of this representation results from the fact that $g(x) \equiv 1$ in equation (5.3.6); the normal density is the unique density for which this is true. For a complete development, see Chapter 6 of Kendall and Stuart (1977).

5.3.3 On computing Gaussian quadrature rules

Now that we know all about the various common families of orthogonal polynomials, we can state the fundamental theorem of Gaussian quadrature:

Theorem 5.3-1. *[Gaussian Quadrature] Let $w(x)$ be an admissible weight function on (a,b), and let $\{p_j(x)\}$, $0 \le j$, be a set of orthonormal polynomials with respect to $w(x)$ on (a,b). Denote the zeros of $p_n(x)$ by $a < x_1 < \ldots < x_n < b$. Then there exist weights $w_1, w_2, \ldots, w_n$ such that*
(a) $\int_a^b f(x)w(x)\,dx = \sum_{i=1}^n w_i f(x_i)$ whenever $f(x)$ is a polynomial of degree at most $2n-1$,
(b) $w_k > 0$, for $1 \le k \le n$,
(c) if $f(x)$ is $2n$-times continuously differentiable, the error satisfies the relationship

$$E_n(f) = \int_a^b f(x)w(x)\,dx - \sum_{i=1}^n w_i f(x_i) = \frac{f^{(2n)}(x^*)}{(2n)!k_n^2}, \qquad (5.3.8)$$

where k_n is the leading coefficient in $p_n(x)$ and x^* is some point on the interior of (a, b), and

(d) the weights satisfy the equation

$$w_j = -\frac{k_{n+1}}{k_n}\frac{1}{p_{n+1}(x_j)p'_n(x_j)}, \tag{5.3.9}$$

where again k_n denotes the leading coefficient in $p_n(x)$.

This and other results are discussed at length in Davis and Rabinowitz (1984). From the theorem it follows that, for each of the weight functions represented in Table 5.3.3, there correspond Gaussian integration rules. The abscissæ and weights for the Gaussian integration rules corresponding to some of those in the table have been well tabulated. The most accessible set of tables is Abramowitz and Stegun (1964), which includes tables for Gauss-Laguerre and Gauss-Hermite integration rules. Also included is a set of formulæ for Gauss-Legendre integration of moments. Stroud and Secrest (1966) also contain integration formulæ for Gauss-Chebyshev rules and for Gaussian integration of low-order moments of functions with respect to the Hermite (normal) and Laguerre (exponential) weight functions. Appendix 4 of Davis and Rabinowitz is a bibliography of tables for use in integration.

Two little-known sets of tables of particular use to statisticians are those of Steen, Byrne, and Gelbard (1969), which give Gaussian quadratures for the integrals $\int_0^\infty f(x)\exp(-x^2)\,dx$ and $\int_0^t f(x)\exp(-x^2)\,dx$, which can be used to obtain half-normal integrals and normal tail probabilities.

Theorem 5.3-1 is also of interest because it serves—theoretically at least—as the basis for constructing Gaussian quadrature rules in nonstandard cases. Rules for situations which are not tabulated can sometimes be generated by computer program. The orthogonal polynomials themselves can in principle be generated from the recurrence relations defined by equations (5.3.5). Given the polynomials, their roots can be evaluated using optimization routines making use of the interleaving property; this gives the abscissæ of the integration rules. Finally, the weights in the rules can be obtained from equation (5.3.9). Unfortunately, this program is not always so easy to carry out.

The sticking point is evaluation of the inner products in (5.3.5) which determine the coefficients in the polynomials. It is generally the case for nonstandard integration problems that the coefficients in the recursion cannot be expressed explicitly. When the moments of $w(x)$ on (a, b) are known, they can be used to generate the orthogonal polynomials by a related recursion involving the moment matrix and its adjoints. Golub and Welsch (1969) give details of this approach, as well as computer programs. The approach via moments often requires solution of ill-conditioned systems.

Golub and Welsch use the Cholesky factorization of the moment matrix instead of the moment matrix itself to reduce this problem.

Davis and Rabinowitz (1984) catalog in Section 2.7 a number of alternative methods for computing Gaussian quadrature rules.

5.3.4 Other Gauss-like integration rules

A number of integration formulæ are similar to Gaussian integration rules in that they determine some (but not all) of the weights and abscissæ by optimization over all possible such values. We list some of the most important here; details are found in Davis and Rabinowitz (1984).

When we are allowed full freedom to select the $2n$ parameters $\{x_i\}$ and $\{w_i\}$ in an n-point rule, the result integrates exactly all polynomials of order at most $2n - 1$. As the number of degrees of freedom is reduced, so to is the degree of the resulting integration rule. Suppose, for instance, that we have already evaluated $f(x)$ at m points, say $y_1, \ldots, y_m$. We wish to avail ourselves of this information in an integration rule using the m predetermined points and n additional points. Such a rule has the form

$$GP(f) = \sum_{i=1}^{m} a_i f(y_i) + \sum_{j=1}^{n} w_j f(x_j), \qquad (5.3.10)$$

where the m parameters $\{a_i\}$ are to be determined in addition to the $2n$ parameters $\{x_i\}$ and $\{w_i\}$. Although one might expect that, if these constants are determined optimally, the resulting rule could always integrate polynomials of order at most $m + 2n - 1$ exactly, this is not always the case. The system of equations which the added $2n$ parameters must satisfy may have complex-valued roots, or roots outside the domain of integration. Optimal extensions (with real roots inside the integration interval) only exist under special circumstances which at present are only poorly understood. General conditions under which such optimal extensions exist are not known, so that the present state of the art consists of a catalog of special cases.

Two specific formulæ of this type are the Radau and Lobatto rules, which respectively include the function values at one end point and at both end points of the interval. These methods are variants of Gauss-Legendre formulæ, and are defined with respect to $w(x) = 1$. The n-point Radau rule has degree $2n - 2$, and has the form $2f(-1)/n^2 + \sum_1^{n-1} w_j f(x_j)$; the rule is closed on the left and open on the right. The n-point Lobatto rule has degree $2n - 3$, and has the form $2[f(-1) + f(1)]/[n(n-1)] + \sum_1^{n-2} w_j f(x_j)$; the rule is a closed one. The abscissæ x_j are roots of functions that can be computed from the Legendre polynomials; and the weights have explicit formulæ as well.

5.3.5 Patterson-Kronrod rules

One of the big advantages to Simpson's rule is that it can be computed in stages, adding more and more points until sufficient accuracy is obtained, retaining the benefit of all previously computed function values. The same is *not* true of Gaussian quadrature rules, since their abscissæ do not form a nested set, so that higher-order Gauss rules cannot be built simply from lower-order results. There *are* situations, however, in which Gauss rules do have optimal extensions in the sense of the preceding section.

Kronrod (1964, 1965) computed the optimal extensions to n-point Gauss-Legendre rules by adding an additional $n+1$ points. It can be shown that these extensions have interleaved abscissæ, and that the weights that result are all positive. Computing G_n, the n-point Gauss rule, requires n function evaluations and application of n weights. After computing G_n, $n + 1$ more function evaluations are needed, followed by the application of $2n + 1$ entirely new weights, to obtain the $(2n + 1)$-point Kronrod rule K_{2n+1}. The rule G_n has degree $2n - 1$, and the rule K_{2n+1} has degree $3n + 1$. Kronrod (1965) contains extensive tables in decimal and octal. Press, Flannery, Teukolsky, and Vetterling (1986) give tables of (G_n, K_{2n+1}) for $n = 7, 10, 15, 20, 25$, and 30.

Patterson (1968a) has shown by example that the Kronrod rules can be extended further, producing rules $P_{2^\nu(n+1)-1}$, although there is no theory currently available which proves that such extensions must exist for all n and ν. The rules which have been constructed to date all have positive weights. The successor to P_m is obtained by adding $m+1$ points, and then redetermining the weights on *all* of the $2m + 1$ points.

QUADPACK uses the Kronrod-Patterson sequence G_{10}, K_{21}, P_{43}, and P_{87}, corresponding to $n = 10$ and $\nu = 0, 1, 2, 3$. Patterson has suggested using the sequence starting with $n = 3$. Tables with 33 significant figures for this sequence have been tabulated by Wichura (1978) for $n = 3$ and $\nu \leq 6$.

Patterson (1968b) has also suggested constructing a sequence of rules in the following manner. Start with a Gauss rule using $2^k + 1$ points. Remove every other point starting with the second point, and call the set of 2^{k-1} removed points S_k. From the $2^{k-1} + 1$ points that remain, remove every second point, and call the set of removed points S_{k-1}. Repeat this process until set S_2 has been obtained, and call the set of three remaining points S_1. The sets S_j are now a sequence of interlaced subsets of the original $2^k + 1$ points. The first rule is the optimal rule on S_1, the second rule is the optimal rule on $S_1 \cup S_2$, and so forth.

Our discussion of optimal extensions has applied to Gauss-Legendre rules only. Extensions of other Gaussian quadrature rules are known to exist in some cases; the integrals of Chebyshev type $(w(x) = (1 - x^2)^\alpha)$ are best understood, with Kronrod extensions with positive weights existing for

$-\frac{1}{2} \le \alpha \le \frac{3}{2}$. Davis and Rabinowitz (1984) conjecture that Kronrod extensions fail to exist for Gauss-Laguerre and Gauss-Hermite rules, "except for isolated small values of n."

EXERCISES

5.26. [25] Using the system of equations (5.3.1), find the coefficients in the 5-point basic closed Newton-Cotes rule.

5.27. [15] Derive the system of equations analogous to (5.3.1) for determining the 2-point Gaussian quadrature rule on the interval $(-1, 1)$ with $w(x) \equiv 1$.

5.28. [11] (Continuation) Deduce that $a_1 = a_2$ in the previous problem.

5.29. [13] (Continuation) Deduce that $a_1 = a_2 = 1$ and that $x_1 = -x_2 = 1/\sqrt{3}$ in the previous problem.

5.30. [18] Express (5.3.2) in terms of an integral on $(-1, 1)$.

5.31. [28] From the facts that $P_k(x)$ is odd (even) whenever k is odd (even), and that $P_k(x)$ has k distinct roots, deduce that the weights $a_i = b_i$, where b_i and a_i correspond respectively to the abscissæ $\pm x_i$.

5.32. [22] Using the results from the previous exercise, derive the Gauss-Legendre rule for $k = 3$ points.

5.33. [01] Let $p_k(x)$ be an arbitrary kth degree polynomial, and let $w(x)$ be an admissible weight function on (a, b). Why is $p_k \in L_w^2$ for all $k \ge 0$?

5.34. [10] Let $f(x) \in L_w^2$, and suppose that $f(x) = \sum_{i=0}^{\infty} a_i p_i(x)$, where $\{p_j(x)\}$ are a set of orthogonal polynomials with respect to w. Show that $a_i = \langle f, p_i \rangle / \langle p_i, p_i \rangle$.

5.35. [26] Using equation (5.3.4) with coefficients $A_j = 0$, $B_j = (2j - 1)/j$, and $C_j = (j-1)/j$, and the starting polynomials $P_{-1} = 0$ and $P_0 = 1$, write a computer program to generate the Legendre polynomials $P_j(x)$ for $j \ge 1$.

5.36. [20] (Continuation) Modify the program of the previous exercise to produce the *Chebyshev polynomials of the first kind*, denoted by $T_j(x)$, which are defined on $(-1, 1)$ with respect to the weight function $w(x) = 1/\sqrt{1 - x^2}$. The recurrence parameters are $A_j = 0$, $B_j = 2$, and $C_j = 1$.

5.37. [18] (Continuation) The *Shifted Chebyshev polynomials of the first kind*, denoted by $T_j^*(x)$, are simply the ordinary Chebyshev polynomials transformed to the interval $(0, 1)$. Show that the weight function corresponding to this transformation is $w^*(x) = 1/\sqrt{x(1 - x)}$, which is the kernel of the "hard" integral (5.2.1).

5.38. [25] (Continuation) Show that the recurrence parameters for $T_j^*(x)$ are given by $A_j^* = -2$, $B_j^* = 4$, and $C_j^* = 1$, and compute the corresponding polynomials.

5.39. [50] Develop general conditions under which optimal extensions to Gaussian integration rules exist.

5.40. [50] Determine whether Kronrod extensions exist for any Gauss-Laguerre or Gauss-Hermite integration rules.

5.4 Automatic and adaptive quadrature

Integrands often behave well over much of the interval of integration, but behave more badly (that is, less like a low-order polynomial) over a restricted range. When this happens, it makes sense to divide the range of integration into two subintervals and to expend most of our function evaluations (hence most of our work) in the subinterval where the integrand is difficult. Whenever the character of the integrand differs markedly in different regions of the integration interval, it is advisable to split the integral up and to apply possibly different integration rules on each piece.

As an example, consider computing by quadrature the normal tail probability $Pr\{Z > \frac{1}{2}\}$. The normal density $\phi(x)$ is well-approximated by a polynomial on the lower end of $(\frac{1}{2}, \infty)$, say on $(\frac{1}{2}, 3)$. For large x, however, $\phi(x)$ behaves like an exponential function. Thus, we would expect to do somewhat better (for a given number of function evaluations) by evaluating $\int_{1/2}^{3} \phi(x)\, dx$ by one method (such as Gauss-Legendre), and the integral $\int_3^\infty \phi(x)\, dx$ by another method (such as Gauss-Legendre after change of parameter), than we would by evaluating the entire integral by a single method, even though the integrand is very smooth and well-behaved. This is in fact the case.

Automatic integration routines are designed to carry out this process automatically, to discover where the integrand is difficult, and to concentrate effort there. Automatic integration programs can be classified as either *adaptive* or *nonadaptive*. Non-adaptive routines follow a fixed scheme of iterative refinement of the integral estimate, proceeding by subdividing the range of integration in a predetermined manner. Adaptive procedures, on the other hand, attempt to discern the behavior of the integrand, and to make decisions about subdivision of the range of integration based on that behavior. The most sophisticated adaptive programs actually use different integration rules on different regions of the integration interval, as in the example of the preceding paragraph. Such an algorithm is called a *composite integration scheme*.

An example of a nonadaptive automatic integrator is given in Algorithm 5.1.2, in which a sequence of Simpson's rules based on increasing numbers of points is used until the estimate changes by less than a pre-specified amount. The sequence of Gauss-Kronrod-Patterson rules of the previous section is another example.

An example of an adaptive automatic integrator is subroutine QAG in the QUADPACK library. This routine is based on a fixed Gauss-Legendre rule of the user's choice, together with that rule's Kronrod extension. The initial interval is evaluated using the Kronrod rule, and the difference between it and the (lower-order) Gauss rule is used as an estimate of the error. If this error is sufficiently small, the program stops. Otherwise, the current interval is bisected and each subinterval evaluated separately, using exactly the same algorithm as before (Kronrod for the integral, |(Kronrod) − (Gauss)| for the error estimate) on each of the subintervals. At any given point in the computation we have a partition P of (a, b) into subintervals J. The current estimate is $\sum_{J \in P} K(J)$, where $K(J)$ is the Kronrod estimate on interval J, and the current error estimate is $\mathcal{E}(P) = \sum_{J \in P} E(J)$, where $E(J)$ is |(Kronrod) − (Gauss)| on interval J. If $\mathcal{E}(P)$ is sufficiently small, we stop, otherwise the interval J for which $E(J)$ is the largest is bisected to form a new partition, and the procedure continues. Routines such as QAG also provide for halting the computation if roundoff error is detected, if the number of subdivisions becomes too large, or if the interval for which $E(J)$ is largest becomes too small. Because $\mathcal{E}(P)$, the error estimate that is reduced at each stage, estimates the error for the entire integral, the strategy described here is sometimes called *globally adaptive*. *Locally adaptive* strategies require that each $E(J)$ be sufficiently small before $K(J)$ is included in the final sum. If $E(J)$ is found to be larger than the error permitted interval J, then J is subdivided, and the error allocated to J is apportioned among the newly subdivided intervals before the process continues.

QUADPACK contains a set of excellent globally adaptive quadrature routines which have options for dealing with a variety of weight functions, infinite intervals, and singularities.

Having sung the praises of automatic quadrature methods, a word of caution is in order. Let λ index a one-parameter family $\{f_\lambda(x)\}$ of integrands that are each well-behaved on $[a, b]$. Denote by $I(\lambda)$ and $Q(\lambda)$ respectively the exact value and the automatic quadrature estimate for $\int_a^b f_\lambda(x) \, dx$. One can then plot $P(\lambda) \equiv |I(\lambda) - Q(\lambda)|$ as a function of λ; Lyness (1983) calls such a plot a *performance profile* for the quadrature routine (relative to the family of test functions). The key observation is that $P(\lambda)$ is generally not smooth even if the test functions are well behaved, since small changes in λ can affect error estimates sufficiently to activate a refinement step in the automatic quadrature program. Lyness notes that if a long series of quadrature values are required for input to other numerical processes, the resulting CPU requirements can be 100 times larger for automatic quadrature than for standard rule-evaluation algorithms, simply because the latter happen to have smooth performance profiles.

EXERCISES

5.41. [32] Let Z be a standard normal random variable, and let $P = Pr\{Z > 0.5\}$. Using Tables 5.3.1 and 5.3.2, investigate two methods for computing this integral. Let $GLT_k(x,\infty)$ denote the k-point Gauss-Legendre rule for (x,∞) using the transformation $t = e^{-z}$. Let $GL_k(x,y)$ denote the usual k-point Gauss-Legendre rule for the interval (x,y). Finally, let $GLC_{2k}(x,y) = GL_k(x,y) + GLT_k(y,\infty)$ denote the composite rule obtained by splitting the infinite interval (x,∞) at y.

Evaluate $GLC_{10}(\frac{1}{2},\frac{3}{2})$ and $GLC_{10}(\frac{1}{2},3)$. Is there an optimal value for y? Can $GLC_{10}(\frac{1}{2},y)$ ever be *worse* than $GLT_5(\frac{1}{2},\infty)$?

5.42. [12] (Continuation) Describe a noncomposite rule $R_{2k}(x,y)$ analogous to $GLC_{2k}(x,y)$ for integration on (x,∞).

5.43. [30] (Continuation) For what values of y is each of the following rules the best: $GLT_{10}(\frac{1}{2},\infty)$, $GLC_{10}(\frac{1}{2},y)$, and $R_{10}(\frac{1}{2},y)$? Why?

5.44. [36] Consider the one-parameter family of scaled Cauchy random variables with density

$$f_\lambda(x) = \frac{1}{\pi}\frac{\sigma}{\sigma^2 + (x-\lambda)^2}.$$

Compute the performance profiles for the automatic integrator of Algorithm 5.1.2 (Simpson's rule) and for a fixed Gauss-Legendre formula. Use $[1,2]$ as the range of integration, and choose a fixed value of σ, such as $\sigma = 0.1$.

5.5 Interpolating splines

Spline functions, often simply called *splines*, are smooth approximating functions that behave very much like polynomials. Their polynomial-like behavior makes them easy to work with, both theoretically and computationally. Spline functions can be used for two rather distinct purposes: to approximate a given function by *interpolation*, and to *smooth* values of a function observed with noise. The theory and computational details in the two cases are somewhat different. Since only *interpolating splines* play much of a role in quadrature, we shall discuss them in this section. *Smoothing splines* occupy a central position in the literature on smoothing and density estimation, so we defer further discussion until Chapter 6.

Spline functions are of greatest use when it is necessary to evaluate an integral of a smooth function which is not easily computed, but values of which are tabulated. In this case, it is relatively easy to construct a spline approximation to the tabled function, and then to integrate the spline exactly on the desired interval. We now turn to a brief examination of interpolating splines and their properties.

COMMENT. The term "spline" is derived from a draftsman's tool called a spline which is a thin beam of flexible material to which weights can be attached. The spline is placed on top of a drawing containing points which are to be connected smoothly. The spline is then bent by hand so as to touch each of the points, and the bend is preserved by attaching weights along the length of the beam. The shape of the spline can then be traced on the drawing. It turns out that, to a first approximation, the actual curve traced out by the beam is a cubic spline function interpolating the points on the drawing.

5.5.1 Characterization of spline functions

Loosely speaking, a spline is a piecewise polynomial function satisfying certain smoothness conditions at the join points. A number of definitions of splines are equivalent. Consider a set of points $K = \{x_1, x_2, \ldots, x_m\}$ with $x_1 < x_2 < \cdots < x_m$. We shall refer to K as the set of *knots*. For notational convenience, we shall set $x_0 = a$ and $x_{m+1} = b$.

Definition. *Let* $x_1 < x_2 < \cdots < x_m$ *and set* $K = \{x_1, x_2, \ldots, x_m\}$. *A function* $s(x)$ *is said to be a spline on* $[a, b]$ *of degree* n *with knots* K *provided that*

(i) $s(x) \in P_n$ *on* (x_i, x_{i+1}) *for* $0 \leq i \leq m$, *and*
(ii) $s(x) \in C^{n-1}[a, b]$, *provided that* $n > 0$.

We denote the class of all splines on $[a, b]$ of degree n with a given set of knots K by $S_n(K)[a, b]$, or by $S_n(K)$ when the interval of interest is the entire real line. To simplify notation, we shall use the symbols $S_n(x_1, \ldots, x_m)$ and $S_n(K)$ interchangeably, depending on whether or not we wish to emphasize the role of the particular knots. Clearly $P_n \subset S_n(K)$ for any knot set K, and if $K_1 \subset K_2$, then $S_n(K_1) \subset S_n(K_2)$. It is also true that $S_n(K) \subset PC^n$.

An alternative characterization of splines is that the nth derivative of a spline of degree n is a step function, with jump points at the knots. Thus, such a spline can be thought of as the n-fold indefinite integral of a step function. Polynomials, of course, are simply n-fold indefinite integrals of constant functions.

5.5.2 Representations for spline functions

An nth-degree polynomial requires $n + 1$ parameters to be determined completely. For instance, one can specify the $n + 1$ coefficients in the polynomial. Alternatively, one could specify the n (possibly complex) roots together with a scale factor. Or one could simply specify the value of the polynomial at $n + 1$ distinct points. The choice of parameterization is largely a matter of convenience. The same is true of splines; alternative representations exist and are useful under different conditions.

5.5.2.1 Truncated power functions

We begin by defining the *plus function*, given by $x_+ = x$ if $x > 0$ and $x_+ = 0$ if $x \leq 0$. Powers of the plus function are sometimes called *truncated power functions*. Theorem 5.5-1 shows how spline functions can be represented in terms of truncated power functions. It's proof is left as an exercise.

Theorem 5.5-1. *If $s(x) \in S_n(K)$, then $s(x)$ has a unique representation of the form*

$$s(x) = p(x) + \sum_{k \in K} c_k \cdot (x - k)_+^n, \qquad (5.5.1)$$

where $p(x) \in P_n$.

A spline so represented can easily be integrated term by term, so that once the spline function has been determined, constructing any number of integrals of the function involves little additional computation.

5.5.2.2 Piecewise-polynomial representations

By expanding the truncated power function terms in (5.5.1), the spline $s(x)$ can be written explicitly as a polynomial in P_n on each interval. Using the canonical ordering of knots x_k, we have

$$s(x) = \begin{cases} p_0(x) = p(x) & x < x_1 \\ p_1(x) & x_1 \leq x < x_1 \\ \cdots & \\ p_m(x) & x_{m-1} \leq x < x_m \\ p_{m+1}(x) & x_m \leq x \end{cases}, \qquad (5.5.2)$$

where each $p_i(x) \in P_n$. Thus, we have $m + 1$ intervals, on each of which the spline can be represented in terms of $n + 1$ coefficients:

$$s(x) = \sum_{j=0}^n a_{ij} x^j \quad \text{provided that } x \in [x_i, x_{i+1}). \qquad (5.5.3)$$

Here, x_0 is taken to be a or $-\infty$, as appropriate. A related and numerically superior representation centers the piecewise polynomials at endpoints of the intervals on which each is used, taking the form

$$s(x) = \sum_{j=0}^n b_{ij} (x - x_i)^j \quad \text{for } x \in [x_i, x_{i+1}). \qquad (5.5.4)$$

Here, we can take $x_0 = a$ or $x_0 = x_1$, whichever is more convenient. Equation (5.5.4) is used to represent splines in the IMSL software library. This representation is also easily integrated exactly.

5.5.2.3 B-splines

A third representation is highly stable numerically and easy both to manipulate and to integrate. In this approach, the spline $s(x)$ is represented as a weighted sum of piecewise-polynomial functions called *B-splines*. B-splines are actually more general than the spline functions with which we have been dealing, since they can be used to form piecewise polynomial functions with different degrees of smoothness at different join points, so that an nth- degree B-spline might be C^{n-1} at some knots and C^j with $j < n - 1$ at others. Ordinary nth-degree splines, by contrast, are C^{n-1} at every knot.

Let $\mathcal{K} = \{k_1, \ldots, k_m\}$ be an ordered set of knots. If we consider ordinary splines $s(x) \in \mathcal{S}_n(\mathcal{K})$, then we can write $s(x)$ in terms of B-splines as

$$s(x) = \sum_{i=1}^{m} w_i B_i(x), \tag{5.5.5}$$

where each B_i is an nth-degree B-spline. The set $\{B_i\}$ depends only on the knot set $\mathcal{K}$ and the degree n. Each B_i has limited support; in particular, $B_i(x) \neq 0$ only if $x \in [k_i, k_{i+n}]$. As a consequence, the value of $s(x)$ at a point depends on at most n nonzero B_i's. The functions B_i have two other nice properties: they are nonnegative, and they sum to one at each x.

Suppose we have a spline $s(x)$ in $\mathcal{S}_n(\mathcal{K})$. What happens when one of the knots in $\mathcal{K}$ is moved toward another knot? We can think of the limit of this process as producing a knot set $\mathcal{K}$ with a *multiple knot*. The limiting function $s(x)$ is no longer a spline, but it *is* a B- spline. The result of the limiting process is to introduce a discontinuity in the $(n-1)$-st derivative of s. Higher-order discontinuities can be introduced by allowing a larger number of knots to coalesce. Thus, one can think of B-splines as an explicit representation for spline functions with multiple knots.

COMMENT. It can be shown that the collection of B-splines of degree k with knots k_1, ..., k_m having C^{ν_j} smoothness at k_j ($\nu_j < k$) spans the linear space of all piecewise-polynomial functions of degree at most k having these breakpoints and satisfying the set of smoothness constraints. As a result, they form a *basis* for this linear space; hence the name *B*-splines.

A full development of B-splines requires the theory of interpolating polynomials and divided differences, which would take us far afield. The interested reader is referred to de Boor (1978) for a full treatment, including discussion of and software for the problem of stable evaluation of B-splines. We discuss the special case of the B-spline representation for cubic interpolating splines in Section 5.5.4.

5.5.3 Choosing an interpolating spline

A consequence of Theorem 5.5-1 is that a spline of degree n with m knots is completely determined by $n + m + 1$ parameters. Thus, when nth-degree splines are used to interpolate m points of a tabulated function, an additional $n+1$ parameters can (and must!) be selected before the spline is determined. In our formulation of spline functions above, we have assumed that the knots are interior points of $[a, b]$. If $[a, b]$ is a finite interval, it is natural to determine the additional $n+1$ parameters by imposing boundary conditions on the spline.

If the spline is being used to approximate a periodic function, examined over a full period $[a, b]$, then it is natural to require that the spline approximation be periodic as well by requiring that $s(x)$ and its derivatives agree at the end points. A spline $s(x)$ is said to be a *periodic spline* if $s^{(k)}(a+) = s^{(k)}(b-)$ for $0 \le l \le n - 1$. The latter condition absorbs all but one of the free parameters, which can be chosen by specifying, say $s(a)$ or by setting $s^{(n)} = 0$.

If the spline is of odd order, say $n = 2r - 1$, then there are $2r$ free parameters. If we require $s(x)$ to agree with a tabulated function at a and b, the remaining $2(r-1)$ parameters can (somewhat arbitrarily) be specified by $s^{(k)}(a+) = s^{(k)}(b-) = 0$, for $r \le k \le n - 1$. This choice is called the *natural spline* of degree n. The cubic natural spline is easily computed and is widely used. Despite the suggestive name, however, the "natural" choice has little to recommend it, particularly for $n > 3$. Rather, it is generally preferable to choose the higher-order derivatives on the boundary in such a way that the lower-order derivatives match known or estimated values.

5.5.4 Computing and evaluating an interpolating spline

For purposes of quadrature from tabled values, one often works with a small number of equally-spaced points. How ought one to go about choosing a representation and evaluating the coefficients? We now present two methods. The first is based on the truncated power function representation (5.5.1). While this representation is very handy for *thinking* about splines, it should be clear from the example that it is dreadful for computing with them. The second method uses the B-spline representation for the cubic splines. It is computationally efficient and stable, although slightly more difficult to think about.

5.5.4.1 Computing with truncated power functions

If the procedure is to be done on a one-time basis, it may be adequate to use the truncated power function representation for the splines and to hoodwink a standard least-squares regression program into evaluating the coefficients, although even for small problems the resulting linear systems may be ill- conditioned and the computations tedious. An example of this approach will make the ideas clear.

Consider the problem of computing the tail area of the normal distribution $1 - \Phi(x) = \int_x^\infty \phi(t)\,dt$ for a value of x that is not in the available tables, say, $\xi = 1.64485$. We have available to us the following values:

x	$\phi(x)$	$1 - \Phi(x)$
1.4	0.149727465635745	0.080756659233771
1.5	0.129517595665892	0.066807201268858
1.6	0.110920834679456	0.054799291699558
1.7	0.094049077376887	0.044565462758543
1.8	0.078950158300894	0.035930319112926
1.9	0.065615814774677	0.028716559816002

Since $1 - \Phi(1.64485) = 1 - \Phi(1.6) - \int_{1.6}^{1.64485} \phi(x)\,dx$, we construct a cubic-spline approximant $s(x)$ which interpolates $\phi(x)$ at the tabled values, which we then use to evaluate the integral. Here $a = 1.4$ and $b = 1.9$. From expression (5.5.1) we have

$$
\int_{1.6}^{1.64485} \phi(x)\,dx \approx \int_{1.6}^{1.64485} \left[p(x) + \sum_{k \in K} c_k \cdot (x - k)_+^3 \right] dx
$$

$$
\equiv \left[P(x) + \sum_{k \in K} \frac{c_k}{4} (x - k)_+^4 \right]_{1.6}^{1.64485} \tag{5.5.6}
$$

Now $n = 3$ and we have $m = 4$ interior knots, so that the interpolating spline is determined by $n + m + 1 = 8$ parameters. Since six function values are specified by the table, two additional constraints remain to be specified. For concreteness, let us compute the natural spline, which takes the second derivatives to be zero at a and b.

A convenient step to take is to reparameterize the polynomial integral in (5.5.6) by writing $p(x)$ in powers of $x - \xi \equiv x - 1.64485$ rather than x, say

$$
p(x) = \alpha_3 (x - \xi)^3 + \alpha_2 (x - \xi)^2 + \alpha_1 (x - \xi) + \alpha_0. \tag{5.5.7}
$$

After doing this, the polynomial corresponding to $P(x)$ needs to be evaluated only once (at 1.6) rather than twice, and the polynomial terms involved in the computations are nearly centered, with a resulting increase in numerical stability.

How do we obtain values for the four α_j coefficients in the polynomial $p(x)$ and the four spline coefficients c_k? For compactness of notation, let an asterisk denote the corresponding value after centering at ξ, so that $a^* = a - \xi$, for example. Also let $\Delta_{ij} \equiv x_i - x_j$. The six interpolation constraints and the two derivative constraints produce the linear system $A\gamma = y$, where $\gamma' = (\alpha_0, \alpha_1, \alpha_2, \alpha_3, c_1, c_2, c_3, c_4)$, $y = (y_1, y_2, y_3, y_4, y_5, y_6, 0, 0)$, and A can be written as

$$
A = \begin{pmatrix}
1 & a^* & (a^*)^2 & (a^*)^3 & 0 & 0 & 0 & 0 \\
1 & x_1^* & (x_1^*)^2 & (x_1^*)^3 & 0 & 0 & 0 & 0 \\
1 & x_2^* & (x_2^*)^2 & (x_2^*)^3 & \Delta_{21}^3 & 0 & 0 & 0 \\
1 & x_3^* & (x_3^*)^2 & (x_3^*)^3 & \Delta_{31}^3 & \Delta_{32}^3 & 0 & 0 \\
1 & x_4^* & (x_4^*)^2 & (x_4^*)^3 & \Delta_{41}^3 & \Delta_{42}^3 & \Delta_{43}^3 & 0 \\
1 & b^* & (b^*)^2 & (b^*)^3 & \Delta_{51}^3 & \Delta_{52}^3 & \Delta_{53}^3 & \Delta_{54}^3 \\
0 & 0 & 2 & 6a^* & 0 & 0 & 0 & 0 \\
0 & 0 & 2 & 6b^* & 6\Delta_{51} & 6\Delta_{52} & 6\Delta_{53} & 6\Delta_{54}
\end{pmatrix}. \tag{5.5.8}
$$

With a modest amount of effort, one can construct the matrices A and y in a statistical package such as Minitab. Having done this, one can then solve for the coefficient vector γ using the regression commands. As with most good statistical packages, Minitab recognizes that this system of equations is nearly singular and complains good-naturedly, after which it omits one or more of the columns from the computation! To avoid this, Minitab allows the user to specify a less stringent tolerance value using a subcommand, which makes it possible to complete the computation. To do so for this relatively well-behaved problem, the tolerance must be set to less than 10^{-7}! The resulting spline approximation to $\phi(x)$ was computed using Minitab on a VAX, with the result

$$
\begin{aligned}
s(x) = {} & 0.104327 - 0.145384(x - 1.64485) \\
& + 0.245265(x - 1.64485)^2 + 0.333899(x - 1.64485)^3 \\
& - 0.390299(x - 1.5)_+^3 + 0.060015(x - 1.6)_+^3 \\
& + 0.086140(x - 1.7)_+^3 - 0.460624(x - 1.8)_+^3.
\end{aligned} \tag{5.5.9}
$$

Although the truncated power functions contribute little to the numerical value of the resulting integral—indeed, the last two contribute factors of precisely zero—they cannot be omitted from the spline computation because doing so would cause the polynomial coefficients to change. The result of integrating (5.5.9) term by term produces the estimate $1 - \Phi(1.64485) \approx 0.0499241$. The correct value is 0.05000037.

5.5.4.2 Cubic splines based on B-splines

After all of the pain of the previous section, there must be an easier and more accurate way to compute cubic interpolating splines. There is. The method we outline here turns out to be equivalent to the B-spline representation for the cubic splines. This representation has an important property that makes it attractive for computation. The linear system which they produce is tridiagonal in the cubic case and totally positive. Such systems of equations are highly stable, and can be solved in $\mathcal{O}(m)$ time. Moreover, the total-positivity property implies that numerical stability can be achieved without pivoting, which makes the algorithm particularly simple.

Consider for a moment interpolating a function $f(t)$ on the unit interval, where $f(0)$ and $f(1)$ are specified. Linear interpolation on $[0, 1]$ simply takes a convex combination of the function values, say $Lin(t) = zf(0) + tf(1)$, where $z + t = 1$. What would a cubic interpolant look like? If we write $s(t) = Lin(t) + Cub(t)$, then we must have $Cub(0) = Cub(1) = 0$, since $s(t)$, too, must agree with $f(t)$ at the endpoints. One such choice is to take

$$s(t) = zf(0) + tf(1) + \frac{z^3 - z}{6}s_0 + \frac{t^3 - t}{6}s_1, \qquad (5.5.10)$$

where s_0 and s_1 are unspecified constants and, as before, $z + t = 1$. It turns out that s_0 and s_1 are equal, respectively, to $s''(0)$ and $s''(1)$. Thus, specifying two function values and two (second) derivatives is enough to determine the cubic interpolating polynomial on the interval.

Return now to the general interpolation setting. Number the interpolation points $a = x_0 < x_1 < \cdots < x_{m+1} = b$ and their corresponding function values y_0 through y_{m+1}. To interpolate on the interval $[x_j, x_{j+1}]$, we write $t = (x - x_j)/(x_{j+1} - x_j)$, we identify $f(0)$ with y_j and $f(1)$ with y_{j+1}, and we have a sequence of constants $s_0, \ldots, s_{m+1}$. All but s_0 and s_{m+1} are determined by the requirement that the first derivatives at $x_1, \ldots, x_m$ must be continuous. Consider, for instance, $s'(1)$. We can derive an expression for $s'(1)$ based on the interpolating function on $[x_0, x_1]$, which depends on s_0 and s_1. We can also differentiate the interpolating function on $[x_1, x_2]$; this expression depends upon s_1 and s_2, and must be equal to the first expression. Subtracting these two expressions and equating to zero produces an equation in three unknowns: s_0, s_1, and s_2. Similarly, $s'(j)$ produces an equation in three unknowns: s_{j-1}, s_j, and s_{j+1}. These m equations in $m + 2$ unknowns can be augmented (at each end) with single equations prescribing values for $s_0 = s''(a)$ and for $s_{m+1} = s''(b)$. This produces a stable, totally positive tridiagonal system of equations in the s_j's.

Once we have the s_j's, further computation is almost trivial. To interpolate at a point x, we need only locate the knots for which $x \in [x_j, x_{j+1}]$, and then using s_j and s_{j+1} evaluate the interpolating cubic. Equation (5.5.10) can easily be integrated exactly, so that once the s_j's have been computed, integration is almost as easy as interpolation.

The method described here is very stable and, for the normal probability example, very accurate. A double-precision implementation based on the code in Press, Flannery, Teukolsky and Vetterling (1986) produces the estimate $1 - \Phi(1.64485) \approx 0.050000084$ for the cubic natural spline, and an estimate of 0.05000037 for the cubic spline with derivative information used at the endpoints.

EXERCISES

5.45. [01] What is $S_n(\emptyset)$, where the symbol $\emptyset$ denotes the empty set?

5.46. [03] Prove that $P_n \subset S_n(K)$ for any knot set K.

5.47. [03] Prove that if $K_1 \subset K_2$, then $S_n(K_1) \subset S_n(K_2)$.

5.48. [17] Show that $s(x) \in S_n(K)$ if and only if $s(x) \in C^{n-1}$ and $s^{(n)}$ is a step function with jumps at points $x \in K$. (Use derivatives from the left when both left and right derivatives exist, but differ.)

5.49. [21] Suppose that $s_i(x) \in S_n(x_i)$ for $i = 1, \ldots, m$, and set $s(x) = \sum_{i=1}^{m} s_i(x_i)$. Show that $s(x) \in S_n(x_1, \ldots, x_m)$.

5.50. [15] Show that, if $s(x) \in S_n(K)$, $n > 0$, then $s'(x) \in S_{n-1}(K)$.

5.51. [12] What is $S_n \cap C^n$?

5.52. [27] Prove Theorem 5.5-1.

5.53. [11] Show how to integrate (5.5.1) exactly.

5.54. [22] What is the relationship between the coefficients a_{ij} of (5.5.3) and the coefficients b_{ij} in (5.5.4)?

5.55. [03] What is the function $P(x)$ in expression (5.5.6)?

5.56. [15] Verify that the matrix A of (5.5.8) is correct.

5.57. [27] Verify the spline coefficient estimates for $1 - \Phi(1.64485)$ given in expression (5.5.9).

5.58. [23] Show that the coefficients s_0 and s_1 in (5.5.10) are the second derivatives of the polynomial $s(t)$ at the endpoints.

5.59. [11] Derive the expression corresponding to (5.5.10) for interpolating in the interval $[x_j, x_{j+1}]$.

5.60. [26] Suppose that the derivative of $f(x)$ were known at x_0. How would you set s_0 to reflect this information?

5.61. [28] Derive the expression corresponding to (5.5.10) for an indefinite integral of $s(x)$ on the interval $[x_j, x_{j+1}]$.

5.62. [33] Verify the estimates given for the cubic spline estimates of $1 - \Phi(1.64485)$ based on the B-spline representation.

5.63. [28] Since we have tabulated values of $1 - \Phi(x)$ in the text, in this case we can fit $1 - \Phi(x)$ directly by spline interpolation. Approximate $1 - \Phi(1.64485)$ using the cubic natural spline and using a cubic spline with derivative conditions at the endpoints.

5.6 Monte Carlo integration

The most common statistical problem requiring integration is that of finding the value for some parameter which describes an aspect of the distribution of a given statistic. While it may be a simple matter to describe the statistic itself, say by providing an algorithm for computing the statistic from a sample, it is generally not simple to write down the density for the statistic at hand. In such cases, *simulation*, also called the *Monte Carlo method*, can save the day. The topics of random-number generation and the proper construction of simulation experiments are worthy of careful and detailed treatment. Space in this volume allows only the most cursory treatment. Fuller discussion of simulation in general and the design of simulation studies is contained in Bratley, Fox, and Schrage (1983), Morgan (1984), Rubinstein (1981), and in the now-classic book by Hammersley and Handscomb (1964). There is also a large literature on methods for generating random numbers on a computer following specified distributions. Bratley, Fox, and Schrage have a good summary of methods useful for generating variates that are uniform on $[0, 1]$ and on generating variates from them with other common distributions. Devroye (1986) is a nearly encyclopedic work on methods for generating variates with specified properties from uniform random deviates. Johnson (1987) is concerned with generating multivariate probability distributions on computers. We shall return to these topics in Chapters 7 and 8. In the meantime, an example will illustrate some of the basic ideas and a few important concepts.

5.6.1 Simple Monte Carlo

Suppose that X is a random variable with density $f(x)$ on $[a, b]$ and that we desire to evaluate the integral

$$\theta = E[g(X)] = \int_a^b g(x) f(x) \, dx. \tag{5.6.1}$$

It may be difficult or impossible to write down $f(x)$, in which case direct quadrature is out of the question, but it may be easy to draw a random sample from the distribution with density $f(x)$. In that case, we can estimate θ using

$$\hat{\theta}_{MC} = \frac{1}{R} \sum_{i=1}^{R} g(X_j), \tag{5.6.2}$$

where the observations $X_1, \ldots, X_R$ are R independent, identically distributed replications of a single random sample from f.

As an example, consider the α-trimmed mean $\overline{X}_\alpha$, in which the mean of the middle $(1 - 2\alpha)$ of the observations is computed. For i.i.d. Gaussian samples, the density $f_\alpha(x)$ of $\overline{X}_\alpha$ can be written down explicitly, although it is not particularly pleasant. With $f_\alpha(x)$ in hand, one could then compute the variance of $\overline{X}_\alpha$ by quadrature. This would correspond to $g(x) = (x - E(\overline{X}_\alpha))^2$ in (5.6.1). For sampling from other distributions, or from non-independent samples, the density is simply too difficult to deal with analytically. Without an explicit form for the density—or at least an algorithm for computing it at a point—it is not possible to use the quadrature methods discussed so far in this chapter.

The idea behind Monte Carlo methods is that the next best thing to knowing the distribution of a statistic—or equivalently, its density—is to have a very large sample from that distribution. Since the empirical cumulative distribution function (ecdf) converges to the true cdf, any function of the ecdf will converge to the corresponding function of the cdf itself. (This statement is correct provided that certain regularity conditions hold, which shall not concern us here.)

In the trimmed-mean example, for instance, it is easy to generate by computer a large number R of samples of a given size n from a specified distribution F. Since $\overline{X}_\alpha$ is location-equivariant, we can take the median of F to be zero without loss of generality. For each such sample, $\overline{X}_\alpha$ is computed and recorded; these give us an empirical cdf based on R observations from the distribution of $\overline{X}_\alpha$. The sample mean (5.6.2) of the R values of $\overline{X}_\alpha^2$ can then be used as an estimate of $\mathrm{Var}(\overline{X}_\alpha)$, say $\hat{\sigma}_\alpha^2$. Virtually the same amount of computing effort is required to get such an estimate for independent normal samples as is needed to obtain an estimate for correlated normals, or independent exponentials, or any of a number of other possibilities. Moreover, the changes to the computer program needed to shift from one to another of these possibilities is small and local. Thus, Monte Carlo quadrature has virtues of flexibility which are generally absent from more direct approaches to quadrature.

The price that one pays for a Monte Carlo answer is imprecision. Because $\hat{\sigma}_\alpha^2$ is based on a sample of size R, different samples will produce different values for $\hat{\sigma}_\alpha^2$. Thus, an important adjunct to a Monte Carlo answer is the corresponding Monte Carlo precision, corresponding to the variability in a sample of size R. In the general case, we require a standard error corresponding to $\hat{\theta}$ of (5.6.2), which we can obtain using

$$s.e.(\hat{\theta}) = \frac{1}{\sqrt{R}} \left(\frac{\sum [g(X_j) - \hat{\theta}]^2}{R - 1} \right)^{1/2} \tag{5.6.3}$$

In the trimmed-mean example, the sample *variance* of the values of $\overline{X}_\alpha^2$ is, say, γ^2. Then $\gamma/\sqrt{R}$ is a standard error for $\hat{\sigma}_\alpha^2$, and we can be reasonably confident that σ_α^2 is within $\pm 3\gamma/\sqrt{R}$.

COMMENT. Except in cases where $g(X)$ has such a long-tailed distribution that its variance is infinite, the central limit theorem assures that the estimate $\hat{\theta}$ has an approximately normal Monte-Carlo distribution. As a consequence, the standard error from (5.6.3) can be used in standard normal-theory confidence intervals for θ, which in turn provide an error estimate for the integral which is largely free of assumptions about the shape of the integrand. Direct quadrature, by contrast, can only supply error estimates which depend upon knowing bounds for higher-order derivatives of the integrand.

5.6.2 Variance reduction

The price of simulation relative to quadrature now becomes clear. Doubling our computational effort by doubling R increases precision in our estimates by only 40%. Thus, four-place accuracy is 10,000 times more expensive that two-place accuracy. In ordinary quadrature, one expects doubling the number of quadrature points to double the precision, and some integration rules do far better than that! Often a great deal of computational effort in simulations can be saved by taking advantage of mathematical structure in the problem. Recognizing that the trimmed mean is location equivariant, for instance, makes it possible to investigate the estimator's behavior at only a single choice for μ rather than having to look at all possible choices for μ.

A more powerful method for increasing precision makes use of the fact that $\overline{X}_\alpha$ and $\overline{X}$ are positively correlated, and that the sampling properties of the latter are well understood, particularly for Gaussian samples. Because the exact answer is known for $\overline{X}$, the observed estimates for $\overline{X}_\alpha$ can be adjusted by an amount that depends upon the degree to which the R realizations of $\overline{X}$ deviate from their expected behavior. In effect, one can observe the pair $(\overline{X}_\alpha^2, \overline{X}^2)$ and estimate the mean of the first component using a regression estimate involving the second. This example illustrates the method of *control variates,* or *regression adjustment.* It is one instance of what are called *variance-reduction methods* for Monte Carlo computations. Details of a regression adjustment can be found in the exercises. There are several such methods for obtaining greater precision from a fixed amount of computation; all involve making explicit use of some known aspect of the mathematical structure of the problem. On occasion one can *introduce* additional special structure into the simulation itself which can be conditioned on to dramatically improve precision. Such methods for variance reduction are aptly called *swindles.* Johnstone and Velleman (1985) employ a powerful swindle that has general applicability in their study of robust regression estimators.

5.6.3 A hybrid method

Siegel and O'Brien (1985) discuss the history of and contribute new results to an intriguing approach which blends the ideas of classical quadrature and integration by Monte Carlo. Siegel and O'Brien consider the case with $f(x) = 1$ and $[a, b] = [-1, 1]$ in (5.6.1). A point of departure for understanding their method is to note two things about formula (5.6.2). First, it provides the exact answer if $g(x)$ is a constant. Second, it is the average of R single-point quadrature formulæ. Having said this much, a generalization of (5.6.2) should be obvious; estimate θ by

$$\hat{\theta}_{SO} = \frac{1}{R} \sum_{j=1}^{R} I(X_{1j}, \ldots, X_{kj}), \qquad (5.6.4)$$

where $I(\cdot)$ is a k-point quadrature estimate for (5.6.1). A natural form for the quadrature estimate is $I(X_{1j}, \ldots, X_{kj}) = a_i(X_{1j}, \ldots, X_{kj}) \cdot g(X_{ij})$. The weights a_i are a function of the k points in $[-1, 1]$ that make up the jth replication, and they are chosen so as to achieve exactness for low-order polynomials. Notice that if the points are equally spaced, then appropriate weights give the Newton-Cotes formulæ. Notice, too, that the Gauss-Legendre rules are also of this form.

Consider now the rules of the form given above for $k = 3$. For every choice of three points in $[-1, 1]$ one can obtain the quadratic interpolant to g, which can then be integrated to give exact results when g happens to be quadratic. By restricting attention to *symmetric* choices of the evaluation points, one can achieve exactness for cubics due to the symmetry. The three design points in the symmetric case have the form $(-\xi, 0, \xi)$. In this case, the coefficients a_i are easily computed, and the integration rule is given by

$$I(-\xi, 0, \xi) = 2g(0) + \frac{g(-\xi) - 2g(0) + g(\xi)}{3\xi^2}. \qquad (5.6.5)$$

A main result of Siegel and O'Brien is that if one chooses ξ to have the appropriate distribution, then (5.6.4) provides an unbiased Monte Carlo estimate for θ regardless of the shape of g. Siegel and O'Brien also discuss error estimates for such schemes, and rules with higher-order exactness.

COMMENT. The combined rule (5.6.4), (5.6.5) can be thought of as a randomized version of Simpson's rule. The randomized version puts substantial weight near the ends of the interval $[-1, 1]$. To obtain unbiasedness for all g, the random variable ξ must have cumulative distribution function $F_\xi(t) = t^3$ on $[0, 1]$. This is the cdf of the maximum of three independent uniforms. Sampling from this "maximum of three" distribution is easy—simply take the cube root of a single uniform.

5.6.4 Number-theoretic methods

Monte Carlo integration on the unit interval can be thought of as a probabilistic version of Newton-Cotes quadrature: the equally-spaced points of the latter are replaced by uniformly distributed points whose *expected values* are equally spaced. An advantage of the probabilistic version over the deterministic version is that the "average equal spacing" property is unaffected by adding a single new observation, whereas the "precisely equal spacing" property of Newton-Cotes cannot be preserved when a single new point is added. Thus, adding a single evaluation point in the latter case entails discarding all of the previously computed information. There is an intermediate ground between random sampling and deterministic equal spacing which is often superior to Monte Carlo in that convergence can be $O(1/n)$ rather than $O(\sqrt{n})$, and superior to Newton-Cotes in that all previously computed values of the integrand can be used when a new point is added. Such methods are based on sequences of points in the unit interval called *quasirandom numbers* selected for their number-theoretic properties, hence the name *number-theoretic integration methods*. The best introduction to quasirandom methods of integration is Niederreiter's lengthy 1978 paper. Chapter 6 of Stroud (1971) discusses these methods with particular emphasis on multivariate integration problems.

Although the details of quasirandom integration are beyond the scope of this volume, we can indicate the basic ideas and their applicability to statistics. In Monte Carlo quadrature, a sequence of n i.i.d. points uniform on $[0, 1]$ are generated. The expected number of such points falling in the subinterval $[a, a+\Delta]$ is $n\Delta$. Of course, some such intervals will contain "too many" points and others "too few;" this is a consequence of independence. By abandoning the requirement that the points appear to be stochastically independent, one can even out the discrepancies in the number of "hits" between different subintervals of the same width Δ, and it turns out that it is desirable to do so if our goal is to reduce the error in quadrature.

Quasirandom number sequences are designed to produce sequences that produce low values of a measure of deviation from equidistribution called the *discrepancy*. This quantity $D(n)$ is the maximum difference over all subintervals of the actual number of points in the interval and the corresponding expected number of points, that is,

$$D(n) = \max_{\Delta} \max_{0 \le a \le 1-\Delta} |\#\{x_i \in [a, a + \Delta]\} - n\Delta|.$$

Definitions of discrepancy for sequences on the unit cube in higher dimensions are defined analogously, and sequences of points achieving low values for these measures can be constructed for higher dimensional integration.

Except for integration on the unit interval (and its generalizations to the unit cube), methods based on quasirandom numbers are not generally available and as a consequence have seen little use in statistical applications. Recent work by J. E. H. Shaw (1986) may change this state of affairs. He shows how to convert common problems in Bayesian statistics such as computing marginal distributions into quadrature problems on the unit cube, and he then constructs quasirandom integration rules that improve on Monte Carlo methods.

EXERCISES

5.64. [07] An estimator $\hat{\mu}(X_1, \ldots, X_n)$ of a parameter μ is said to be *location equivariant* if $\hat{\mu}(X_1 - c, \ldots, X_n - c) = \hat{\mu}(X_1, \ldots, X_n) - c$ for all real c. Show that $\overline{X}_\alpha$ is location equivariant.

5.65. [11] Why can the sample mean of the R values of $\overline{X}_\alpha^2$ be used as an estimate of $\mathrm{Var}(\overline{X}_\alpha)$?

5.66. [01] The text claims that doubling R increases precision by only 40%. Where does the "40%" figure come from?

5.67. [26] (Regression adjustment) Suppose that one observes independent, identically distributed pairs (Y_i, X_i) from a distribution with mean (μ, θ), variances(σ_y, σ_x), and correlation ρ. Suppose that $E(Y \mid X)$ is approximately linear, and that one knows the value of θ. How would you estimate μ given this information, and how would you assess the precision of this estimate?

5.68. [35] (continued) Use the ideas of the previous exercise to carry out a simulation study of the trimmed mean for various degrees α of trimming and a few choices of n. How much does regression adjustment increase precision over unadjusted Monte Carlo?

5.69. [19] (Siegel and O'Brien, 1985) Prove that the hybrid quadrature rule (5.6.4), (5.6.5) is unbiased for any integrand $g(x)$ on $[-1, 1]$ provided that ξ has cumulative distribution function $F_\xi(t) = t^3$ on $[0, 1]$.

5.70. [24] Let U_0, U_1, U_2, U_3 be independent, identically distributed $U[0, 1]$ random variables. Show that $V = \sqrt[3]{U_0}$ and $W = \max(U_1, U_2, U_3)$ have the same distribution.

5.71. [45] Extend the hybrid quadrature ideas of Siegel and O'Brien to integrals of functions of normal random variables, providing a randomized quadrature rule corresponding to Gauss-Hermite rules.

5.7 Multiple integrals

Virtually every treatment of multiple integration begins with the lament that quadrature in higher dimensions is hard. Indeed, the general problem of integrating an arbitrary function over a region of arbitrary shape in p dimensions is extraordinarily difficult to attack by general methods. Instead, the literature on multiple integration is a collection of results and methods for specific subproblems that are constructed about properties of the integrand or of the regions of integration. For some of these subproblems—integration on the cube, for example—a body of theory can be worked out, but for most problems the state of knowledge remains simply a collection of special techniques for highly specialized situations, coupled with a handful of general approaches such as Monte Carlo integration.

Even though the problem is hard, we still have to do something about it. In this section we shall give an idea of some of the difficulties and some approaches that can be used to solve problems that arise in statistical applications.

The best single reference on multiple integration remains A. H. Stroud's 1971 book entitled, *Approximate Calculation of Multiple Integrals*. Abramowitz and Stegun (1964) give some additional useful formulæ for low-dimensional integration. Haber (1970) gives an expository survey of the most commonly used methods for multiple integration. Davis and Rabinowitz (1984) bring the area nearly up to date.

It is instructive to consider the contents of Stroud's book to get an initial feel for the problem. The book is divided into two parts of roughly equal size, entitled "Theory" and "Tables." The theory section can be divided into two parts, dealing respectively with special regions such as the cube, simplex, and sphere, and general regions. Product formulæ can be developed for the special regions, whereas there are no general techniques with a satisfactory theory for dealing with more general regions. The tables section is an encyclopedic compendium of formulæ for the various simple regions and rules.

Why is integration in two or more variables harder than univariate quadrature? Stroud notes two facts. First, the geometry of one-dimensional space is much simpler than that of higher-dimensional space. Consider, if we identify sets that are equivalent under affine transformation, there is essentially just one bounded connected region in $\mathcal{R}^1$, whereas in higher dimensions, there are infinitely many (non-equivalent) bounded connected regions. The integration formulæ for these different regions are essentially different from one another, so that a single unified theory is not possible. Second, as we have seen in our treatment of Gaussian quadrature, the theory of integration is intimately connected with that of orthogonal polynomials. Unfortunately, the theory for dimensions $p > 1$ is exceedingly complicated and poorly understood.

5.7.1 Iterated integrals and product rules

When the region of integration is a rectangle $R \subset \mathcal{R}^p$, then a natural approach is to write the integral in iterated form. For example in $\mathcal{R}^2$, if $R = R_1 \times R_2$, where R_1 and R_2 are each intervals, then we write

$$\int_R f(x_1, x_2)\, d(x_1, x_2) = \int_{R_1} \int_{R_2} f(x_1, x_2)\, dx_2 dx_1$$

$$= \int_{R_1} g(x_1)\, dx_1, \tag{5.7.1}$$

where

$$g(t) = \int_{R_2} f(t, x_2) dx_2. \tag{5.7.2}$$

If we knew $g(\cdot)$ then we could approximate (5.7.1) by univariate quadrature over R_1. Of course, we don't know $g(\cdot)$, but according to (5.7.2) we can approximate *it* by quadrature over R_2! Using 10-point Gaussian quadrature as our basic integration rule, we would need to evaluate $g(\cdot)$ ten times, and for each of these ten evaluations we would have to perform a separate 10-point Gaussian integration. Thus, in all, we could approximate (5.7.1) using 100 evaluations of $f(\cdot)$.

In general, when the integration region R is the Cartesian product of rectangles on $\mathcal{R}^p$, an integration rule using k^p function evaluations can be constructed from univariate k-point quadrature rules applied to the iterated integrals. Such multivariate methods are called *Cartesian-product quadrature rules*. In their Section 5.6, Davis and Rabinowitz (1984) discusses such rules at length.

COMMENT. It is worth noting that the n-dimensional Laplace transform

$$\int_0^\infty \cdots \int_0^\infty \exp(-\sum t_i x_i) f(x_1, \ldots, x_n)\, dx_1 \cdots dx_n$$

can be written as a Cartesian product of 1-dimensional Gauss-Laguerre formulæ.

Suppose that X_1 and X_2 are positive random variables with joint density $f(x_1, x_2)$. How could we compute $\Pr\{X_2 < X_1\}$? An obvious approach that extends the idea of the Cartesian-product rule is to write

$$\int_{x_2 < x_1} f(x_1, x_2)\, dx_1 x_2 = \int_0^\infty dx_1 \int_0^{x_1} f(x_1, x_2)\, dx_2. \tag{5.7.3}$$

Integrals such as (5.7.3) that can be written in the iterated form

$$\int_{a_1}^{b_1} dx_1 \int_{a_2(x_1)}^{b_2(x_1)} dx_2 \cdots \int_{a_p(x_1,\ldots,x_{p-1})}^{b_p(x_1,\ldots,x_{p-1})} f(x_1, \ldots, x_p)\, dx_p \tag{5.7.4}$$

can be approximated in the obvious way by using univariate rules "from the inside out." Such rules are called *generalized product rules;* the weights in these rules are simply the products of the weights obtained from their component univariate rules.

What is nice about iterated integrals is that there is such an obvious and straightforward way to obtain an estimate for them using general product rules. The obvious way is not always a good way, however, but it is a starting point. In equation (5.7.3) for instance, if f is fairly flat near the origin then it is overkill to use 10-point Gaussian integration on the interior integral for small values of x_1. One might prefer to use more function evaluations for large x_1 values and fewer function evaluations for small x_1. On the other hand, if most of the probability mass is near the origin, then the reverse approach might be preferable.

Generally, iterated integrals can be written in more than one way. Equation (5.7.3), for instance, could also be written as

$$\int_0^\infty dx_2 \int_{x_2}^\infty f(x_1, x_2)\, dx_1. \tag{5.7.5}$$

The product rule for (5.7.3) would involve a product of a method for a semi-infinite interval and one for a finite interval. A product rule for (5.7.5) would involve a product of two rules for semi-infinite intervals. When an integral is expressed as an iterated integral, the numerical accuracy of the result can be dramatically affected by the order of integration. In this case, (5.7.3) would be the preferable formulation provided that the integrand is well-behaved near the x_2 axis.

When iterated integrals can be constructed, different integration rules can be used for the different component quadratures. For example, integral (5.7.3) could use Gauss-Laguerre for the outer quadrature (over a semi-infinite range), and Gauss-Legendre or Simpson's Rule for each of the inner integrals.

5.7.2 General multivariate regions

There are really two kinds of "general" regions: those that cannot be built up simply as unions of simpler regions, and those that can. Many statistical problems, particularly those involving computation of a probability, fall into the former category; we discuss such problems briefly at the end of this section. In the latter case, however, if the region is defined explicitly, and if it is a simple matter to "tile" the region with elementary regions for which quadrature rules exist, then the problem is little different in principle from constructing extended Newton-Cotes methods.

One difficulty in passing from $\mathcal{R}^1$ to higher dimensions is that the one-dimensional interval can be generalized in so many ways. Consider the plane, for instance. Natural analogs of the one-dimensional finite interval are the square and the disk in two dimensions. For specific regions such as the square, the triangle, and the disk, bivariate quadrature rules have been constructed, so that regions which can be represented as the disjoint union of such elementary regions can be integrated relatively efficiently. Because they can be used to tile the plane, square and triangular regions are particularly interesting as building blocks for general regions. The disk is of interest only for integrating regions with circular boundaries, disjoint unions of disks being rarely encountered integration regions.

As in univariate problems, one can often convert a given region R of integration into a more standard region R' (such as the disk or the square) by an appropriate change of variables.

Kahaner and Rechard (1987) have constructed a useful and efficient general-purpose FORTRAN routine for bivariate integration over regions that can be represented as the union of triangles. At each step the program approximates the integral and obtains an error estimate within each of the triangles. The triangle with the largest error is subdivided into subtriangles, within each of which the algorithm is applied. This process continues until either a prespecified global error tolerance is achieved, or one of several stopping criteria suggesting lack of progress is encountered. This algorithm is a globally adaptive method similar in design to the univariate procedure QAG discussed in Section 5.4.

Of some statistical interest is the development, given in Stroud (1971), of product rules of a given order d for the p-dimensional simplex as a product of p univariate rules of order d. Thus, if an order-d rule on $\mathcal{R}^1$ requires M points, a dth order rule for the simplex requires M^p points. Stroud also gives product rules for p-dimensional cones and for the p-sphere (by writing in spherical coordinates).

Orthogonal polynomials in p-dimensions exist for some common regions, and have been tabulated, at least for low-order methods (say $d \leq 5$). For some regions, particularly the cube and $\mathcal{R}^n$ weighted by $\exp(-\sum x_i^2)$, the orthogonal polynomials are simply products of the corresponding univariate orthogonal polynomials.

Just as in the univariate case, multivariate quadrature rules are generally constructed so as to be exact for polynomial integrands of low-order. While this approach works well in the univariate case even for functions that do not much look like polynomials, it does not generally work so well in higher dimensions. It is not hard to see why this is the case.

Consider a univariate integral evaluated using Simpson's rule and 100 points. The estimate is made up of the sum of the integral over 50 subintervals, on each of which a different quadratic approximation to the integrand is used. Thus, the "quadratic assumption" is really being used only very locally—each approximation applies only over 2% of the range of integration. For the method to work well, an integrand need only look approximately quadratic on small subintervals of the range of integration. Now consider a bivariate integral on a square, and suppose that we wish to use 100 points again. Four-point rules can achieve exactness for quadratic monomials, so that we can subdivide our region of integration into no more than 25 regions, each covering about 4% of the total area and having sides covering about 20% of the sides of the square integration region. Integrands for which the bivariate rule is exact must be precisely quadratic for long distances at a time in both the x and the y directions—for as much as a fifth of the length of a side of the integration region. Unfortunately, such integrands are hard to come by. To achieve the same degree of insensitivity to the "quadratic assumption", one would have to square the number of points when going to the bivariate rule, which is often just not possible. This *curse of dimensionality* plainly gets worse in higher dimensions.

We now return briefly to the problem of integrals defined over implicitly-defined regions. As a simple example, consider a random sample X_1, ..., X_n of $\mathcal{N}(\mu, 1)$ random variables, and let $Z_k = \sqrt{k} \cdot \overline{X}_k$, where $\overline{X}_k$ is the mean of the first k observations. What is $\text{Prob}\{|Z_k| < |Z_n|\}$ for different choices of k and μ? It is extremely difficult to write the probability as an iterated integral because the boundary of the region is so complicated. One obvious idea to apply is to write loosely

$$\text{Prob}\{|Z_k| < |Z_n|\} = \int_{\mathcal{R}^n} I(|z_k| < |z_n|) f(x) \, dx,$$

where the right-hand side can be evaluated as a Cartesian product. At each evaluation point x, if $|z_k| < |z_n|$ then the integrand's value is just the value of the density at that point, otherwise the value of the integrand is zero. Unfortunately, this integrand is *highly* discontinuous, and methods of quadrature designed to be exact for polynomials fail miserably at integrating functions such as indicators with interior discontinuities. Monte Carlo methods, on the other hand, are particularly well-suited to such integrands.

5.7.3 Adaptive partitioning methods

The problem with multivariate quadrature is that the analogs of univariate integration rules require so many function evaluations. When the integrand is fairly flat over a large portion of the range of integration, however, we only need a handful of function values in order to pin down the contribution of that portion to the total integral. In an area where the integrand is changing rapidly, we need many function evaluations to get a good estimate of the integral. As we evaluate the function in different parts of the range, we can learn about the shape of the integrand in the various subregions, and perhaps can decide where it is most productive to expend additional effort, that is, to compute additional function evaluations. Thought of in this way, *adaptive quadrature rules* are algorithms for sequential experimental designs.

Friedman and Wright (1981a,b) consider integration on the p-dimensional cube in this light. Their method adaptively subdivides the cube into hyperrectangles so as to minimize the variation of function values within each of the subdivisions. After having established a partition of the cube, either Monte Carlo or quasirandom sampling is used within each partition. Friedman and Wright's approach is a particularly interesting hybrid of ideas from optimization and quadrature. The measure of "variation" on a rectangle R is the product of its volume $v(R)$ and what we might call its "height," $h(R) = \max_{x \in R} f(x) - \min_{x \in R} f(x)$. The two components of $h(R)$ are determined by constrained optimization. Since derivative information is not generally available, Friedman and Wright employ a quasi-Newton method for the optimization. At a given stage, the rectangle R whose variation $s(R) = v(R) \cdot h(R)$ is largest is subdivided; this partitioning is carried out recursively until all of the values of $s(R)$ are sufficiently small. What is remarkable is that for most integrands it is well worth spending some function evaluations on optimization in order to achieve a good partition for stratified Monte Carlo sampling.

5.7.4 Monte Carlo methods

Most simulations are implicitly evaluating integrals over regions in high-dimensional space. Indeed, the Monte Carlo approach is often the only alternative left when multivariate quadrature becomes intractable.

An interesting compromise between full product-Gaussian quadrature and pure Monte Carlo integration has been studied in some detail by Evans and Swartz (1986), based on a suggestion by Hammersley. In $\mathcal{R}^p$ an n-point product-Gaussian rule requires n^p function evaluations. One useful property of Gaussian integration rules is that the weights are all nonnegative, so that after normalization they form a probability mass function. We can write the rule as

$$\int_R f(x)\, dx \approx C \sum w_\alpha f(x_\alpha), \qquad (5.7.6)$$

where $\sum w_\alpha = 1$. By sampling x_α with probability w_α repeatedly, we can obtain an unbiased estimate of the product-Gaussian rule's value. Evans and Swartz show that this approach to Monte Carlo integration can produce substantial variance reduction.

5.7.5 Gaussian orthant probabilities

Probably the most natural multivariate integration problem that arises in statistics is that of computing probabilities for the multivariate normal distribution over orthants (or more generally, over rectangular regions). Even such a simply stated problem is quite difficult. Consider the probability $P_k(\mu, \Sigma) = \text{Prob}\{\mathbf{x} \geq \mathbf{0}\}$, where $\mathbf{x} \sim \mathcal{N}(\mu, \Sigma)$ is the k-variate normal distribution with mean vector μ and variance matrix Σ, and the inequality is taken to mean that all components simultaneously satisfy the inequality. This is the probability that $\mathbf{x}$ lies in the *positive orthant*. If $\mu = 0$ this is simply a standard normal orthant probability. Unless $k \leq 3$ or $\mu = 0$ and $\Sigma \propto I$, no closed-form expression for $P_k(\mu, \Sigma)$ is known.

Dutt and Lin (1975) give tables for computing orthant probabilities for $k = 4$ and $k = 5$. Gehrlein (1979) shows how to reduce the problem with $k = 4$ to a problem involving three univariate quadratures. Farebrother (1986) discusses methods for computing orthant probabilities in the context of linear regression with inequality constraints, in which the null distribution of a test statistic for constrained regression models are weighted sums of chi-squared random variables, the weights being normal orthant probabilities.

Evans and Swartz (1986) discuss evaluating multivariate normal probabilities by Monte Carlo methods. The "hit-or-miss" method generates a sequence of independent realizations from $\mathcal{N}(\mu, \Sigma)$ and counts the proportion that fall into the positive orthant. Evans and Swartz examine several alternatives to the hit-or-miss method for evaluating $P_k(0, \Sigma)$, and they show that each of the three algorithms they propose is optimal for particular choices of Σ. They also show that for some choices of Σ, the hit-or-miss method can be superior to some of the more sophisticated estimators. Evans and Swartz also briefly discuss the general problem of estimating $P_k(\mu, \Sigma)$ for general μ and Σ, which is equivalent to estimating the multivariate normal cumulative distribution function.

Related to the orthant problem is the problem of computing normal probabilities for arbitrary regions. Some recent work illustrates the current state of the art. DiDonato, Jarnagin, and Hageman (1980) proposed a method for computing bivariate normal probabilities over convex polygons; this work was extended to arbitrary polygons by DiDonato and Hageman (1982). Bohrer and Schervish (1981) have constructed algorithms for normal probabilities over rectangular regions for which error bounds can be computed. Schervish (1984) has done further work for multivariate normal probabilities with error bounds.

EXERCISES

5.72. [11] Let $x \sim \mathcal{N}(0, \Sigma)$, and set $p = \text{Pr}\{x_1 \circ_1 0, x_2 \circ_2 0, \ldots, x_p \circ_p 0\}$, where each $\circ_i$ denotes either $\leq$ or $\geq$. Show that p can be written in terms of a *positive* normal orthant probability.

5.73. [42] Compare alternative methods for approximating the normal orthant probability $P_k = \text{Prob}\{\mathbf{x} \geq \mathbf{0}\}$, where $\mathbf{x} \sim \mathcal{N}(0, I_k)$ is the k-variate standard normal, and the inequality is taken to mean that all components simultaneously satisfy the inequality. Some candidates for comparison include the method of Friedman and Wright, quasirandom number methods, product-Gaussian quadrature, and simple Monte Carlo methods.

5.74. [48] Devise an algorithm for computing the positive orthant probability exactly for $x \sim \mathcal{N}_p(0, \Sigma)$, for $p \geq 4$.

5.75. [50] Devise an algorithm for computing general Gaussian orthant probabilities for $x \sim \mathcal{N}_p(\mu, \Sigma)$, with $\mu \neq 0$.

5.8 Bayesian computations

Bayesian methods of inference are based on the posterior distribution of a parameter θ given the observed data X. If the observables have joint density $f(x \mid \theta)$, then as in Chapter 4 we define the likelihood function to be $L_X(\theta) = p(X \mid \theta)$, the joint density considered as a function of θ with the values of x being fixed at the observed values. Given a prior density $\pi(\theta)$ on the parameter θ, the *posterior density* of θ given X is given, using Bayes's Theorem by

$$p(\theta \mid X) = \frac{p(X \mid \theta)\pi(\theta)}{\int p(X \mid \theta)\pi(\theta)\, d\theta} = \frac{L_X(\theta)\pi(\theta)}{\int L_X(\theta)\pi(\theta)\, d\theta}. \qquad (5.8.1)$$

When θ is multidimensional, additional interest centers on the conditional and marginal posterior distributions of elements of the θ vector. If $\theta = (\theta_1, \theta_2)$, then the posterior marginal density of θ_1 is given by

$$p(\theta_1 \mid X) = \int p(\theta_1, \theta_2 \mid X)\, d\theta_2, \qquad (5.8.2)$$

and the posterior conditional density of θ_1 given θ_2 is

$$p(\theta_1 \mid \theta_2, X) = \frac{p(\theta_1, \theta_2 \mid X)}{p(\theta_2 \mid X)}. \qquad (5.8.3)$$

Of additional interest is the *predictive density* of a new observation given the observed information about θ. The predictive density is

$$f(z) = f(z \mid X) = E[f(z \mid \theta)]$$
$$= \int f(z \mid \theta)p(\theta \mid X)\, d\theta, \qquad (5.8.4)$$

where the expectation in the first line is with respect to the posterior distribution of θ.

Practical Bayesian inference depends upon being able to examine features of posterior and predictive distributions such as means, variances, densities, and percentiles. When the prior and likelihood are conjugate to one another, this enterprise is easily carried out. However in general, one must resort to numerical approximations. The most common approximation, discussed below, is to take the posterior distribution to be normal and to approximate only the mean and variance matrix of this normal distribution. Unfortunately, the normal approximation is too often inadequate, in the sense that the posterior density may not be nearly normal in shape.

5.8.1 Exploring a multivariate posterior density

How can we understand the structure of a multivariate distribution? What aspects of it are important? Since it is hard to visualize multivariate distributions, it is useful to have ways of "deconstructing" a joint density into smaller, more easily digestible pieces. The two building blocks with which we are most familiar from which multivariate distributions can be constructed are conditional distributions and marginal distributions. In effect, these distributions are lower-dimensional aspects of the higher-dimensional object in which we have interest. Marginals and conditionals can be looked at and comprehended in one or two dimensions. But in order to look at a marginal distribution, for instance, one must be able to integrate out the other variables, that is, one must be able to do fairly high-dimensional quadrature on an integrand that is generally not well-understood. Approximation by a multivariate Gaussian distribution makes the computation of marginals straightforward, but it is often the case that the Gaussian approximation is too poor to use. In order to assess whether a Gaussian approximation is adequate one must be able to get a handle on some of the marginal distributions—which puts us back at square one.

5.8.2 Some computational approaches

We now turn to several computational methods for examining multivariate posterior distributions. We begin with the standard normal approximation, and then we examine three computationally-intensive approaches that can be used to obtain more refined estimates.

5.8.2.1 Laplace's method

The normal approximation used in Bayesian analysis generally works like this. Suppose that we need to compute the posterior mean of $g(\theta)$, that is,

$$E[\,g(\theta)\mid X\,] = \frac{\int g(\theta)e^{\ell(\theta)}\pi(\theta)\,d\theta}{\int e^{\ell(\theta)}\pi(\theta)\,d\theta}, \qquad (5.8.5)$$

where $\ell(\theta)$ is the log likelihood function and $\pi(\theta)$ gives the prior density. Set $L(\theta) = \ell(\theta) + \log(\pi(\theta))$, so that $\exp(L(\theta))$ is the integrand in the denominator of (5.8.5). We can think of $L(\theta)$ as a modified likelihood function.

If $\hat{\theta}$ is the mode of L, and if we set $\sigma^2 = -1/L''(\hat{\theta})$, then approximately

$$\int e^{L(\theta)} \, d\theta \approx \int e^{L(\hat{\theta}) - (\theta - \hat{\theta})^2 / (2\sigma^2)} \, d\theta$$

$$= \sqrt{2\pi}\sigma e^{L(\hat{\theta})}.$$

(5.8.6)

This approximation is called *Laplace's method* for estimating an integral, and is based on the idea that, if the integrand is very peaked near its mode, then the integral only depends on the behavior of the integrand in the vicinity of the mode. The usual normal approximation uses the same approximation $L(\theta) \approx L(\hat{\theta}) - (\theta - \hat{\theta})^2 / (2\sigma^2)$ in both the numerator and denominator integrals of (5.8.5).

5.8.2.2 Gauss-Hermite quadrature

Naylor and Smith (1982) describe the use of Gauss-Hermite quadrature to evaluate posterior expectations of functions $g(\theta)$. Since they use product Gaussian quadrature rules in the multivariate case, the Naylor-Smith approach can be used to produce marginal posterior moments automatically. Naylor and Smith also describe an iterative orthogonalization method which improves convergence when, as an approximation in the multivariate case, conditional moments are replaced by marginal moments. The method depends upon the posterior density being approximately equal to the product of a normal density and a polynomial; it appears to be tractable computationally for problems of dimension less than eight. Naylor and Smith demonstrate their methods on a collection of data sets. Their examples illustrate the fact that posterior distributions based on data sets of small or moderate size are often not very close to normal in shape.

5.8.2.3 The Tanner-Wong method of data augmentation

A promising idea is to develop some means of generating samples from specified posterior marginals. Provided that that can be done efficiently, then in principle any aspect of the posterior marginal distributions can be inspected and assessed by Monte Carlo. This idea has been developed by Tanner and Wong (1987), and they have proposed a powerful method based on it.

The main insight on which the Tanner-Wong method is based is similar to the main idea underlying the EM algorithm (see Section 4.7). In the EM algorithm an incomplete-data problem is augmented by estimates of (unobservable) latent variables to produce the so-called "complete-data problem," which is constructed so as to have a straightforward solution. This solution is then used to improve the estimates of the latent variables, and the process is repeated.

Suppose that we wish to learn about the posterior density $p(\theta \mid X)$, and suppose further that a modified posterior based on additional data $p(\theta \mid Y)$

has a known form, where $Y = (X, Z)$. Here, Z represents the additional (unobserved) data. We can then write

$$p(\theta \mid X) = \int p(\theta \mid z, X)p(z \mid X) \, dz. \qquad (5.8.7)$$

Note that $p(z \mid X)$ is the predictive density of the latent data given the observed data. We can write this predictive density as

$$p(z \mid X) = \int_{\Theta} p(z \mid \phi, X)p(\phi \mid X) \, d\phi, \qquad (5.8.8)$$

where the variable ϕ is a dummy variable of integration over the parameter space Θ. Now use expression (5.8.8) for the predictive density in (5.8.7) to obtain

$$
\begin{aligned}
p(\theta \mid X) &= \int_{\Theta} p(\phi \mid X) \, d\phi \int p(\theta \mid z, X)p(z \mid \phi, X) \, dz \\
&= \int_{\Theta} K_X(\theta, \phi)p(\phi \mid X) \, d\phi.
\end{aligned}
\qquad (5.8.9)
$$

Thus, $p(\theta \mid X)$ is a solution to the integral equation

$$g(\theta) = \int_{\Theta} K(\theta, \phi)g(\phi) \, d\phi, \qquad (5.8.10)$$

which suggests an iterative algorithm for obtaining a solution. Start with an initial guess $g^{(0)}(\theta)$, and then repeat the iteration

$$g^{(i+1)}(\theta) = \int_{\Theta} K_X(\theta, \phi)g^{(i)}(\phi) \, d\phi. \qquad (5.8.11)$$

In general, the kernel K in expression (5.8.11) cannot be written down explicitly. However when the "complete-data" posterior $p(\theta|Z, X)$ is easy to write down, then iteration (5.8.11) can be approximated using the following Monte Carlo idea. From (5.8.8) it is clear that one can sample from the predictive distribution for Z by first generating a parameter value ϕ from the "current" posterior $p(\cdot | X)$, and then generating Z from the conditional density $p(z | \phi, X)$. Thus, we have a method for sampling from our (current approximation to) the predictive distribution. We can approximate (5.8.7), however, by Monte Carlo as

$$p(\theta \mid X) \approx \frac{1}{R} \sum_{i=1}^{R} p(\theta \mid Z_i, X), \qquad (5.8.12)$$

where the Z_i's are independent draws from the predictive density. The right-hand side of (5.8.12) gives a new estimate for the posterior density, which can then be used to repeat the process.

The method depends critically on being able to compute $p(\theta \mid Z_i, X)$ explicitly. This will generally be the case for situations in which the maximum likelihood problem could be solved using the EM algorithm and in which conjugate priors are employed. This is a sufficiently rich class of problems to make the Tanner-Wong method of general interest.

5.8.2.4 The Tierney-Kadane-Laplace method

Tierney and Kadane (1986) describe methods for approximating posterior means and variances for positive functions of parameters in $\mathcal{R}^p$, as well as methods for obtaining marginal posterior densities of general parameters; their approach is applicable provided that the posterior distribution is unimodal. Tierney and Kadane's method can also be used to construct an approximate predictive density.

The Tierney-Kadane method for approximating posterior moments such as (5.8.5) is based on applying Laplace's method twice—both to the denominator and to the numerator integrals of (5.8.5). The method can be shown to be highly accurate for approximating posterior means and variances for positive functions of parameters in $\mathcal{R}^p$. Tierney and Kadane also show how to adapt this method to obtain marginal posterior densities of general (not necessarily positive) parameters. This approximation requires the same amount of effort as two maximum likelihood computations (with Hessians). The resulting estimator is of the form

$$\hat{E}(g) = (\sigma^*/\sigma) \exp[L^*(\hat{\theta}^*) - L(\hat{\theta})],$$

where $L^*(\theta) = L(\theta) + \log(g(\theta))$, and $\hat{\theta}^*$ and σ^* are defined analogously to their unstarred counterparts. In the multivariate case, the ratio of posterior standard deviations would be replaced by $\sqrt{\det \Sigma^* / \det \Sigma}$, where Σ and Σ^* are the negative inverse Hessian matrices of L and L^*, respectively, evaluated at their modes.

The Tierney-Kadane approach can also be used to obtain values of predictive and marginal posterior densities. Each of these computations requires one application of Laplace's method for each point at which the density is to be evaluated. Tierney and Kadane applied their method to one of the examples considered by Naylor and Smith (1982), who used Gauss-Hermite quadrature to solve the problem. The Tierney-Kadane method was about 20 times faster than Gauss-Hermite quadrature, and the two approximations produced comparable accuracy.

EXERCISES

5.76. [23] How are the two steps of expression (5.8.5) derived?

5.9 General approximation methods

Theoretical statistical computations frequently involve values of functions such as tail areas from common probability distributions such as the normal, chi-squared, and Student distributions, and their inverses (*percent points* or *quantiles*). With few exceptions, these functions do not have closed-form representations, so that one must make do with approximations that can be simply expressed. For the most common distributions there are many standard approximations now in use.

Closely related to the problem of approximating probabilities and percent points is the more general problem of approximating the distribution of a function of a random variable X whose distribution is known. In this section, we discuss in barest outline a few general methods that can be used to obtain approximations that can be converted to useful computing formulæ. Our focus will be on general aspects of computing tail areas and their inverses. Section 5.10 will then deal with specific algorithms for a few of the most common distributions. For a more complete discussion, including a useful collection of algorithms, one should consult Kennedy and Gentle (1980).

In the remainder of this section, we shall employ the following notation. The random variable X will be assumed to have a cumulative distribution function (cdf) given by $F_X(t) = \Pr\{X \leq t\}$, with a density function $f_X(t)$ with respect to Lebesgue measure. Of interest will be approximations to cdf $F_X(t)$, and its inverse, $F_X^{-1}(p)$, for $0 < p < 1$. When we wish to obtain either $F_X(t) = p$ for small values of t (so that p is close to zero), or values of the complementary cdf $G_X(t) \equiv 1 - F_X(t) = \int_t^\infty f_X(x)\,dx$ for large t, we speak of evaluating the *tail area* of F_X, and we treat this problem separately.

Kennedy and Gentle (1980) quite correctly point out that cdf's, tail areas, and percent points are often used as intermediate quantities in computations, and thus may require high accuracy. A very simple example comes from the literature of ranking and selection. Suppose that X_1 and X_2 are independent, with $X_1 \sim \mathcal{N}(\mu, 1)$ and $X_2 \sim \mathcal{N}(0, 1)$. Let $Y = \max(X_1, X_2)$. The density of Y is given by $p(y) = \phi(y-\mu)\Phi(y) + \phi(y)\Phi(y-\mu)$, where $\phi(t)$ denotes the standard normal density and $\Phi(t)$ the corresponding distribution function. Computing $E(Y) = \int yp(y)\,dy$ to six decimal places requires evaluation of the integrand to *at least* six digits. Thus, for theoretical computations it is important to have highly accurate approximations for such fundamental building blocks as cdf's and tail areas, even though for such tasks as computing p-values in applied work much cruder approximations will suffice.

5.9.1 Cumulative distribution functions

The distribution function $F_X(t) = \Pr\{X \leq t\}$ can be written as the integral

$$F_X(t) = \int_{-\infty}^{t} f_X(x)\, dx \qquad (5.9.1)$$

which suggests that quadrature methods can be used to approximate the integral. Indeed, numerical quadrature can often be a good choice, although for specific common distribution functions alternative approximations can be derived which either require less computational effort than quadrature of similar accuracy, or which have guaranteed (small) error bounds, which are generally not obtainable from the general quadrature theory.

Some general methods for obtaining approximations to integrals involve series expansions, continued fractions, and rational approximations. These methods are discussed in Section 5.9.4.

5.9.2 Tail areas

At first glance it would seem that computing the tail area $G(t)$ for large t would be easy; given an approximation for $F(t)$ one could simply evaluate $1 - F(t)$. Excellent approximations exist, for instance, for the cdf of a normal random variable, so that normal tail areas would seem to be trivial to obtain. Unfortunately, life is not so simple. Consider single-precision floating-point approximation of $1 - \Phi(4) \approx 3.16713 \times 10^{-5}$, in a format that supplies five decimal digits. Then an accurate approximation to $\Phi(4)$ is 0.99997, the complement of which, 3×10^{-5}, is only good to a single decimal place. Arguments much larger than four produce answers with no significant digits!

5.9.3 Percent points

Let p be a probability in $(0, 1)$, and let $F(x)$ denote the cumulative distribution function of a random variable X. A point x_p for which $F(x_p) = p$ is called a pth quantile or fractile of F. If F is a continuous monotone increasing function, then $x_p = F^{-1}(p)$.

A common approach for evaluating x_p is to solve for x in the nonlinear system $F(x) = p$ using any of the methods of Chapter 4. For many standard distributions the density $f(x) = F'(x)$ is available in closed form, and adequate approximations to F itself are also at hand. In this case, Newton's method can be used effectively, provided only that a starting value sufficiently close to the answer can be obtained.

COMMENT. For small values of p, computing F^{-1} is an ill-conditioned problem. This makes it very difficult to compute F^{-1} to a fixed absolute accuracy. On the other hand, it is less difficult to compute percent points to a given *percentage* accuracy.

Sometimes it is possible to evaluate F itself accurately, but accurate closed-form approximations to F^{-1} are not available. If a crude, easily computed approximation $\phi(p)$ to $F^{-1}(p)$ exists, there is a simple iteration, ascribed by Gentleman and Jenkins (1968) to John Tukey, that can be used to obtain an improved solution $\phi^*(p)$, namely,

$$
\begin{aligned}
\phi^*(p) &= \phi[2p - F(\phi(p))] \\
&= \phi[p - \{F(\phi(p)) - p\}].
\end{aligned}
\tag{5.9.2}
$$

This iteration can be use recursively. The derivation of (5.9.2) is actually fairly simple. Let $x_p = F^{-1}(p)$. Suppose that near x_p we have that $F(x) \approx \phi^{-1}(x) + c$. Evaluating this approximation at $x = x_p$ and at $x = \phi(p)$ and subtracting the two results gives the result. A second derivation is to construct a simple iteration as in Section 4.2.1; this method also produces information about the convergence of the algorithm. The details are left as an exercise.

5.9.4 Methods of approximation

It is useful to know some of the main types of approximations that are used in computing formulæ, so we mention the most common here. Approximation theory and asymptotic expansions form the basis for these approaches, and are unfortunately beyond the scope of this book. The reader interested in the mechanics of constructing approximations should consult Kennedy and Gentle (1980) for important examples relevant to statistical computing, Powell (1981) for an introduction to the modern theory of approximation theory and methods, and Daniels (1987) for asymptotic approximations useful in constructing tail-area approximations.

5.9.4.1 Series approximation

The simplest method for approximating an integral such as (5.9.1) is to expand the function in a Taylor series about a convenient point ξ. Generally, the resulting series will be convergent only in some region about ξ, and convergence may be slow unless t is near ξ.

As an example, the normal integral can be written as

$$
\Phi(t) = \frac{1}{2} + \phi(t) \sum_{k=0}^{\infty} \frac{t^{2k+1}}{(2k+1)!!},
\tag{5.9.3}
$$

where $k!!$ denotes $k \cdot (k-2) \cdot \ldots \cdot 3 \cdot 1$. To obtain (5.9.3), the proof of which is left as an exercise, one must use Taylor's theorem in a somewhat unconventional way; expand $\Phi(x)$ about the point t and evaluate the resulting expression at $x = 0$. Clearly, a truncated version of (5.9.3) may be an adequate approximation to $\Phi(t)$ only near $t = 0$.

Sometimes it is possible to expand an integrand in a Taylor series and then to integrate the resulting expression term by term. This method only occasionally produces results that are computationally competitive with other approximations, but as is the case with the trapezoidal rule, even crude, easily derived methods can be valuable for understanding a problem quickly. As an example, consider $\Phi(t)$ once more. Writing $\Phi(t) = \frac{1}{2} + [\Phi(t) - \Phi(0)]$, we have

$$
\begin{aligned}
\Phi(t) &= \frac{1}{2} + \int_0^t \phi(x)\,dx \\
&= \frac{1}{2} + \frac{1}{\sqrt{2\pi}} \int_0^t e^{-x^2/2}\,dx \\
&= \frac{1}{2} + \frac{1}{\sqrt{2\pi}} \sum_{k=0}^{\infty} \frac{(-1)^k x^{2k+1}}{k!\,2^k(2k+1)} ,
\end{aligned}
\tag{5.9.4}
$$

where the last step is obtained by writing $\exp(-x^2/2)$ in a power series, and integrating term by term. This expression converges fairly rapidly near $t = 0$; at $t = 0.1$ it is accurate to four decimal places after one iteration, seven places after two iterations, and almost ten places after three iterations.

For values of t much larger than 2.0, neither (5.9.3) nor (5.9.4) converges very quickly. At $t = 3$ for instance, the estimates based on (5.9.4) oscillate wildly at first, producing early iterates ranging from 2.0 to -0.27. Indeed, every even-numbered iterate before the fourteenth is greater than one! Still, the iteration converges fairly quickly after that, with 5.5-place accuracy at iteration 18, and 9.7-place accuracy at iteration 24. Although (5.9.3) produces estimates that are always less than one, at $t = 3$ its convergence is only slightly more rapid; iteration 18 has 6.5 places of accuracy, and iteration 24 has 10.7-place accuracy.

5.9.4.2 Continued fractions

A more general relative of the power series expansion is the continued fraction expansion of a function. Like power series, continued fractions have a radius of convergence within which they are convergent. One can construct continued fractions from power series, and *vice versa;* the advantage of one over the other is in terms of the speed and radius of convergence. The continued-fraction representation of a function may converge in a different domain from that of the series; often the fraction behaves well where the series behaves badly. Continued-fraction expansions for the normal cdf, for instance, can converge quite rapidly in the tails, where power series do badly. It is sometimes even the case that divergent asymptotic series can be converted to globally convergent continued fractions, although there is no general theory. A number of methods exist for converting a series to a continued fraction; the interested reader should consult Jones and Thron (1980) for complete details.

COMMENT. Euler showed in 1748 how to construct a continued fraction whose approximants coincided with the partial sums of a given infinite series, and vice versa. When the method is applied to the partial sums of a power series, the resulting continued fraction has exactly the same convergence properties as the power series, since by construction the partial sums of the series and the approximants of the fraction coincide.

To fix ideas, consider a pair of sequences $b_0, b_1, \ldots$, and $a_1, a_2, \ldots$, with all $a_k \neq 0$. Define the sequence of values f_n by the finite continued fraction

$$f_n = b_0 + \cfrac{a_1}{b_1 + \cfrac{a_2}{b_2 + \cfrac{a_3}{b_3 + \cdots \cfrac{}{+ \cfrac{a_n}{b_n}}}}}. \tag{5.9.5}$$

This continued fraction is often written more compactly as

$$f_n = b_0 + \left(\frac{a_1}{b_1} \ + \ \frac{a_2}{b_2} \ + \ \frac{a_3}{b_3} \ + \ \cdots \ + \ \frac{a_n}{b_n} \right).$$

If the sequences $\{a_k\}$ and $\{b_k\}$ are infinite sequences and if the sequence of *approximants* $\{f_n\}$ has a limit, then the continued fraction is said to converge to that limiting value. The approximants are also called *convergents* of the fraction.

Given n, one can compute f_n by a backward recurrence, starting at the bottom of expression (5.9.5). This method requires only n multiplications or divisions. This algorithm is generally quite stable numerically. Unfortunately, the computation for f_n cannot make use of any previously computed values of f_k for $k < n$, so that computing the *sequence* of values $f_1, f_2, \ldots, f_n$ requires $O(n^2)$ operations.

Alternatively, continued fractions can be evaluated using a pair of three-term forward recurrence relations. The nth approximant can be written as $f_n = A_n/B_n$, where A_n and B_n are given by

$$\begin{aligned} A_n &= a_n A_{n-2} + b_n A_{n-1}, & n \geq 1, \\ B_n &= a_n B_{n-2} + b_n B_{n-1}, & n \geq 1, \end{aligned} \tag{5.9.6}$$

where $A_{-1} = 1$, $A_0 = b_0$, $B_{-1} = 0$, and $B_0 = 1$. This method is less stable numerically than the backward recurrence, but has the advantage that the entire sequence f_1, f_2, ..., f_n can be computed in $O(n)$ multiplications. Successive values of f_k can then be compared to test for convergence.

Spurred by advances in computational hardware, the area of continued fractions has seen a resurgence of interest in the last three decades. For computing purpose, of course, an infinite continued fraction is replaced by one of its approximants. The truncation error so introduced as been carefully studied only recently, the most important results having only been discovered since 1965. Virtually the only modern treatment of both theoretical and computational issues in continued fractions is the comprehensive work of Jones and Thron (1980).

As an example, we shall mention two continued-fraction representations for normal tail areas. The ratio of the normal tail area to the density is called *Mills's ratio*, for which a famous continued-fraction expansion is available. Let $G(x) = 1 - \Phi(x)$. Then for $x > 0$ we have the continued fractions

$$\frac{G(x)}{\phi(x)} \approx \left(\frac{1}{x} + \frac{1}{x} + \frac{2}{x} + \frac{3}{x} + \frac{4}{x} + \cdots \right), \qquad (5.9.7)$$

and

$$G(x) \approx \frac{1}{2} - \phi(x) \left(\frac{x}{1} + \frac{-x^2}{3} + \frac{2x^2}{5} + \frac{-3x^2}{7} + \frac{4x^2}{9} + \cdots \right). \qquad (5.9.8)$$

These expressions can be found in Abramowitz and Stegun (1964), and can also be derived from classical series for the error function and its complement, as found, for example, in Jones and Thron (1980). The fraction in (5.9.7) converges quite slowly for small x; at $x = 1.64485$, 22 terms are required to obtain five significant figures, and even at $x = 3$, 10 terms are required. The fraction in (5.9.8), by contrast converges much more rapidly for small x, achieving five-place accuracy at $x = 1.64485$ in just 10 iterations. Unfortunately, for larger x, it takes some time before it settles down to convergence; at $x = 3$, five-place accuracy requires 18 iterations.

The fraction in (5.9.7) is an example of an *S-fraction*, a special class of continued fractions whose convergence properties are particularly well understood. Expression (5.9.7) converges for all positive x. Because it is an S-fraction, (5.9.7) has errors whose magnitude alternate in sign and which are monotone decreasing in magnitude, so that the difference in successive terms can be used reliably to monitor convergence. It is also possible to construct what is called the *even part* of the fraction, a new continued fraction whose approximants are equal to the even approximants of (5.9.7). Thus, for computational purposes it is possible to reduce by half the number of terms which require evaluation. In practice, the terms in the even part require about twice as much effort to compute as those in the original fraction, so computational savings are often minor. For example,

a_0	0.6556782	a_4	0.6629607
a_1	1.3772310	a_5	−0.6078250
a_2	−0.2634126	a_6	−0.1589534
a_3	−0.4245714	a_7	1.5683740

TABLE 5.9.1 *Coefficients for the polynomial* $P(w) = \sum a_k w^k$ *in the approximation* $1 - \Phi(x) \approx \phi(x)P(\frac{1-x}{2(1+x)})$. *The approximation produces five or more significant digits in the range* $0 < x \leq 10$.

the even part of the fraction in (5.9.7) is given by

$$\frac{1}{x}\left(1 - \frac{1}{x^2 + 3} + \frac{-2 \cdot 3}{x^2 + y} + \frac{-4 \cdot 5}{x^2 + 11} + \frac{-6 \cdot 7}{x^2 + 15} + \cdots\right). \tag{5.9.9}$$

5.9.4.3 Polynomial approximation

In dealing with distribution functions, their complements, and percent points, polynomial approximation is generally not helpful except for the very limited purpose of interpolating tabulated values. Continuous distribution functions must have a first derivative that goes to zero as x approaches infinity; polynomials have first—and higher-order—derivatives that explode as x goes to infinity. As a consequence, polynomials are generally better suited to interpolation than extrapolation. Press, Flannery, Teukolsky, and Vetterling (1986) provide a good discussion of the pitfalls in polynomial approximation, and they also give FORTRAN programs for computing the unique interpolating polynomial of degree $m - 1$ passing through a given set of m points.

Polynomial approximation is sometimes useful after a suitable change of variables to make the tail behavior of the function more nearly polynomial. As an example, consider evaluating the normal tail probability $G(x) = 1 - \Phi(x)$ for $x > 0$. Since the normal density drops off quite rapidly in the tails, it is helpful to consider instead Mills's ratio $G(x)/\phi(x)$. The change of variables $t = t(x) = 1/(1+x)$ converts the range from $(0, \infty)$ to $(0, 1)$. Write $H(t) = G(x(t))/\phi(x(t))$. A plot of $H(t)$ against t appears to the eye to be very nearly linear in t. Indeed, this function can be well approximated by a low-order polynomial in t, which is dominated by the quadratic and linear components. Thus we can write

$$G(x) = 1 - \Phi(x) \approx \phi(x)P\left(\frac{1-x}{2(1+x)}\right), \tag{5.9.10}$$

where $P(w)$ is a polynomial in w, and we have set $w = t - \frac{1}{2} = (1-x)/2(1+x)$ for reasons of numerical stability. Table 5.9.1 gives a set of coefficients for a seventh-degree polynomial $P(w)$ that produces a good approximation to $G(x)$ for $0 < x \leq 10$.

For a given function $f(x)$ on a finite interval $[a, b]$ there is no uniquely best approximating polynomial. One criterion for choosing among competing polynomials is the *minimax criterion*. The minimax approximating polynomial of degree p is that pth-order polynomial $P(x)$ for which $\max_{[a,b]} |P(x) - f(x)|$ is minimized. Unfortunately, the minimax polynomial is generally very difficult to compute. An excellent approximation to the minimax polynomial, however, is quite easy to compute. The Chebyshev polynomials of the first type form a basis for continuous functions on $[-1, 1]$; they can be extended to arbitrary finite intervals $[a, b]$. The *Chebyshev approximation* to $f(x)$ of order p is simply the projection of f onto the span of the first p Chebyshev polynomials. Details, as well as software, can be found in Section 5.6 of Press, Flannery, Teukolsky, and Vetterling (1986).

5.9.4.4 Rational approximation

Whereas polynomial approximation is rarely useful in approximating percent points and distribution functions, approximation by rational functions is often very helpful. Consider a function

$$R(x) = \frac{P(x)}{Q(x)} = \frac{a_0 + a_1 x + a_2 x^2 + \cdots + a_p x^p}{b_0 + b_1 x + b_2 x^2 + \cdots + b_q x^q}. \qquad (5.9.11)$$

We say that $R(x)$ is a rational function of order (p, q). Given m points, one can generally find a rational function of this form that interpolates these points provided that $m = p + q + 2$. Press, Flannery, Teukolsky, and Vetterling (1986) give a FORTRAN program for computing the *diagonal rational interpolant* to a set of m points, the function of form (5.9.11) for which $p = q$ if m is even, and $p = q - 1$ if m is odd.

Rational approximations are often used to compute normal probabilities and percent points. For computing the normal integral (or equivalently the error function $\mathrm{erf}(x) = (2/\sqrt{\pi}) \int_0^x \exp(-t^2)\, dt$), decent methods using rational approximations exist. Kennedy and Gentle (1980) discuss one such approximation. Rational approximations for normal percent points due to Hastings (1955) are given in Abramowitz and Stegun (1964). They are well-known but not very accurate.

Given a power series representation for a function $f(x)$, it is possible to construct from it the *Padé approximation* to $f(x)$. The Padé approximation is a rational function whose first $p + q$ derivatives coincide with those of f. There is a close connection between Padé approximants and continued fractions. When the a_k's and b_k's in the finite continued fraction (5.9.5) are all polynomial functions of x, then the approximant $f_n(x)$ is a rational function. The *quotient-difference algorithm* of Rutishauser (1954) is the most important tool for converting power series, rational function approximations, and continued fraction approximations from one to another. Jones and Thron (1980) give an excellent account, which has the added virtue of being written in English.

EXERCISES

5.77. [26] Show that, if $X_1 \sim \mathcal{N}(\mu, 1)$, $X_2 \sim \mathcal{N}(0, 1)$ independently of X_1, and $Y = \max(X_1, X_2)$, then the density of Y is given by $p(y) = \phi(y - \mu)\Phi(y) + \phi(y)\Phi(y - \mu)$.

5.78. [23] (Continuation) Show that $p(y)$ from the previous exercise is actually a probability density.

5.79. [22] (Continuation) Suppose that $X_1 \sim \mathcal{N}(\mu, 1)$, that $X_i \sim \mathcal{N}(0, 1)$ for $i = 2, \ldots, k$, and that the X_i's are independent. Derive the density for $Y = \max(X_1, \ldots, X_k)$. (This distribution arises in ranking and selection problems.)

5.80. [14] Let $F(x)$ be a continuous, strictly increasing function. Show that $dF^{-1}(x)/dx = 1/F'(F^{-1}(x))$.

5.81. [29] Show that computing $F^{-1}(p)$ becomes ill-conditioned as p approaches zero when F is a continuous, strictly increasing function with infinite support on the left. Deduce a similar result for right-hand tail areas.

5.82. [19] Derive expression (5.9.2) by assuming that, near x_p, $F(x) \approx \phi^{-1}(x) + c$ and then proceed as suggested in the text.

5.83. [27] Derive (5.9.2) as a simple iteration as described in Section 4.2.1.

5.84. [28] State conditions under which the iteration implied by expression (5.9.2) will converge.

5.85. [32] Prove (5.9.3) using the Taylor series argument of Section 5.9.4.1.

5.86. [27] (Continuation) Prove (5.9.3) using integration by parts. Compare this method to that of the previous exercise.

5.87. [19] Fill in the details of equation (5.9.4).

5.88. [12] Why must $\Phi(t)$ be rewritten as $\frac{1}{2} + [\Phi(t) - \Phi(0)]$ in the first step of (5.9.4)?

5.89. [05] What does it mean to have "6.6-place accuracy"?

5.90. [27] Compare the behavior of (5.9.3) and (5.9.4) for estimating $\Phi(t)$ at $t = 0.1$, 0.5, 1.0, 2.0, 3.0, 4.0, and 5.0.

5.91. [02] Show that only n multiplications or divisions are required to compute f_n in expression (5.9.5) by backward recurrence.

5.92. [26] Prove that the approximants f_n of equation (5.9.5) are given by A_n/B_n, where A_n and B_n satisfy the relations of (5.9.6).

5.93. [20] A continued fraction is said to be *regular* if $a_k = 1$ for all $k \geq 1$. The first few values of b_k in the regular continued-fraction representation of π are 3, 7, 15, 1, 292, 1, 1, 2, ... Compute f_k for $0 \leq k \leq 4$ and examine the accuracy of the approximation.

5.94. [29] Write computer programs to compute the sequence of approximants to an arbitrary continued fraction using (a) the forward recurrence and (b) the backward recurrence.

5.95. [38] Carry out the program of developing an approximation to the normal tail area function $G(x) = 1 - \Phi(x)$ for $x > 0$ as outlined in the text, using a Chebyshev approximation in $t = 1/(1+x)$ to $\log(G(x)/t) + x^2/2$. How does the Chebyshev approximation compare to that of Table 5.9.1? Is it better to approximate $\log(G(x)/[t\phi(x)])$ or $G(x)/\phi(x)$ using Chebyshev polynomials? Are there better choices for the transformation t?

5.10 Tail-areas and inverse cdf's for common distributions

In this section I list some useful approximations that are frequently needed in practical work. For greater detail, for distributions that are omitted here, or for computing routines— generally of greater complexity—with higher accuracy, the reader is referred to Kennedy and Gentle (1980), Griffiths and Hill (1985), and to Press, Flannery, Teukolsky, and Vetterling (1986).

General methods, on which many of the results cited in this section are based, produce approximations for families of related distributions. The cube-root approximation of Wilson and Hilferty (1931) and the approximations of Peizer and Pratt (1968) are noteworthy in this regard. Recent work by Skates (1987) has produced excellent approximations for such distributions as t, χ^2, and F that provide considerable improvement in accuracy over previous methods at a relatively low cost in additional complexity.

Of particular note are the rough methods of Andrews (1973), which give simply-computed tail-area approximations for the normal, t, χ^2, and F distributions with a relative error of no more than five percent for $p \leq 0.05$. Let $f(x)$ denote the density function, $g(x) = f'(x)/f(x)$ denote its logarithmic derivative, and let $h(x) = g'(x)/g^2(x)$. Andrews approximates $G(x) = \int_x^\infty f(x)\,dx$ by

$$G(x) \approx \frac{1}{K-1}\frac{f(x)}{g(x)}\left[1 + \frac{1}{2}(h(x) - K)\right], \qquad (5.10.1)$$

where $K = \lim_{x \to \infty} h(x)$. Andrews (1973) provides a table of g, g', and K for the normal, t, χ^2, and F distributions, and he indicates that the method also works well for other distributions such as the log-normal, logistic, and extreme value distributions.

5.10.1 The normal distribution

The normal distribution is naturally the one for which approximations are in the greatest demand, not only because the distribution is common and important in its own right, but also because of its central role in statistical inference. What is more, many approximations for other distributions are expressed in terms of an equivalent normal deviate, so that accurate approximations to these distributions require accurate approximations to the normal as well.

5.10.1.1 Normal tail areas

Andrews's (1973) method gives a rough, 1.5–2 significant-digit tail area for $p \leq 0.05$ using the formula

$$G(x) = 1 - \Phi(x) \approx \frac{\phi(x)}{x}\left(1 - \frac{1}{2x^2}\right). \tag{5.10.2}$$

The handiest method with good accuracy for obtaining the Gaussian cdf is based on an algorithm for approximating the complementary error function given in Press, Flannery, Teukolsky, and Vetterling (1986):

$$\mathrm{erfc}(x) \approx \left(\frac{1}{1+x/2}\right)\exp\left(-x^2 + P\left(\frac{1}{1+x/2}\right)\right), \tag{5.10.3}$$

for $x > 0$, where $P(t)$ is a ninth-degree polynomial with coefficients given in Table 5.10.1. This particular polynomial is the ninth-degree Chebyshev approximation to what Press, *et al*, refer to as "an inspired guess as to the functional form." The approximation has a *relative error* of less than 1.2×10^{-7} for all $x > 0$.

The relationship between $\mathrm{erfc}(x) = (2/\sqrt{\pi})\int_x^\infty \exp(-t^2)\,dt$ with $x \geq 0$ and the normal integral is given by

$$\Phi(x) = \begin{cases} \frac{1}{2}\,\mathrm{erfc}(-x/\sqrt{2}) & x \leq 0, \\ 1 - \frac{1}{2}\,\mathrm{erfc}(x/\sqrt{2}) & x > 0 \end{cases} \tag{5.10.4}$$

and

$$\mathrm{erfc}(u) = \begin{cases} 2\Phi(-u\sqrt{2}) & u \leq 0, \\ 2G(u\sqrt{2}) & u > 0 \end{cases}. \tag{5.10.5}$$

Another excellent approximation to the normal integral, with FORTRAN code, is that of Cooper (1968), which uses a series expansion for small x and a continued fraction expansion for large x. Hill (1985) gives a short FORTRAN code that produces nine-figure accuracy.

a_0	-1.26551223	a_5	0.27886807
a_1	1.00002368	a_6	-1.13520398
a_2	0.37409196	a_7	1.48851587
a_3	0.09678418	a_8	-0.82215223
a_4	-0.18628806	a_9	0.17087277

TABLE 5.10.1 *Coefficients for the polynomial* $P(t) = \sum a_k t^k$ *of equation (5.10.3), from an algorithm of Press, Flannery, Teukolsky, and Vetterling (1986).*

5.10.1.2 Normal quantiles

We now turn to the problem of computing *quantiles*, or percent points, of the normal distribution. We seek to obtain values z_p for which $\Phi(z_p) = p$. The inverse of the complementary cdf, which in many respects is a more natural quantity, is simply $-z_p$.

For obtaining normal percent points the algorithm of Beasley and Springer (1985) produces results that are guaranteed to be good up to an absolute perturbation of 2^{-31} in the *argument*. Unfortunately, $p = 2^{-31}$ corresponds to a z value of only about -6.5. In fact, their algorithm is rather better than they claim; unpublished work of Michael Wichura shows that the Beasley-Springer algorithm gives seven significant figures for $|z| \le 3.5$, dropping to six figures at $|z| = 4$, five figures at $|z| = 5.5$, and four figures at $|z| = 9.5$. Their algorithm is based on a pair of rational-fraction approximations, one used for p near 0.5 and the other used for p near zero or one.

COMMENT. Recall that percent-point problems are generally ill-conditioned with respect to absolute error in the extreme tail. As a result it might seem sensible to describe error behavior in terms of perturbations to the input rather than to the output of the algorithm, as indeed Beasley and Springer have done. However, for *relative errors* quite the opposite is the case. Indeed, at a particular value for z_p with p near zero or one, the relative error in z is *smaller* than the relative error in p by the multiplicative factor of about z^{-2}.

Noting that Φ^{-1} is really not badly conditioned in terms of relative error over useful ranges of p, Wichura (1987) has developed an algorithm for normal quantiles that produces sixteen significant digits for $|z| \le 37.5$. It is worth noting that $\Phi(-37.5) < 10^{-307}$, which should suffice for most purposes. Wichura's method is a modification of that of Beasley and Springer. The range of p is divided into three segments, one for p near 0.5, one for p near one (or zero), and the last segment for middle values of p. On each segment, a rational approximation is used to the inverse function.

Wichura has also developed a family of approximations for $G^{-1}(p)$ valid for small values of p, say $|z| \ge 1$. One particular choice from this family is given as Algorithm 5.10.1. For all $p < 0.1$, the method gives at least two significant figures, and the accuracy improves as $p \to 0$. For $|z| > 2$, the method gives four or more significant figures.

Algorithm 5.10.1 (Wichura's method for normal quantiles)

{This algorithm returns $z = G^{-1}(p)$ for $0 < p < 0.1$}
$v := -2\log(p)$
$x := \log(2\pi v)$
$\theta := x/v + (2 - x)/v^2 + (-14 + 6x - x^2)/2v^3$
$z := \sqrt{v(1 - \theta)}$

Huh (1986) describes the following approach for computing normal percent points, given an accurate approximation to the cdf, that converges cubicly. It is based upon Halley's method, which is described by the following iteration. Let x_0 be greater than x_p, and $p > 1/2$. Then iteratively set

$$x_{i+1} = x_i + \frac{p - \Phi(x_i)}{\phi(x_i) - \frac{1}{2}(p - \Phi(x_i))x_i}. \qquad (5.10.6)$$

5.10.2 The χ^2 distribution

For approximating the distribution of a χ^2 random variate on m degrees of freedom, the transformation of Wilson and Hilferty (1931) can be used to approximate the variate in terms of a normal random variable. Let X denote the χ^2_m random variable and Z denote a standard normal deviate. Then

$$X \approx m \left\{ 1 - \frac{2}{9m} + Z\sqrt{\frac{2}{9m}} \right\}^3. \qquad (5.10.7)$$

This can be inverted to obtain the more familiar formula

$$\frac{\left(\chi^2_m/m\right)^{1/3} - (1 - 2/9m)}{\sqrt{2/9m}} \approx \mathcal{N}(0, 1). \qquad (5.10.8)$$

The Wilson-Hilferty approximation is particularly good in the right-hand tail, where it is adequate even for degrees of freedom as low as $m = 3$. It may not be adequate for the extreme left tail (near zero) for small degrees of freedom.

Although it is slightly more complex than the Wilson-Hilferty approximation, the approximation of Peizer and Pratt (1968) is substantially more accurate, particularly in the tails. The equivalent normal deviate for χ^2_m

is given by

$$Z \approx \frac{\chi^2 - m + \frac{2}{3} - \frac{0.08}{m}}{|\chi^2 - m + 1|} \sqrt{(m-1)\log\left(\frac{m-1}{\chi^2}\right) + \chi^2 - (m-1)}. \quad (5.10.9)$$

COMMENT. Two special cases bear mention. For a χ^2 random variable on a single degree of freedom, the probability integral and its inverse can be deduced from the corresponding functions of the normal distribution, since $\chi_1^2 \sim Z^2$, where $Z \sim \mathcal{N}(0,1)$. For two degrees of freedom, one can use the relationship to the exponential distribution to obtain the cdf $F(x) = 1 - \exp(-x/2)$.

Except for small m, or very extreme probabilities p, the Wilson-Hilferty approximation is used as a starting point for a Taylor series expansion of the inverse χ^2 distribution in Best and Roberts's (1985) algorithm for percent points of the χ^2 distribution.

5.10.3 The F distribution

The F distribution has a normalizing transformation that is closely related to the Wilson-Hilferty transformation for the χ^2 distribution. Paulson (1942) gives the following generalization of Wilson-Hilferty for $F \sim F_{m,n}$ in terms of a standard normal variate Z:

$$Z = \frac{\left(1 - \frac{2}{9n}\right) F^{\frac{1}{3}} - \left(1 - \frac{2}{9m}\right)}{\sqrt{\left[\frac{2}{9n} F^{\frac{2}{3}} + \frac{2}{9m}\right]}}. \quad (5.10.10)$$

This somewhat obscure-looking formula can be derived from the cube-root transformation of (5.10.8). Define the *mean-squared distribution* on k degrees of freedom to be the distribution χ_k^2/k. An F random variable on (m, n) degrees of freedom is the same as the that of the ratio of two independent mean-squared random variables on m and n degrees of freedom, respectively. What equation (5.10.8) asserts is that the cube root of a mean-squared random variable has approximately a normal distribution. We can obtain an equivalent normal deviate for an F tail area by letting $U \sim \chi_m^2$ and $V \sim \chi_n^2$, and then writing

$$\Pr\left(\frac{U}{V} > F\right) = \Pr\left(\sqrt[3]{\frac{U}{V}} > \sqrt[3]{F}\right)$$

$$= \Pr\left(U^{\frac{1}{3}} - F^{\frac{1}{3}} V^{\frac{1}{3}} > 0\right). \quad (5.10.11)$$

Equation (5.10.10) is easily deduced from (5.10.11); the details of this calculation are left as an exercise.

The Peizer-Pratt approximation to the F distribution is similar to that for χ^2 given in (5.10.9). Define $S = n-1$, $T = m-1$, $N = S+T = m+n-2$, $p = n/(mF + n)$, and $q = 1 - p$. Then the Peizer-Pratt approximation to $F_{m,n}$ is given by

$$Z = \frac{n - \frac{2}{3} - (N + \frac{2}{3})p + c}{|S - Np|} \sqrt{\frac{3N}{3N + 1} \left(S \log \frac{S}{Np} + T \log \frac{T}{Nq} \right)}, \quad (5.10.12)$$

where the adjustment constant c is given by

$$c = 0.08 \left(\frac{q}{n} - \frac{p}{m} + \frac{q - \frac{1}{2}}{n + m} \right). \quad (5.10.13)$$

Kennedy and Gentle (1980) describe more accurate algorithms in some detail.

5.10.4 Student's t distribution

The t distribution has some elegant approximations. The approximation due to Peizer and Pratt (1968) for t on n degrees of freedom is

$$Z = \pm \left(n - \frac{2}{3} + \frac{1}{10n} \right) \sqrt{\frac{1}{n - \frac{5}{6}} \log \left(1 + \frac{t^2}{n} \right)}, \quad (5.10.14)$$

where the sign of Z is chosen so as to agree with that of t. An approximation of Wallace (1959) performs as well as (5.10.14) for $n \geq 2$ and better for smaller n is the slightly more complicated formula

$$Z = \left(1 - \frac{2\sqrt{1 - \exp(-s^2)}}{8n + 3} \right) \sqrt{n \log(1 + t^2/n)}, \quad (5.10.15)$$

where

$$s = \frac{0.368(8n + 3)}{2\sqrt{n^2 \log(1 + t^2/n)}}. \quad (5.10.16)$$

The most accurate algorithms for tail areas of the Student distribution are hybrid methods, the particular choice of algorithm depending upon the sizes of t and n. Both Cooper's (1985) algorithm and Hill's (1970a) method have this property.

Percent points of the t distribution require even greater care. Hill (1970b) has constructed such an algorithm for obtaining percent points. Computer programs for Cooper's and for both of Hill's algorithms are available. Kennedy and Gentle (1980) examine both of Hill's (1970a,b) algorithms in considerable detail. Readers who wish to understand the difficulties and subtleties required in constructing good approximations should consult Kennedy and Gentle's treatment of the t distribution.

5.10.5 Other distributions

Press, Flannery, Teukolsky, and Vetterling (1986) give approximations to and FORTRAN subroutines for the incomplete beta and incomplete gamma functions, from which tail areas of the χ^2, F, t, Poisson, binomial, and negative binomial distributions can be obtained. Their algorithms provide a relative error in the incomplete beta function, for instance, of less than 10^{-6}. Griffiths and Hill (1985) are another source of FORTRAN subroutines for probability distributions and percent points.

Of particular value in the Griffiths and Hill collection is an index to the first 207 algorithms published in *Applied Statistics*, together with cross references to related algorithms and comments on the algorithms. The *Applied Statistics* algorithms include methods for computing the distributions of a wide variety of statistics, including those of the largest multinomial frequency, linear combinations of χ^2 random variables, and the Studentized range.

EXERCISES

5.96. [14] Derive expression (5.10.2).

5.97. [21] Compare Andrews's (1973) normal tail-area approximation of (5.10.2) to the first three approximants of the continued-fraction expansion (5.9.7).

5.98. [30] How might one be led to the "inspired guess" from which expression (5.10.3) results?

5.99. [31] Write a computer program that computes both normal tail probabilities using (5.10.3) and normal percent points using expression (5.10.6). How does the latter compare with Algorithm 5.10.1?

5.100. [42] (Continuation) Compare the method of (5.10.6), Tukey's method of (5.9.2), the rational approximations of Hastings [found in Abramowitz and Stegun (1964)], and Wichura's rough method of Algorithm 5.10.1 for computing percent points of the normal distribution.

5.101. [26] (Michael Wichura) Set $p = \Phi(z)$. Show that computing z from p is badly ill-conditioned in terms of absolute error, but is well-conditioned in terms of relative error for p near zero or one.

5.102. [21] Show how to obtain tail probabilities and percent points for the χ^2 distribution, given acceptable approximations for normal tail areas and percent points.

5.103. [19] Derive equation (5.10.11) from (5.10.10) and (5.10.8).

5.104. [21] How does one use (5.10.10) to obtain tail areas of the $F_{m,n}$ distribution?

5.105. [25] (Continuation) How can (5.10.10) be used to obtain percent points of the $F_{m,n}$ distribution?

5.106. [43] Conduct a study of alternative methods for computing tail areas from the t and F distributions, including both direct methods and methods based on the incomplete beta function.

6 SMOOTHING AND DENSITY ESTIMATION

One definition of statistics is that it is the study of discerning pattern against a background of variability. Simple linear regression of Y on X, for instance, can be viewed as the superposition of a simple structure (a line) on a variable background (a scatterplot). It can also be viewed in terms of estimating the conditional expectation function $f(x) = E[Y \mid X = x]$ under the assumption that $f(x)$ is linear. Whether or not one is interested in estimating a conditional mean function, a suitable line superimposed on a scatterplot can be an invaluable aid to perception.

COMMENT. Some might argue that the line superimposed on the scatterplot can even help us to perceive what is not there! Whenever we employ smoothing or fitting methods, we are superimposing structure on the data that the data themselves may not support. Smoothing can help to reveal structure against a background of randomness; it can also disguise the fact that there is little structure other than randomness.

Of course, patterns encountered in real life are often nonlinear, so it is natural to seek methods for approximating or even discovering nonlinear patterns in data. Many of these patterns are functional relationships between variables, and we quite naturally seek *methods for estimating functions* on the basis of data.

Consider for a moment a bivariate relationship between a response variable Y and a covariate X. We are interested in whether the conditional distribution of Y given X changes in some easily understood way as a function of X. In particular, denote by $M(x)$ the "middle" of the distribution of $Y \mid X = x$, say the conditional mean or median. The smoother M is, the more easily we can understand the relationship between X and Y.

How might we estimate $M(x_0)$ based on a set of observed (X, Y) pairs? Since we expect that the middle of Y's conditional distribution varies smoothly with x, our guesses for $M(\cdot)$ should be smooth, too. If this is true, then when x is *near* to x_0, M(x) is very nearly equal to $M(x_0)$, as well, so that the middle of the Y values for points with X near x_0 should provide a good estimate of $M(x_0)$. By the same reasoning, unless $M(\cdot)$ has very special structure, we generally expect observations with X values far from x_0 to give us little information about the conditional distribution of Y at x_0, and hence little information about $M(x_0)$. Thus, one is led naturally to smoothing algorithms which take *local* averages of observations.

For purposes of exposition, the terms "local," "average," and "near" are being used here in a loose sense, and each needs to be taken with a grain of salt. By a *local smoothing method,* we mean one whose value $\hat{M}(x)$ depends most heavily on points with X near x. A (locally) weighted regression, for instance, would be a local smoother if the weights were close to zero for points with X far from x. (Robustness considerations might allow for a small weight for an occasional nearby point, however.) The term "average" is meant to be any function of a set of observations that ends up in the "middle" of the data set. Such functions are often, but not always, location equivariant. Examples include the sample mean, the sample median, and, say, the mean of the medians of the lowest three and highest three of five observations.

There is a tradeoff between fidelity to the data and smoothness. A curve forced to pass through all of the data points is rarely smooth. At the other extreme, the curve $M(x) \equiv \overline{Y}$, while very smooth, rarely captures all of the important features of the data. All smoothers must effect a compromise between these two extremes. In estimation terms, we can identify fidelity to the data with low bias in estimating the conditional mean, and smoothness with low variability. The fewer points used in a local average, the lower the bias and the higher the variability.

Roughly speaking, a smoothing algorithm is defined by the meaning given to the terms "average," "local," and "near," and—most critically— by the particular way in which smoothness is traded off against fidelity to the data.

Density estimation is closely linked computationally to smoothing, in the following sense. If we have a random sample from some distribution with observed values $x_1, x_2, \ldots, x_n$, the natural estimate of the distribution function is the empirical cdf (ecdf), which is an average of n point masses at the observations. While the true cdf is often continuous and differentiable, the ecdf is neither. If we let $\delta(x)$ denote the Dirac delta function at x, that is a (generalized) function that is zero everywhere except at x, where it has an infinite point mass integrating to unity, then we can represent the "natural" density estimate as $\sum \delta(x_i)/n$. Although adequate for some purposes, this formulation is not very good for such purposes as estimating the value of the density at a point.

What we generally do believe is that regions in which we have observations are likely to have greater density than regions in which no observations are seen. It is natural, then, to "smear out" the observed point masses over larger regions which include them, and to base estimates of the density on this smeared probability mass. This process of redistributing the observed mass over a larger region corresponds directly to the local averaging process in scatterplot smoothing. Smaller regions achieve greater fidelity to the data, whereas larger regions produce estimates of the density which are relatively smoother. Because density estimation methods can often be thought of as applying smoothing techniques to a basic estimator such as the histogram, the purely computational aspects of smoothing and density estimaition are similar (Titterington, 1985).

In the remainder of this chapter we shall focus on some of the computational aspects of density estimation and smoothing algorithms. We begin with some of the basic concepts from density estimation, then we proceed to examine some recent developments in smoothing. Readers interested in a treatment of those aspects of density estimation not covered here should consult Silverman's (1986) book and the survey of methods by Wegman (1972a,b). Devroye and Györfi (1985) develop an approach to density estimation based on L_1 considerations. For the area of smoothing Titterington's survey of smoothing techniques (1985) goes well beyond the topics examined here. A recent report by Buja, Hastie, and Tibshirani (1987) does provide a unifying description of linear smoothers and their role in additive regression models. Silverman (1985) gives an excellent overview of the use of spline smoothing.

6.1 Histograms and related density estimators

The histogram is the first, simplest, and most familiar example of a density estimator. It is a good context in which to fix some of the important ideas and to raise some important computational issues.

6.1.1 The simple histogram

Suppose we have a collection of real-valued observations $x_1, \ldots, x_n$ on the interval $[a, b]$. Break the interval into m *bins* of equal width h, with boundary points or *cutpoints* $a + ih$, for $0 \leq i \leq m$. For notational convenience, set $a_j = a + jh$. Denote by n_j the number of the x_i's falling in the jth bin, say $[a_{j-1}, a_j]$. The histogram for the data set is then

$$\hat{f}(x) = \frac{1}{n} \sum_{j=1}^{m} \frac{n_j}{h} \cdot I_{[a_{j-1}, a_j]}(x), \qquad (6.1.1)$$

where $I_A(x)$ denotes the indicator function of the set A.

In terms of our introductory comments, when x is in the jth bin, the histogram estimator can be written as

$$\hat{f}(x) = \frac{1}{nh} \int_{a+(j-1)h}^{a+jh} \sum_{i=1}^{n} \delta(x - x_i) \, dx. \qquad (6.1.2)$$

In this form, it can be seen that the histogram estimator is a smoothed version of the unsatisfactory weighted sum of Dirac delta functions. For histograms the term "local" means "in the same bin as," and "average" means just the usual mean. The tradeoff between smoothness and closeness to the data is determined by the bin width h.

COMMENT. To some degree one can effect this tradeoff by allowing bins of different width and by selecting cutpoints judiciously. In particular, one can start with a small trial bin width, and then amalgamate neighboring bins as seems necessary. Other methods such as the kernel methods discussed below are more convenient, however, both in terms of computing and in communicating just what was done.

COMMENT. In our development we have ignored the possibility that an observation x_i might fall exactly on a bin boundary. If the data are a random sample from a continuous distribution, then such an event has infinitesimal probability. In practice, however, it is important to define the bins to be half open rather than closed, because most digital computers generally represent data to a fixed number of places.

6.1.2 A naive density estimator

The shape of the histogram, especially for small n, depends in large measure on the location of the cutpoints—in effect, the choice of a and b when the actual potential range of the data is unbounded on either or both ends. A way to avoid this problem is to redefine the histogram estimator in a slightly different way, by allowing the bin used to estimate $f(x)$ not to be fixed, but rather to be centered on x. This idea produces what Silverman (1986) calls the *naive density estimator*

$$\hat{f}(x) = \frac{1}{n}\sum_{i=1}^{n}\frac{1}{h}w\left(\frac{x - x_i}{h}\right),\qquad (6.1.3)$$

where

$$w(x) = \begin{cases} 1/2 & |x| < 1 \\ 0 & \text{otherwise.} \end{cases}\qquad (6.1.4)$$

The function $w(\cdot)$ here is simply the uniform density on $[-h, h]$. The density estimate is constructed from a moving *window* of fixed width $2h$, within which delta functions are averaged uniformly. We might well call (6.1.3) a *moving-window histogram*.

How does the naive estimator of (6.1.3) differ from the histogram of (6.1.2)? The meaning of "average" hasn't changed, nor has the parameter controlling the tradeoff between smoothness and data-tracking. What has changed slightly is the meaning of "local." For the histogram, x is "near" x_i if they occupy the same fixed bin of width h; for the naive estimator

x and x_i are "near" if their Euclidean distance from one another is no greater than h. Thus the histogram defines nearness in terms of common set membership, while the moving-window histogram defines it in terms of Euclidean distance.

6.1.3 Kernel estimators

From a practical standpoint, the naive estimate is somewhat better than the histogram, in that the notion of nearness that it embodies is more sensible. Even though the former can appear to be less ragged than the latter, neither estimate is very smooth. The jumps in the naive estimator occur at the observation points. As the moving window of width $2h$ moves to the right, it "picks up" larger observations lying to the right of the window, and it "drops off" observations on the left of the window. This is a consequence of the finite support of the uniform density $w(\cdot)$. By modifying the weight function $w(\cdot)$ in (6.1.3) to be continuous with support on the real line, the raggedness of the naive estimator can be overcome.

We call a density estimate of the form

$$\hat{f}(x) = \frac{1}{n} \sum_{i=1}^{n} \frac{1}{h} K\left(\frac{x - x_i}{h}\right), \qquad (6.1.5)$$

where

$$\int_{-\infty}^{+\infty} K(t)\, dt = 1,$$

a *kernel estimate*, and the function $K(t)$ is said to be the *kernel* of the estimator. Generally kernels are both positive and symmetric about zero, but it is occasionally useful to relax these conditions. The quantity h is called the *bandwidth* or (half) *window width* of the estimator; as before, it controls the degree of smoothness that the resulting function exhibits.

COMMENT. The infinite-support requirement can be replaced by one of compact support if the kernel drops to zero smoothly. An example is Tukey's biweight kernel

$$K(x) = \frac{15}{16}(1 - x^2)_+^2, \qquad (6.1.6)$$

which is symmetric and unimodal with support on $[-1, 1]$, but which is continuously differentiable.

All kernel estimators are based on the notion of Euclidean distance as a measure of the nearness of two points.

6.1.4 Nearest-neighbor estimates

Yet another basic notion of "nearness" begins with a distance measure between points. Given a distance measure and any point x_0, the points in a data set can be ranked in terms of their distances from x_0.

In one dimension the computation of nearest neighbors can be simplified by ordering the points in a sample (often along the x-axis), and then simply noting the order in which the points appear. The nearness of two points is then measured in terms of the number of other points between them in the ordering. Thus, if x and x_0 are adjacent in the ordering, then x is said to be a *1-neighbor* of x_0. If a single other point is between them, then x is said to be a *2-neighbor*. If $k - 1$ points intervene, then x is a *k-neighbor* of x_0.

The *nearest-neighbor* density estimators are based on averages of the k nearest neighbors in the sample to the point x_0 at which the density is being evaluated. It is instructive to examine the moving-window histogram estimator of (6.1.3) to see how it might plausibly be modified to employ this modified notion of nearness. If the density $f(\cdot)$ is nearly flat near x, then the expected number of points in an interval of width $2h$ is $k \approx 2hf(x)$. Here we have thought of h as fixed. Suppose instead that we fix k and ask how wide an interval $2h$ we need to encompass k points. The answer is approximately inversely proportional to the density $f(x)$. If we denote by $h_k(x)$ the half-width of the smallest interval centered at x that contains exactly k of the data points, then we have approximately $f(x) \approx k/2nh_k(x)$. Corresponding to (6.1.3) and (6.1.5) we have

$$\hat{f}(x) = \frac{1}{n} \sum_{i=1}^{n} \frac{1}{h_k(x)} K\left(\frac{x - x_i}{h_k(x)}\right). \qquad (6.1.7)$$

Quite evidently, we have exchanged a fixed-width window containing a variable number of observations for a variable-width window containing a fixed number of observations. The nearest-neighbor approach, while often adequate for smoothing, does not produce satisfactory results for density estimation, as Silverman (1986) discusses at length in his Chapter 5.

6.1.5 Computational considerations

The most common applications for density estimation require more than just the estimate of the density at a point. Rather, the estimates are required at a grid of points so that, for instance, the density can be plotted. When selecting the window width, it is always useful to plot the density estimates for various values of h, and to inspect the results visually rather than relying on numerical summaries (or theory) alone. Thus, we are led to consider efficient means for evaluating a density estimate at some set of k points.

Despite the superficial similarities between the histogram estimator and the naive uniform kernel estimator, they are quite different *computationally*. The histogram is quite easy to compute, since the bins are fixed at the outset. Only a single pass through the data is needed in order to compute the histogram estimator for any number of ordinates x, since all that is really required is to compute the quantities n_j, as in Algorithm 6.1.1. This can be done once and for all and the work required is independent of h.

Algorithm 6.1.1 (Simple histogram bin counts)

$\quad$ **for** $j := 1$ **to** m **do** $n_j := 0$
$\quad$ **for** $i := 1$ **to** n **do**
$\quad\quad k := \lceil (x_i - a)/h \rceil$
$\quad\quad n_k := n_k + 1$

COMMENT. Only slightly more complex is an algorithm that can be used with bins of unequal width, when the cutpoints have been specified in advance; the construction of such an algorithm is left as an exercise.

By contrast, there is no simple method for setting up the computation of the estimator in equation (6.1.3). The most straightforward method begins by sorting the data set into ascending order. For notational convenience, let us denote the ordered data by $z_1 \le z_2 \le \cdots \le z_n$. To compute $\hat{f}(\cdot)$ at a single point x, one finds a z_j within h of x by a binary search in the z-array, and then searches linearly up and down from z_j to identify points on either side that are also within h of x. The sorting step requires $O(n \log n)$ operations, as does the binary search (Knuth, 1973). The expected number of subsequent comparisons depends both upon h and the actual form of the underlying density near x. Since this can be at most linear in n, the overall computational cost is dominated by the $n \log n$ term. The expected cost to evaluate $\hat{f}(x)$ a total of k times is $O(kn \log n)$, compared to $O(\min(k, n))$ for the simple histogram estimator.

Everything that has just been said applies equally to any kernel estimator. The direct method for evaluating a kernel estimate at, say, k points is of order $O(kn)$. Silverman (1986) notes that the direct method can be improved upon greatly if the evaluation points are equally spaced by noting that the right-hand side of equation (6.1.5) is the convolution of the empirical distribution of the data with the kernel $K(\cdot)$. As a result, the convolution can be done efficiently using Fourier transform methods, which are to be treated in Chapter 11. A FORTRAN algorithm for computing kernel density estimates using the Fast Fourier Transform is given in Silverman (1982). For computational efficiency Silverman uses the Gaussian kernel, as its Fourier transform is particularly simple. It is also possible to compute other quantities such as $\hat{f}''(x)$ at little added expense (Schiffelbein, 1986). The second derivative of $\hat{f}$ is useful for diagnostic purposes; Silverman's method of *test graphs* (1978) uses $\hat{f}''$ to help evaluate choices for h.

The nearest-neighbor method presents some computational simplifications in special cases, but in general is more difficult to work with than either of the two approaches discussed up to this point. When the kernel $K(x)$ is piecewise constant over intervals of fixed width (as is (6.1.4), for instance) then it is easy to move this fixed-width window to the right and to update the computation of $H_k(x)$ at the same time. Suppose, for instance, that the kth-nearest neighbor of x_0 is at $x_0 - h_k(x_0)$. As the window moves to the right, $H_k(x_0 + t)$ increases linearly in t, until a new observation is encountered, say for $x_1 = x_0 + t_0$. As the window moves further to the right, the old kth-nearest point drops out of the set of neighbors, and $h_k(x_1 + t)$ decreases linearly in t, until a point drops out on the left. When $K(\cdot)$ is constant within the window, then simple updating formulæ can be used to introduce or to delete a point. Unfortunately, unless the kernel has very simple structure, updating formulæ do not suffice, and Fourier methods are not available either. Thus, the only practicable alternative requires direct computation based on the definition (6.1.7).

When x is not univariate, the problem of finding nearest neighbors is more difficult. Fast algorithms for near-neighbor computations have been developed by Friedman, Baskett, and Shustek (1977), and by Friedman, Bentley, and Finkel (1977). Murtagh (1984) reviews computationally efficient methods for computing nearest neighbors.

EXERCISES

6.1. [14] After using Algorithm 6.1.1 to obtain $\{n_j\}$, how would you evaluate the histogram estimator $f(x)$ in expression (6.1.2)?

6.2. [10] Prove that the constant $15/16$ is correct in (6.1.6).

6.3. [13] Prove that Tukey's biweight kernel (6.1.6) is continuously differentiable.

6.4. [22] Write an algorithm to replace Algorithm 6.1.1 for use with bins of unequal size.

6.2 Linear smoothers

We now turn our attention to smooth estimates $\hat{M}$ of a regression function

$$M(x) = E[Y \mid X]. \tag{6.2.1}$$

It is well known that the fitted values from ordinary linear regression can be expressed as linear combinations of the observed values as

$$\hat{y} = Hy, \tag{6.2.2}$$

where $H = X(X'X)^{-1}X'$ is the so-called hat matrix that "puts the hat" on y. Note that H depends only on the configuration of the X matrix. What is less well-known is that many popular methods for smoothing, although they produce fitted values $\hat{y} = \hat{M}(x)$ that are nonlinear functions in X, are linear in the sense that the observed fitted values can be written in the form

$$\hat{y} = Sy, \qquad (6.2.3)$$

where S is an $n \times n$ matrix that once again depends only on the matrix X. Any smoothing algorithm that has this property we shall refer to as a *linear smoother*. Common linear smoothers include running means, running lines, kernel smoothers, polynomial regression, segmented regressions (regression splines), and smoothing splines. Although this linear structure provides little in the way of direct *computational* advantage, it does make it possible to understand the properties of linear smoothers fairly thoroughly. Buja, Hastie, and Tibshirani (1987) examine in detail the theoretical properties of linear smoothers, particularly in the context of the additive regression model, which is discussed in Section 4.6.

What if we wish to evaluate $\hat{M}(x)$ at some point other than the observed x_i's? If x lies between two of the design points, we can define $\hat{M}(x)$ by linear (or higher-order) interpolation; if x is outside the range of the observed data, we can use linear extrapolation based on the two nearest points, subject to all of the caveats and disclaimers that always should accompany prediction in areas where we have little information. If we think of the elements y_j in (6.2.3) as being ordered by increasing values of x_j, then for $x_i < x < x_{i+1}$, we have $\hat{M}(x) = s'y$, where s is simply a convex combination of the ith and $(i+1)$st rows of S.

We now turn to some important examples of smoothers with linear structure.

6.2.1 Running means

The simplest of the smoothers is the *running-mean smoother*, or the *running-average smoother*, which computes $\hat{y}_i$ by averaging y_j's for which x_j falls in a neighborhood N_i of x_i. As we discussed for density estimation, there are several definitions of neighborliness that could be used; by far the simplest choice from the computational standpoint is to use symmetric nearest neighborhoods, which define N_i to be those points x_j for which $|i - j| \leq k$. Such a neighborhood contains k points to the left and k points to the right of x_i, unless x_i is too close to the ends of the range, where the neighborhoods are truncated. The parameter k controls the degree of smoothing, and is often more conveniently expressed in terms of the approximate fraction f of the sample contained in the middle N_i's, namely $k = (\lfloor fn \rfloor - 1)/2$. The fraction f is called the *span* of the smoother.

The running-means smoother is easy to implement. The data are sorted in ascending order by x value, and then the data values are stepped through in order, updating the current mean and the neighborhood size, if necessary. Algorithm 6.2.1 illustrates the details.

Algorithm 6.2.1 (Running means)

Select span $0 < f < 1$
Sort data pairs (x_i, y_i) so that $x_1 \leq \cdots \leq x_n$
$k := \lfloor (\lfloor fn \rfloor - 1)/2 \rfloor$
$T := 0$
for $j := 1$ **to** $k + 1$ **do** $T := T + y_j$
$l := 1$
$u := 1 + k$
for $i := 1$ **to** n
$\quad \hat{y}_i := T/(u - l + 1)$
$\quad$ **if** $i > k$ **then**
$\quad\quad T := T - y_l$
$\quad\quad l := l + 1$
$\quad$ **if** $u < n$ **then**
$\quad\quad u := u + 1$
$\quad\quad T := T + y_u$

This smoother is obviously easy to compute, requiring $O(n \log n)$ operations for the sorting, followed by $O(n)$ additional computations for the smoothing itself. The price that one pays is that the resulting estimates are not actually very smooth. In addition, the estimates are generally biased near the end points.

6.2.2 Kernel smoothers

The product of a running-mean smoother is usually quite unsmooth. This is due to the fact that all points in the neighborhood N_i, even those quite far from the target point x_i, get equal weight in fitting $\hat{y}_i$. By using weighted means instead, with the greatest weights reserved for points closest to x_i, we might well do better. The *kernel smoother* with kernel K and window width $2h$ uses

$$\hat{y}_i = \sum_{j \in N_i} w_{ij} y_j, \qquad (6.2.4)$$

where

$$w_{ij} = K\left(\frac{x_i - x_j}{h}\right) \Big/ \sum_{j \in N_I} K\left(\frac{x_i - x_j}{h}\right). \qquad (6.2.5)$$

If the kernel is smooth, then the resulting output will also be smooth. This is easily seen from (6.2.4), which can be interpreted as a weighted sum of the (smooth) kernels, where the y_j's are the weights.

The problem of bias in the corners does not go away with estimators such as (6.2.4); the problem is inherent in nearly all methods using only locally weighted means. Let us see why that might be the case. Suppose that $M(x)$ is monotone increasing near the left boundary. Thus, $M(x_1) < M(x_i)$ for $i > 1$, so that any convex combination of y_i's that fall within the range of monotonicity must overestimate y_1. The only possible way out of this difficulty is to allow some of the weights to be negative.

How does one go about computing kernel smoothers? Unlike kernel density estimators, the kernel smoothers cannot be constructed using the fast Fourier transform since (6.2.4) does not have the form of a convolution, so that the best general methods amount to computing (6.2.4) directly from the definition, which is an $O(n^2)$ proposition. Moreover, unless the kernel has bounded support, the smoother matrix S in (6.2.3) is not a band matrix as it is in the case of the running-mean smoother. Partly for this reason, kernels with compact support are frequently used in practice. This does result in a band matrix, but the bandwidth of the matrix depends upon the particular data set. When n is large and the kernel is moderately expensive to compute—as is the Gaussian kernel $K(t) = \exp(t^2/2)$, for example—the computational burden associated with kernel smoothers can become onerous.

6.2.3 Running lines

The primary advantage of the running-mean smoother is that the estimate for $\hat{M}(x_{i+1})$ is easily obtained by a simple update of $\hat{M}(x_i)$. The disadvantages accrue from using a very crude model of the underlying regression function $M(x)$; in each neighborhood we act as if $M(x)$ were constant.

With only a little bit more computational effort—still only $O(n)$—we can replace the locally-constant model by a locally-linear model for $M(x)$, thus reducing bias near the endpoints, to the extent that $M(x)$ is approximately linear there. This is accomplished using formulæ for updating linear regressions rather than means, as follows.

Using standard regression results, the estimate $\hat{y}_i$ is simply

$$\hat{y}_i = \overline{y}_i + \hat{\beta}_i(x_i - \overline{x}_i), \tag{6.2.6}$$

where $\overline{y}_i$ and $\overline{x}_i$ are the means of the y_j's and x_j's, respectively, in the neighborhood N_i. The local regression coefficient $\hat{\beta}_i$ can be obtained from the updated values for $\overline{x}_i, \overline{y}_i$, the updated cross-product $\sum_{N_i}(x_j - \overline{x}_i)(y_j - \overline{y}_i)$, and the corresponding sum of squared deviations for the x_j's. Algorithms based on (6.2.6) are called *running-line smoothers* or *local regression* smoothers.

Stable updating algorithms for these quantities are given in Chan, Golub, and LeVeque (1983), and are discussed in Chapter 2. Serviceable updating formulæ are easily derived. If we set $C = \sum_{N_i}(x_j - \bar{x}_i)(y_j - \bar{y}_i)$ and $V = \sum_{N_i}(x_j - \bar{x}_i)^2$, then we can adding the point (x, y) to the neighborhood N_i can be accomplished by computing the quantities

$$
\begin{aligned}
\bar{x}'_i &= (x + n_i\bar{x}_i)/(n_i + 1)\\
\bar{y}'_i &= (y + n_i\bar{y}_i)/(n_i + 1)\\
C' &= C + (n_i + 1)(x - \bar{x}'_i)(y - \bar{y}'_i)/n_i\\
V' &= V + (n_i + 1)(x - \bar{x}'_i)^2/n_i.
\end{aligned}
\tag{6.2.7}
$$

Removing a point from the neighborhood N_i uses the formulæ

$$
\begin{aligned}
\bar{x}'_i &= (n_i\bar{x}_i - x)/(n_i - 1)\\
\bar{y}'_i &= (n_i\bar{y}_i - y)/(n_i - 1)\\
C' &= C - n_i(x - \bar{x}_i)(y - \bar{y}_i)/(n_i - 1)\\
V' &= V - n_i(x - \bar{x}_i)^2/(n_i - 1).
\end{aligned}
\tag{6.2.8}
$$

Running-line smoothers, while better than running averages, still appear ragged to the eye. At the cost of increased computation, the jaggedness can be alleviated by using *weighted* local regression in a neighborhood of x_i, where the weight accorded to each point in N_i is a decreasing function of its distance from x_i. Local weighted-line smoothers bear the same relationship to kernel smoothers that running lines bear to running means. This is an important element of Cleveland's (1979, 1981) locally-weighted regression method (LOWESS). Because the weights depend on actual distance between points and not simply their relative order, it is no longer possible to apply updating formulæ to obtain the local regression coefficients $\hat{\beta}_i$. Instead, the weights must be recomputed for each neighborhood and the usual regression computations done. This results in computing time that is $O(n^2)$.

6.2.4 General linear smoothers

Many other smoothing algorithms—spline smoothing and bin smoothing among them—produce linear smoothers. Smoothing splines are important in their own right and are considered in the next section.

The advantages of using linear smoothers is theoretical, not computational, although for *particular* linear smoothers there are sometimes special structures that can be exploited to good computational advantage. We can deduce much about the behavior of linear smoothers *a priori*, due purely to the linearity of the theory. As an example, consider the problem of *repeated smoothing*. If an initial result produced by running means or lines is too jagged, for instance, a natural remedy would be to smooth (again) the fitted values from the previous attempt. For symmetric linear smoothers, the theory is straightforward. If a smoother produces $\hat{y} = Sy$, then m repeated applications produces the result $\hat{y} = S^m y$, so that $\hat{y}$ will look more and more like a multiple of the eigenvector of S corresponding to the largest eigenvalue λ_1 (see Section 3.9.1).

COMMENT. It is instructive to examine the behavior of some simple smoothers under repeated application. As an example, consider the running-mean smoother with local row weights $(1/2, 1/2)$. Iteration of this smoother produces very nearly Gaussian weights very quickly.

If $\lambda_1 < 1$, then repeated smoothing will shrink the resulting values toward zero; if $\lambda_1 > 1$, then the smoothing process will actually diverge. The S matrix for locally-weighted running lines, for instance, often has a largest eigenvalue in excess of unity. It is easy to see that the hat matrix $H = X(X'X)^{-1}X'$ from linear regression has eigenvectors equal to the columns of X, with eigenvalues of unity. Indeed, H is a projection; repeated applications leave the estimates $\hat{y}$ unchanged.

EXERCISES

6.5. [17] Write out the 10×10 smoothing matrix S of equation (6.2.3) for the running-mean smoother with $n = 10$ and span $f = 0.5$, assuming that the x values appear in increasing order.

6.6. [10] Why is the running-mean smoother of Algorithm 6.2.1 often biased near the end points?

6.7. [16] What is the difference between the running-mean smoother of Section 6.2.1 and the kernel smoother with a uniform kernel $K(t) = 1/2$ for $|t| < 1$ and $K(t) = 0$ elsewhere?

6.8. [09] What advantages do the running-line smoothers of Section 6.2.3 enjoy over the corresponding running-mean smoothers from Section 6.2.1 when the x_j's are equally spaced?

6.9. [33] Write a computer program to implement running-line smoothing as discussed in Section 6.2.3.

6.10. [14] Why is weighted local-regression smoothing an $O(n^2)$ computation?

6.11. [40] Write computer programs for a running-mean smoother, a running-line smoother, and a kernel smoother using the Gaussian kernel. Compare running times and performance for a variety of input data sets of differing size and smoothness.

6.12. [27] Prove equations (6.2.7) and (6.2.8).

6.13. [31] Consider the equally-spaced time series (i, y_i) for $i = 1, \ldots, n$. Show that, except for end effects, iteration of the running-mean smoother with row weights $(1/2, 1/2)$ is nearly equivalent to using a kernel smoother with appropriate Gaussian weights.

6.14. [24] (Continuation) Suppose that weights of $(p, 1 - p)$, $0 < p < 1$, were used in the previous exercise. Does repeated iteration still produce Gaussian weights?

6.3 Spline smoothing

Of all of the linear smoothing algorithms, spline smoothers are perhaps the most attractive. Among other properties, the smoothing matrix S is symmetric, has eigenvalues no greater than unity, and reproduces linear functions. Spline methods are particularly appealing from the statistical standpoint because they arise as the solutions to optimization problems closely related to least squares and maximum likelihood. In addition, they can be viewed in Bayesian terms as the solution to an estimation problem. As a result, the posterior distribution of $\hat{M}(\cdot)$ can be used as the basis for inference about even nonlinear aspects of $M(\cdot)$, such as its maximum slope.

In Section 5.5 we developed interpolating splines, piecewise polynomial functions with $m - 1$ continuous derivatives, for some m. If you haven't done so already, you should glance over that section to become familiar with splines and their basic properties. The rest of this section makes use of the material in Section 5.5.

For a broader overview and more detailed bibliography of splines in statistics, one should see Wegman and Wright (1983). Any serious user of smoothing splines in practice should examine Silverman (1985), which distills a great amount of theoretical knowledge and practical advice. Wahba (1987) contains an overview of theoretical aspects of spline smoothing.

6.3.1 Smoothing splines

As in Section 6.2, we are interested in estimating $M(x)$, where

$$y_i = M(x_i) + \epsilon_i, \tag{6.3.1}$$

for $1 \leq i \leq n$, and with $\{\epsilon_i\}$ being uncorrelated, zero-mean random variables. As in the previous section, we shall assumed that the design points are written in ascending order. We shall also assume that the design points all fall in a known finite interval $[a, b]$. The problem with fitting $M(x)$ using simple linear least squares is that since straight lines have no curvature, even the best straight line may not come very close to some of the data points if $M(\cdot)$ is in fact curved. On the other hand, if a fitted function is allowed that is arbitrarily wiggly, then we can interpolate every point, provided that there are no duplicates among the x_i's. A compromise between these extremes is to select $\hat{M}$ so as to minimize the sum of squared errors *plus* a penalty for curvature or roughness in $\hat{M}$, namely

$$S_\lambda(M) = \frac{1}{n} \sum_{i=1}^{n} [y_i - M(x_i)]^2 + \lambda \int_a^b [M''(t)]^2 \, dt, \qquad (6.3.2)$$

where $\lambda \geq 0$, and $M \in C^3$. The second term on the right-hand side of (6.3.2) is often called the *roughness penalty*. As $\lambda \to \infty$, the minimizing function M is the least-squares regression line. As $\lambda \to 0$, the solution is the cubic natural interpolating spline with knots located at the interior data points.

COMMENT. Methods which minimize functions such as (6.3.2) are called *penalized least-squares methods*. The penalty function in (6.3.2) minimized a measure of overall curvature. Note that linear functions automatically receive zero penalty. In general, penalty functions may involve higher-order derivates. If we use $\int [M^{(k)}]^2 \, dt$ as the penalty term, then the minimizers of the penalized least-squares problem are $(2k-1)$st order splines, again with knots at the interior design points. Cubic splines are most commonly used in data analysis.

There is another useful interpretation of expression (6.3.2). If the errors in (6.3.1) have a normal distribution, then the first term on the right of (6.3.2) is the negative log-likelihood function for the parameter M. The penalty term, then, can be viewed as the negative logarithm of a prior density for M, so that the minimizer of (6.3.2) can be thought of as a posterior mean value for the regression function. We interpret this prior to put mass only on cubic splines with knots set $\{x_i\}$ and satisfying the natural boundary conditions. In this case, each possible M can be written as a linear combination of B-splines, $M(x) = \sum_1^n \gamma_j B_j(x)$ (see Section 5.5.2.3). Silverman (1985) shows how this particular prior distribution can be thought of as a prior multivariate normal distribution on the coefficients γ_j.

A third useful interpretation for smoothing splines is that they are the conditional expectation of a stochastic process observed with noise. The cubic spline solutions on $[0,1]$ arise, for instance, when the $\{\epsilon_i\}$ are independent with $\epsilon_i \sim \mathcal{N}(0, \sigma^2)$, with a diffuse prior on $M(0)$ and $M'(0)$, and with $M(t)$ generated by the stochastic differential equation

$$\frac{d^2}{dt^2} M(t) = \sqrt{\frac{n\sigma^2}{\lambda}} \frac{dW(t)}{dt}, \qquad (6.3.3)$$

where $W(t)$ is a Wiener process with mean zero, $W(0) = 0$ and $\mathrm{Var}(W(1)) = 1$, independent of the ϵ_i's (Wahba, 1978). The second-order differential operator on the left side of (6.3.3) can be replaced by a general mth-degree linear differential operator, and the resulting conditional expectations are more general spline functions.

Given a set of data, how can we compute a smoothing-spline estimate for the regression function? One approach is to select a value for λ in (6.3.2), and then to solve the minimization problem. By investigating the results for a range of λ's, the data analyst can settle on a visually acceptable choice for the smoothing parameter.

COMMENT. An alternative way of proceeding is to (somehow) construct a spline with knot set $\{x_i\}_2^{n-1}$, to compute λ for this fit, and then to decide whether this value of λ is acceptable. Although the approach that computes the fit directly from λ is generally adopted, there are situations in which that is not desirable. De Boor (1978) constructs his cubic smoothing splines by specifying the maximum allowable residual sum of squares, the first term on the right of (6.3.2), instead of λ. Woltring (1986) describes a FORTRAN package (GCVSPL) that allows its user to specify λ directly, to specify a number of "degrees of freedom" (corresponding roughly to the number of estimated parameters in a linear regression), or to have the smoothing parameter chosen automatically by the method of generalized cross validation, which is discussed at somewhat greater length in Section 6.5.

For a given choice of λ, the conditions on the cubic smoothing spline can be expressed in terms of a tridiagonal linear system, just as was the case for cubic interpolating splines of Section 5.5.4.2. Details, together with FORTRAN code, can be found in de Boor (1978). The method is numerically stable, and because of the tridiagonality of the linear systems, quite fast. Silverman (1985) reports a modification to de Boor's algorithm that is well-suited to constructing fits for several values of λ. The first such fit requires $35n$ flops, and subsequent calls require only $25n$ flops.

De Hoog and Hutchinson (1987) describe an efficient and highly stable algorithm for computing polynomial smoothing splines. Of particular note is the algorithm's ability to compute the trace of the smoothing matrix S in $O(n)$ flops; this quantity is the hard part to get in computing the generalized cross-validation score. This algorithm is incorporated in GCVSPL.

Silverman (1985) also describes a fast method for Monte Carlo sampling from the posterior distribution of $\hat{M}(x)$ to obtain estimates and precisions for nonlinear functionals of $M(\cdot)$.

The interpretation of optimal smoothing splines as conditional expectations of stochastic processes can also be exploited computationally. Viewed in this way, spline smoothing can be seen to be equivalent to a signal extraction problem, which has long been studied by electrical engineers. The Kalman filter, discussed in Section 4.8.4, is the basic tool. A number of authors have studied the problem from this perspective over the past five years or so. Wecker and Ansley (1983) show how to obtain the joint posterior distribution for $M(x)$ and its derivatives as byproducts of the filtering method. Kohn and Ansley (1987) have developed an efficient algorithm for spline smoothing based on modified Kalman filtering techniques that is somewhat more efficient than the Wecker and Ansley algorithm. Both methods allow for arbitrary mth-degree linear differential operators in the penalty function and quite general priors on the derivatives of M.

Although the Kohn and Ansley algorithm is efficient and numerically stable, in the case of the cubic (natural) smoothing spline, there is essentially no computational advantage over the de Hoog and Hutchinson algorithm.

6.3.2 Regression splines

Two objections to using smoothing splines are that they require special software, and that they must estimate more parameters—albeit subject to a smoothness constraint—than there are observations. If the goal is to capture some obvious (or suspected) features of curvature in a data set, these objections can be overcome using so-called *regression splines*.

The regression splines, sometimes called least-squares splines, are based upon the truncated power-series representation of equation (5.5.1). In the cubic case, (5.5.1) asserts that $M(x)$ can be represented as a linear combination of $n + 2$ basis functions, namely the monomials 1, x, x^2, and x^3, and then $n - 2$ truncated power functions of the form $(x - x_i)_+^3$. The cubic smoothing spline estimates all $n + 2$ coefficients, using the smoothness constraint to avoid singularity. The regression-spline approach has the data analyst set most of these coefficients to zero *a priori* and then estimate the remaining coefficients by least squares. To do this, the statistician selects a small number $m << n$ of design points for the knots, selected so that areas of apparent positive curvature are separated from areas of negative curvature by one or more knots. The handful of truncated power functions can then be constructed and used as regressor variables in standard regression programs.

COMMENT. When using regression splines, there is no explicit parameterization of the tradeoff between smoothness and goodness of fit. Rather, the data analyst can only compare goodness of fit between alternative smooth estimates by reference to the residual sum of squares or, perhaps better, the residual mean square. At the same time, the degree of smoothness in the resulting fits can be gauged only by eye. Of course it is possible to evaluate $\int [\hat{M}'']^2$ given the truncated power-series representation, but once again that would have to be done by hand when the main fitting is accomplished using a standard statistical computer package.

As we mentioned in the context of interpolating splines, the truncated power-function representation can produce numerically ill-conditioned systems. The regression-spline method of smoothing can be implemented using B-splines, with a great improvement in numerical behavior. Unfortunately, this approach cannot be combined easily with standard regression packages.

There is nothing special about cubic regression splines, of course. One can use the same method for splines of any order. In particular, this method is useful for introducing a change-point (elbow) into a linear regression problem.

6.3.3 Multivariate spline smoothing

There are several nonequivalent generalizations of the univariate spline functions, only one of which appears to have the desirable properties of being the solution to a natural variational problem and possessing a Bayesian interpretation. The *thin-plate splines* are the solution to the same penalized least-squares problem (6.3.2) as the univariate smoothing splines that we have considered, except that the roughness penalty $\int f^2(t)\,dt$ is replaced by $\int \|\nabla^2 f(t)\|^2\,dt$, where $\|A\|^2 = \sum a_{ij}^2$.

The special structure of univariate smoothing splines that makes it possible to compute them efficiently does not carry over to the multivariate case. Software for computing thin-plate splines with automatic choice of smoothing parameters using generalized cross validation (GCVPACK) is described in Bates, Lindstrom, Wahba, and Yandell (1987); the FORTRAN subroutines are available.

EXERCISES

6.15. [23] Prove that the minimizer of (6.3.2) tends toward the least-squares line as $\lambda \to \infty$.

6.16. [18] Prove that the minimizer of (6.3.2), in the limit as $\lambda \to 0$, is the cubic natural interpolating spline with knots at the design points $\{x_i\}$.

6.17. [11] Describe how to fit a continuous, piecewise-linear regression curve that has two linear pieces joined at $x = a$.

6.4 Nonlinear smoothers

The smoothing algorithms that we have discussed up to this point are all linear smoothers, in the sense that the fitted values they produce have the form $\hat{y} = Sy$ for some matrix S that depends only on the design points and not the responses y. We have seen that the choice of smoothing parameter, and hence the particular matrix S used to smooth, is a free parameter at the data analyst's command. It is often useful to have the data help in choosing the degree of smoothing; doing so makes the resulting smoother nonlinear. Just as the nonlinearity of stepwise regression methods affect regression computations very little, this kind of nonlinearity has little computational impact on constructing smoothers. In the remainder of this section we shall briefly describe some useful smoothing methods that are intrinsically nonlinear.

6.4.1 LOWESS

We have already mentioned Cleveland's method of locally weighted linear fits called LOWESS (1979, 1981). However, we neglected to mention that Cleveland also proposed downweighting apparent outliers in the local weighted regressions. This has the effect of making the smoother much less sensitive to outliers, although the nonlinearity so introduced makes it harder to examine the properties of the smoother theoretically. The practical gains, however, seem to far outweigh any loss of theoretical tractability.

6.4.2 Super smoother

Throughout our discussions above, we have tacitly assumed that the bandwidth or span of the smoother is constant for all values of x. There is no reason why that should be the case. Indeed, where $M(x)$ is flat a large span is appropriate, since variance can be reduced with little increase in bias by enlarging the span. Similarly, in areas where $M(x)$ has high curvature a small span is desirable, since a large span would obscure features of the regression function and pay a heavy price in terms of squared bias.

Once we have an initial smooth estimate, we can use it to help us decide where the span might be increased and where decreased. This is the idea that underlies an algorithm developed by Friedman (1984). This smoother, dubbed *Super Smoother*, is essentially a running-lines smoothing algorithm with variable spans. The optimal span locally is estimated by (local) cross validation. A FORTRAN subroutine implementing the method is available from Friedman.

6.4.3 Running medians

Most of our discussion has centered on estimating a conditional mean function $M(x)$. However, we have some concern for robustness (see Section 6.6) of the resulting estimate, and we have indicated in our introduction to the topic a willingness to consider other measures of the middle of a conditional distribution than the mean.

An approach that is almost better suited to hand computation than to machine computation is the method of *running medians*. As the name suggests, instead of averaging points in, say, a symmetric nearest neighborhood, we take the median of the responses there. At each step of the smoothing process, one datum is added and one deleted from the neighborhood.

Unlike running means or running lines, there are no closed-form updating formulæ for medians. There are some computational tricks that can be employed, however. If both the entering point and the departing point are on the same side of the current median, then the median remains unchanged, so that roughly half the time no further update is required. For small bandwidths k, say $k \leq 15$, one cannot do much better than simply to sort the responses and then to extract the middle observation each time an actual update is required. More sophisticated methods (using AVL trees, for instance) have better asymptotic performance, but have fixed overhead that makes them uncompetitive for the fairly small values of k useful in practice.

6.4.4 Other methods

Tukey (1977) begins his chapter on smoothing with running medians of 3 ($k = 1$). This operation can be repeated until the results converge. Tukey proposes a method he calls splitting to file down some of the jaggedness in the result. These methods, combined with short weighted-mean smoothers—"Hanning," for instance, is a linear smoother with local row weights $(\frac{1}{4}, \frac{1}{2}, \frac{1}{4})$—produce remarkably attractive smooth estimates. They can be easily computed by hand.

6.5 Choosing the smoothing parameter

In all of the methods for smooth-function estimation that we have discussed up to this point, there has been a user-selected parameter λ that controls the degree of tradeoff between low variability (smoothness) and low bias (closeness to the data). We call λ the *smoothing parameter*. Many statisticians have an abiding distrust of estimation methods that require subjective input from the data analyst. For other statisticians, all statistical practice requires choice and the exercise of judgment. It is not surprising, then, that much work has focused on developing guidelines for making the choice of λ less subjective, and on constructing data-based methods for estimating λ. Those methods which use the data to help determine the degree of smoothness are said to be *adaptive*, while data-free choices are said to be *nonadaptive*.

In density estimation, the degree of smoothness is determined by the window width h, or the number of near neighbors k. There has been considerable theoretical work on the problem of choosing these parameters well, both adaptively and nonadaptively.

Scott (1979) suggests that for data whose density is nearly Gaussian, one should use a bin size of approximately $h = (7/2)\hat{\sigma}\sqrt[3]{n}$. Terrell and Scott (1985) show that, under mild smoothness conditions, the optimal histogram for samples of size n requires something more than $m = \sqrt[3]{2n}$ bins. This suggests setting h to be equal to the range of the data divided by m or $m + 1$. This suggestion works well for a variety of densities and estimation methods.

Chow, Geman, and Wu (1983) examine the consistency properties of cross-validatory choice of h for kernel density estimation. The interested reader should also see Silverman (1978, 1986) and Scott (1986) for more on selecting a window width.

COMMENT. While computational issues naturally arise in choosing the smoothing parameter adaptively, few such issues concern the less critical matter of selecting a kernel. A word or two might still be said. As we noted in Section 6.1 for nearest-neighbor estimates, the choice of kernel can have an effect on computational feasibility since updating formulæ that simplify computation can be used for very simple kernels. Singpurwalla and Wong (1983) note that kernels that are negative in spots can produce estimates with desirable asymptotic properties; such kernels can, however, produce negative estimates for the density in spots, too, in finite samples. The interested reader should consult Silverman (1986) for analysis and advice.

Adaptive methods for deciding on the degree of smoothing are often based on the notion that, with respect to some criterion, there is an optimal degree of smoothing. The notions of "optimality" on which particular algorithms are based depend heavily on the criteria selected, which amount to implicit or explicit quantifications of the variance/bias tradeoff. Perhaps the most appealing criteria for regression smoothing are based on *predictive mean squared error* (PMSE). At a single point x, the PMSE is simply $\text{PMSE}(x, \lambda) = E[s_\lambda(x) - E(Y \mid x)]^2$. Here we have explicitly represented the dependence of the smoother s on the smoothing parameter λ. To obtain an overall figure of merit, the $\text{PMSE}(x, \lambda)$ is usually averaged over some distribution on the x's. A common choice is to assume that future values of X will have the same distribution as those observed in the sample, so that the figure of merit is taken either to be the average of $\text{PMSE}(x, \lambda)$ over the observed x points, or over a probability distribution with parameters estimated from the observed points.

Cross-validation (CV) estimates are based on ideas of *predictive sample reuse* pioneered by Stone (1974) and Geisser (1975), and developed extensively in the smoothing context by Grace Wahba and her colleagues. The idea is to hold out one observation at a time, say the kth observation, and then to construct a smoother based upon the remaining $n - 1$ observations.

Call this smoother $s_{-k,\lambda}(x)$. The cross-validation function

$$CV(\lambda) = \frac{1}{n} \sum_{i=1}^{n} [s_{-i,\lambda}(x_i) - y_i]^2. \qquad (6.5.1)$$

Ordinary cross validation leaving out one point at a time is designed to obtain an unbiased estimate of the PMSE. For a given value of λ, the cross-validation function simply measures the average squared error when each point is predicted using only the remaining points in the sample. Simple cross-validation chooses the value of λ which minimizes $CV(\lambda)$.

The method of *generalized cross validation* (GCV) minimizes instead a weighted sum of the terms in (6.5.1); these weights are chosen to eliminate the artificial effects of unequally-spaced design points and choice of coordinate systems. Generalized cross validation finds λ so as to minimize an estimate of

$$GCV(\lambda) = \sum_{i=1}^{n} \text{PMSE}(x_i, \lambda)/n. \qquad (6.5.2)$$

This value of λ approximates the optimal choice, and is asymptotically correct as well, in the sense that the GCV estimate for λ achieves the minimum average PMSE achievable by any estimator in the class of s_λ's. In problems with linear structure, it turns out that GCV simplifies to produce a function that is computationally much simpler than CV, and as a consequence is more useful and more widely used in practice. More details can be found in Wegman and Wright (1983).

COMMENT. Ridge regression estimators, like splines, also compromise between goodness of fit and some measure of "roughness". In the case of ridge regression, what is penalized is the sum of squared regression coefficients, $|\hat{\beta}|^2$. This corresponds in the spline case to defining roughness in terms of the 0th derivative of the underlying function rather than in terms of the second derivative (curvature). For an example of the use of generalized cross validation to choose a tradeoff parameter in the context of ridge regression, see Golub, Heath and Wahba (1979).

Cross-validatory methods for choosing the degree of smoothing can also be applied in the density-estimation context. The most common figure of merit for density estimation is the integrated mean squared error

$$IMSE(\lambda) = E \int_{-\infty}^{\infty} [\hat{f}_\lambda(x) - f(x)]^2 \, dx. \qquad (6.5.3)$$

Scott and Terrell (1986) suggest cross-validation methods for kernel and histogram estimators that correspond to GCV procedures for orthogonal-series estimators and regression estimators. They call their proposal "biased cross validation" because their criterion function is a slightly biased estimate of IMSE. Scott (1986) reviews several different approaches to this problem, including cross validation, for histogram, orthogonal-series, and kernel density estimators. His conclusion is that, "for large enough n, empirical evidence suggests the biased cross-validation algorithm works almost without fail." Marron (1987) also compares cross-validation techniques in density estimation.

6.6 Applications and extensions

In this section we briefly outline a few of the many ways smoothing methods have found applications in statistics. In addition, a number of the basic smoothing algorithms have been modified or extended to provide a rich collection of tools for data analysis.

We have already discussed at some length additive nonparametric and semiparametric regression models in Section 4.6. In view of the interest in these models, methods of nonparametric function estimation gain renewed importance. The topics of projection pursuit, generalized additive models, and partial spline models are all important, and all rely on one or more of the methods discussed in this chapter.

6.6.1 Robust smoothing

An approach to making least-squares fits more resistant to occasional outliers is to use weighted least squares in which the weights assigned to discrepant points are substantially lower than the weights given to most of the points. Robust regression methods were discussed in Section 3.12.2. Linear smoothers in general, and spline smoothers in particular, can be easily modified to allow differential weights that will downweight possible outliers. This is particularly easy to do for regression splines, since standard robust-regression packages can be employed.

Lenth (1977) and Utreras (1981) have developed some methods for making smoothing splines more robust in this sense.

Wong (1987) notes that there are really two distinct aspects of robustness in nonparametric functions estimation. One can make the estimators less sensitive to outliers (hence relaxing the definition of "fidelity to the data"). This corresponds to making the least-squares term more robust in the penalized least-squares formulation of (6.3.2), for instance. But one can also imagine making the roughness penalty in (6.3.2) more robust as well. This could mean, for example, allowing possibly very large values of $[M''(t)]^2$ over very short intervals.

6.6.2 Smoothing on circles and spheres

For directional data Wahba (1981) has developed methods for spline smoothing on spheres and circles. These methods have been used in such fields as meteorology, hydrology, and geophysics. Dierckx (1984) gives algorithms for smoothing on the sphere, using tensor-product splines with adaptively chosen knots.

6.6.3 Smoothing periodic time series

McDonald (1986) has developed a clever method for estimating a smooth function M in the model

$$y_i = M(\omega t_i) + \epsilon_i, \qquad (6.6.1)$$

where M is a periodic function with period 1, ω is a fixed frequency possibly estimated from the data, and t_i is the time at which y_i is observed. The basic idea is to smooth the y_i's as a function of (ωt_i) mod 1, the fractional part of ωt_i. In effect, the scatterplot of y against t is folded over and the panels of unit length are superimposed. McDonald has developed software for drawing such scatterplots on a computer display as the frequency parameter ω is varied by using a mouse or trackball. As the displayed scatterplot's value of ω approaches a frequency at which the underlying mean function is periodic, the plot can change strikingly from an apparently random sprinkling of points into a clearly discernible functional pattern. Situations that ordinarily cause great difficulty in time-series estimation such as missing data, unequally-spaced points, or possible outliers can easily be handled using standard smoothing algorithms. Using projection-pursuit ideas, McDonald shows how to estimate a pseudo-periodogram of the form

$$y_i \approx \sum M_k(\omega_k t_i). \qquad (6.6.2)$$

6.6.4 Estimating functions with discontinuities

Occasionally it is useful to fit mean functions that for the most part are continuous and smooth, but with a handful of discontinuities in either the regression function or its derivatives. McDonald and Owen (1986) have developed a family of smoothing algorithms that allow for discontinuous output. As a result, this approach can be used for edge detection; McDonald and Owen illustrate their method on a measurements of ocean-surface temperature, where the discontinuities are the result of currents.

Wong (1987) notes that another way in which the same result can be achieved is by making the penalty term in (6.3.2) more robust, as discussed in Section 6.6.1.

6.6.5 Hazard estimation

The it hazard function is a quantity of primary interest in survival analysis. In effect, the hazard function is the instantaneous failure rate associated with a survival distribution. As such, it is closely related to the density function. If X has density $f(x)$ and cdf $F(x)$, then the we write the hazard function as $h(x) = f(x)/[1 - F(x)]$. A nonparametric estimate of the hazard function can be obtained by methods similar to those used for density estimation. Tanner (1984) gives an algorithm for hazard estimation that uses cross validation to select the bandwidth for a kernel estimator.

ANSWERS TO SELECTED EXERCISES

2.1. The bits that are given specify the first two (of six) hexadecimal digits, namely, 35_{16}. Since the first digit is nonzero, the number represented is normalized. Thus, on an IBM machine, as many as three leading *bits* can be zero in a properly normalized floating-point number.

2.2. VAXen.

2.3. The floating-point sum 105+(3.14+2.47) is (111,3) using the model of arithmetic used in Exercise 2.5.

2.4. What must be shown is that the relative error $(f(x^*) - f(x))/f(x)$ is small; let's examine the denominator first. From the analysis of basic floating-point computations, we know that $f(x) = 1 \oslash x = (1/x)(1 + \delta)$, where $|\delta| \leq \beta^{-t}$. Similarly $f(x^*) = (1/x^*)(1 + \eta)$, where $|\eta| \leq \beta^{-t}$. Finally, from the well-conditioning of real reciprocals, we know that $1/x^* = 1/[x(1+\epsilon)] \approx (1/x)(1 - \epsilon)$, where $|\epsilon| \leq \beta^{-t}$. Pasting things together shows that the relative error is smaller than $|\eta - \epsilon - \delta|$.

2.5. We must show that $f^*(x) = (1.01 \oplus x) \oslash (1.01 \ominus x) \approx f(\tilde{x})$, where $\tilde{x}$ is close to x and where the circled operators denote the obvious floating-point operations. For notational convenience, write $a = 1.01$. Using the error bounds at the end of section 2.1, we can write $f^*(x) = f(x)[(1 + \epsilon)/(1 + \eta)](1 + \delta) \approx f(x)(1 + \gamma)$, where $|\gamma| \leq 3\beta^{-t}$. This, in turn, equals $(a + \tilde{x})/(a - \tilde{x})$, where

$$\tilde{x} = a \left[\frac{(a + x)\gamma + 2x}{(a + x)\gamma + 2a} \right].$$

Note that $\tilde{x}$ need not be a representable floating-point number. From this expression, we obtain the relative error in $\tilde{x}$ to be

$$\frac{1}{x} \left[\frac{(x^2 - a^2)\gamma}{(a + x)\gamma + 2a} \right].$$

As long as x is not too close to zero, the magnitude of this quantity will be bounded by a small constant times β^{-t}.

2.6. Expand $f(\tilde{x})$ in a Taylor series about x to get $f(\tilde{x}) = f(x) + \epsilon x f'(x) + \epsilon^2 x^2 f''(x^*)/2$, where x^* is a point on the line segment connecting x to $\tilde{x}$.

2.7. Hint: the "extra" term in equation (2.3.1) results from dividing the sum by n.

2.8. Expression (2.3.2) assumes that $x_{\max}$ is the term with largest magnitude (not largest signed magnitude). The magnitude of the arithmetic mean cannot exceed this value. (If you are having trouble getting the constant to come out right, note the range of summation.)

2.9. Here is a fairly crude approach: Suppose that n numbers $x_1, \ldots, x_n$ are to be added. Initialize an array $B[j]$ of $k = \lceil \log_2 n \rceil$ bits and an array $S[j]$ of k real numbers to zero. Before the ith data point is read in, the bits in $B[]$ are to contain the binary representation of $i - 1$. When the ith data point is read in, it is immediately added to the current contents of $S[0]$. Then a single bit is added to the binary representation of the index stored in $B[]$, emulating binary addition. Thus, if $B[0] = 0$, when a bit is added, the result is $B[0] = 1$. If $B[0] = 1$, however, a carry takes place, and a bit is added to $B[1]$ and $B[0]$ is reset to zero. The carry into $B[1]$ may itself produce a carry into $B[2]$, and so forth. Each time a carry takes place, say from $B[j]$ into $B[j + 1]$, *also* add $S[j]$ to the current contents of $S[j + 1]$ and reset $S[j]$ to zero.

2.10. Algebraically, the result of the second summation is exactly zero. Numerically it may not be, due to loss of less significant digits in the accumulation of the first sum.

2.11. The last step adds (990,1) to (893,0), that is 9.90 and 0.893, to obtain 10.8 (108,2). [This is a tricky calculation!] The standard calculation would give 9.90, which would be in error by slightly more than 9%.

2.12. The last step adds (109,2) to (-990,-1), that is 10.9 and -0.099, to obtain 10.8 (108,2). The standard algorithm would have produced 10.9, which is correct to three places. The "correction" applied by algorithm 2.3.1 actually makes the answer worse, although the result is still in error by less than 1%. This illustrates that the usual algorithm does quite well when the data are sorted into increasing order, but may do very badly when the ordering is reversed. Algorithm 2.3.1, on the other hand, produces a consistent result which is never very much in error.

2.15. Note that pairwise summation can be applied *twice* in this algorithm, once in the computation of the mean ($xbar$) and again in the summation of the squared deviations from the mean.

2.20. Hint: Note that $\min X_i \le G \le \max X_i$, so that $(\overline{X} - G)^2 \le \sum (\overline{X} - X_i)^2$.

2.24. Symbolic algebra packages such as MACSYMA or REDUCE may be better suited to this problem than Fortran or Pascal.

2.28. Let X, Y, and Z be centered respectively about G, H, and J. Let S_x denote $\sum (X_i - G)$, S_{xy} denote $\sum (X_i - G)(Y_i - H)$, S_{xyz} denote $\sum (X_i - G)(Y_i - H)(Z_i - J)$, and similarly for other subscripts. Then the required sum can be written as

$$S_{xyz} + (G - \overline{X})S_{yz} + (H - \overline{Y})S_{xz} + (J - \overline{Z})S_{xy}$$
$$- 2n(G - \overline{X})(H - \overline{Y})(J - \overline{Z}).$$

This means that it is necessary to retain sums of lower-order cross products in order to reconstruct the desired quantity from the arbitrarily centered version.

2.29. Let $\tilde{X}_i = X_i - G$ and $\tilde{Y}_i = Y_i - H$. Then

$$\sum w_i(X_i - \overline{X}_w)(Y_i - \overline{Y}_w)$$

$$= \sum w_i(\tilde{X}_i + G - \overline{X}_w)(\tilde{Y}_i + H - \overline{Y}_w)$$

$$= \sum w_i \tilde{X}_i \tilde{Y}_i + (G - \overline{X}_w)\sum w_i \tilde{Y}_i + (H - \overline{Y}_w)\sum w_i \tilde{X}_i$$

$$+ (\sum w_i)(G - \overline{X}_w)(H - \overline{Y}_w).$$

Noting, for instance, that $\sum w_i \tilde{Y}_i = (\sum w_i)(\overline{Y}_w - H)$, the result follows.

2.30. $(x - \overline{x}e)'Q(y - \overline{y}e) = (x - Ge)'Q(y - He) - (\overline{x} - G)(\overline{y} - H)e'Qe$, provided that $\overline{X}$ is defined by $\overline{X} = x'Qe/e'Qe$.

3.1. $\hat{\beta} \sim \mathcal{N}(0, \sigma^2(X'X)^{-1})$.

3.2. Squaring both sides of (3.1.4) and writing in terms of an inner product, we have $x'(Q'Q)x = x'x$. Equivalently, $x'(Q'Q - I)x \equiv x'Ax = 0$ for all x. Note that A is symmetric. By setting x to e_j, the vector with 1 in the jth position and zero elsewhere, we see that $a_{jj} = 0$ for all j. By then setting x to e_{ij}, the vector with 1 in the ith and jth positions and zero elsewhere, and using symmetry, we then see that $a_{ij} = 0$ as well. The second part of the proof is to show that $Q'Q = I$ implies that $QQ' = I$ as well. Since (3.1.4) holds for all $x \in R^n$, the range space of Q must be R^n. Let y be an arbitrary point in R^n and let x be the pre-image of y under Q, that is, $y = Qx$. Then $Q'y = Q'Qx = x$, so that $QQ'y = Qx = y$, for all $y \in R^n$. Thus, $QQ' = I$, the identity transformation. (We have just established a special case of the uniqueness of the inverse matrix.)

3.4. From (3.1.10), $RSS = |Y_1^* - X_1^*(X_1^*)^{-1}Y_1^*|^2 + |Y_2^*|^2$.

3.5. By construction, $X_2^* = 0$.

3.7. From (3.1.14), it follows that $Hw = -w$ if w is proportional to u, and $Hw = w$ if w is orthogonal to u. Hence H *reflects* vectors in the plane S orthogonal to u.

3.12. Apply H_j again, since it is its own inverse.

3.14. From (3.1.8) we have that $y_p^p = x_p^p \beta_p + \epsilon_p$, so that $\hat{\beta}_p = y_p^p/x_p^p$, $\text{Var}(\hat{\beta}_p) = \sigma^2/|x_p^p|^2$, and $RSS = \sum_{j=p+1}^{n}(y_p^j)^2$ is independent of $\hat{\beta}_p$.

3.15. The matrix G has the form $\begin{pmatrix} c & s \\ -s & c \end{pmatrix}$, where $c = \cos(\theta)$ and $s = \sin(\theta)$. Thus, $c^2 + s^2 = 1$, and we must have $-a_{11}s + a_{21}c = 0$. Solving for s and c gives $c = a_{11}/\sqrt{a_{11}^2 + a_{21}^2}$ and $s = a_{21}/\sqrt{a_{11}^2 + a_{21}^2}$.

3.20. Symmetry follows immediately from (3.2.10). For the second part, note that

$$P^2 = Q^{(1)'}(Q^{(1)}Q^{(1)'})Q^{(1)}$$
$$= Q^{(1)'}Q(1)$$
$$= P,$$

since $Q^{(1)}Q^{(1)'} = I_p$.

3.21. Since $\hat{Y} = PY$, $\text{Var}(\hat{Y}) = \sigma^2 PP' = \sigma^2 P$, so that $\text{Var}(\hat{y}_i) = \sigma^2 p_{ii}$. Similarly, $\text{Var}(R) = \text{Var}(Y - \hat{Y}) = \text{Var}((I-P)Y) = \sigma^2(I-P)(I-P)' = \sigma^2(I-P)$, so that $\text{Var}(r_i) = \sigma^2(1 - p_{ii})$.

3.23. Let $\tilde{X}$ represent the X matrix augmented by Y; that is, $\tilde{X} = [X\,Y]$. For any v, $v'Sv = v'\tilde{X}'\tilde{X}v = (\tilde{X}v)'(\tilde{X}v) = |\tilde{X}v|^2 \geq 0$.

3.24. Apply the answer to the previous exercise replacing $\tilde{X}$ by the first j columns of the $\tilde{X}$ matrix, where $\tilde{X}$ is defined in the previous answer.

3.25. $(AX_1^*)'(AX_1^*) = (X_1^*)'A'AX_1^* = (X_1^*)'X_1^* = S$, since $A'A = I$ by orthogonality of A.

3.26. If $\text{Cov}(Y)=\Sigma$, then $v'\Sigma v = \text{Var}(v'Y) \geq 0$, the latter since variances are non-negative. If $\Sigma \geq 0$, then there exists a matrix square root $\Sigma^{1/2}$ of Σ. Let Z be a random vector with independent standard Gaussian components. Then $Y = (\Sigma^{1/2})'Z$ has the desired covariance.

3.28. Show by direct computation that $RSWEEP[k]\,SWEEP[k]S = S$.

3.29. Hint: use induction.

3.30. Let I denote a set of indices, and let j be an index not contained in I. Then we must show that $SWEEP[I\,j]\,(\equiv SWEEP[j]\,SWEEP[I]) = SWEEP[j\,I]\,(\equiv SWEEP[I]\,SWEEP[j])$. This, together with an induction argument on the size of I establishes the result.

3.33. All of the diagonals start out greater than zero, since they are sums of squares. Sweeping on other columns replaces a particular diagonal by a residual sum of squares, which must remain positive. Sweeping on a particular diagonal element changes its sign, and subsequent adjustments involve subtracting positive quantities from something that is already negative.

3.34. All are monotone functions of $s_{iy}/\sqrt{s_{ii}}$.

3.36. After sweeping columns 0 through p of S, the upper left-hand block of the S array contains S_{XX}^{-1}. Similarly, S_{xx}^{-1} is the result of sweeping on the first p columns of CSSCP, which is itself obtained after first having swept on column zero of S.

3.37. Part a) follows immediately from equation (3.4.6). To obtain parts b) and c), write S_{22} of (3.4.6) as $S_{22} = \begin{pmatrix} Y'Y & Y' \\ Y & I \end{pmatrix}$; S_{11} as $X'X$, and S_{12} similarly. Upon writing $H = X(X'X)^{-1}X'$ (the so-called *hat matrix*), the lower right submatrix of (3.4.6) becomes $\begin{pmatrix} Y'(I-H)Y & Y'(I-H) \\ (I-H)Y & (I-H) \end{pmatrix}$. The residual sum of squares is just $Y'(I-H)Y$, which establishes b). Since $HY = \hat{Y}$, we must have the vector of residuals $Y - \hat{Y} = (I-H)Y$, which gives part c).

3.38. First, verify that adding a dummy variable with one in the ith position and zero elsewhere makes the ith fitted value exactly equal the observed value (so that its residual is zero). But sweeping on row $p+1+i$ of the augmented matrix is equivalent to adding the $p+1+i$th column of the augmented matrix to the regression equation, and this column is just such a dummy variable! Note that the element in the $p+1+i$th position of column $p+1$ is now the regression coefficient of the dummy variable, which is to say, the residual of the ith observation from the regression based on the remaining $n-1$ observations.

3.40. See equation (3.4.11).

3.41. $U\mathcal{D}V' = U(U'XV)V' = (UU')X(VV') = X$, since U and V are orthogonal.

3.42. $Xv_i = U\mathcal{D}V'v_i = U\mathcal{D}e_i = U(d_iE_i) = d_iu_i$, where e_i is the p-vector with one in the ith component and zero elsewhere, and E_i is the n-vector with one in the ith component and zero elsewhere.

3.43. Let $p = 1$, and let $X = (-1)$. Then $\mathcal{D} = (+1)$, (since $d_1 \geq 0$), so that $U = -V$.

3.45. This exercise actually corrects the statement of Theorem 8.3-1 in Golub and Van Loan (1983), the proof of the erroneous part of which is left as an exercise for the reader!

3.46. If X is an $n \times p$ matrix with orthonormal columns, then let U be an orthogonal matrix whose first p columns equal those of X. Then $X = U\mathcal{D}V'$ with $V = I_p$ and $\mathcal{D} = \begin{pmatrix} I \\ 0 \end{pmatrix}$.

3.47. The rank of a matrix is equal to the number of linearly independent columns (or rows) of the matrix; these in turn can be considered to be a collection of points (vectors) in $\mathcal{R}^n$ (or $\mathcal{R}^p$). Linear independence is unaffected by a change of basis. $U'X$ is a re-expression of the columns of X in terms of a new basis (the columns of U). Similarly, the rows can be re-expressed in terms of a new p-dimensional basis by post-multiplying by V. Thus $U'XV = \mathcal{D}$ has the same rank as X, and the latter clearly has as many linearly independent columns as non-zero singular values.

3.49. Any least-squares solution satisfies $\mathcal{D}\theta = Y_1^*$, from the SVD. Since $|\beta|^2 = |V\theta|^2 = |\theta^2| = (y_1^1/d_1)^2 + \ldots + (y_k^1/d_k)^2 + |\theta_*|^2$, where $\theta_* = (\theta_{k+1}, \ldots, \theta_p)$, the minimum norm solution has $|\theta_*|^2 = 0$, which is the Moore-Penrose solution.

3.52. By the SVD (or Householder computations), X can be decomposed into $X = QR$, where R has the form required by the right-hand side of (3.5.12).

3.54. $(X'X) = VD'DV' = VD^2V'$, so that $(X'X)^+ = V(D^+)^2V'$. Thus, we have that $(X'X)^+X' = V(D^+)^2V'V(D \quad 0)U' = V(D^+ \quad 0)U' = X^+$, from which the result follows.

3.55. $V(\hat{\beta}) = V(X^+Y) = (X^+)(X^+)' = VD^+U'UD^+V' = V(D^+)^2V' = (X'X)^+$.

3.60.

$$(n - p - 1)s^2(i) = \sum_{j \neq i}(y_j - x_j\hat{\beta}(i))^2$$

$$= \sum_{j=1}^{n}\left(e_j + \frac{p_{ij}e_i}{1 - p_{ii}}\right)^2 - \frac{e_i^2}{(1 - p_{ii})^2}$$

$$= (n - p)s^2 + \frac{2e_i}{1 - p_{ii}}\sum_{1}^{n}e_jp_{ij} + \frac{e_i^2}{(1 - p_{ii})^2}\sum_{1}^{n}p_{ij}^2 - \frac{e_i^2}{(1 - p_{ii})^2}$$

$$= (n - p)s^2 - \frac{e_i^2}{1 - p_{ii}},$$

using the facts that $Pe = 0$ and that $\sum_j p_{ij}^2 = p_{ii}$. This proof is due to Peters (1980).

3.58. The numerator and the denominator are not independent, although if $n - p$ is large, the dependence will be quite small.

3.61. $DFFIT_i = \hat{y}_i - \hat{y}_i(i) = x_i(\hat{\beta} - \hat{\beta}(i)) = x_i(X'X)^{-1}x_i'e_i/(1 - p_{ii}) = p_{ii}e_i/(1 - p_{ii})$.

3.62. Let $Z = XA$. Note that the hat matrix is unchanged, since $P = X(X'X)^{-1}X' = XAA^{-1}(X'X)^{-1}(A')^{-1}A'X' = Z(Z'Z)^{-1}Z'$. Thus, fitted values $\hat{Y} = PY$ and residuals are unchanged, so that the components of (3.6.5) are all unaltered by change of basis.

3.63. $DFFIT_{ij} = x_j(\hat{\beta} - \hat{\beta}(i)) = p_{ij}e_i/(i - p_{ii})$. Note that computing this quantity requires that the entire hat matrix P be available.

3.64. $e_i(i) = y_i - x_i\hat{\beta}(i) = y_i - x_i(\hat{\beta} - e_i(X'X)^{-1}x_i'/(1 - p_{ii})) = e_i/(1 - p_{ii})$. (Cook and Weisberg, 1982).

3.65. "Efficient" here means at least a constant factor better than that obtainable by examining each subset directly. Branch-and-bound methods such as those used in all-subsets regression may be adaptable to this problem.

3.66. Alternatively, one could disprove the conjecture, preferably by constructing an appropriate computer program.

3.67. Premultiply the right-hand side of $(3.7.2)$ by $(A + uv')$ to obtain

$$
AA^{-1} - AA^{-1}u(I + v'A^{-1}u)^{-1}v'A^{-1} + uv'A^{-1}
$$
$$
- uv'A^{-1}u(I + v'A^{-1}u)^{-1}v'A^{-1}
$$
$$
= I - u[(I + v'A^{-1}u)^{-1} - I + v'A^{-1}u(I + v'A^{-1}u)^{-1}]v'A^{-1}
$$
$$
= I - u[I - (I + v'A^{-1}u) + v'A^{-1}u](I + v'A^{-1}u)^{-1}v'A^{-1}
$$
$$
= I.
$$

3.69. The matrix $(A - \lambda I)$ has zero determinant if, and only if, some nontrivial linear combination of its columns is zero. If $|A - \lambda I| = 0$, then let v contain the coefficients in such a linear combination, that is $(A - \lambda I)v = 0$, so that $Av = \lambda v$. By the same token, if v is an eigenvector of A corresponding to λ, then $(A - \lambda I)v = 0$, and $(A - \lambda I)$ must have zero determinant.

3.70. . Since A is positive semidefinite, we can write $A = B'B$, so that $v'Av = v'B'Bv = |Bv|^2$. Now $|Bv|^2 = 0$ implies that $Bv = 0$, which in turn implies that $B'Bv = Av = 0$.

3.71. Let

$$
A = \begin{pmatrix} 0 & 1 \\ 1 & 0 \end{pmatrix} \quad \text{and} \quad v = \begin{pmatrix} 1 \\ 0 \end{pmatrix}.
$$

3.72. Let $\lambda = \max_v v'Av/v'v$, and denote by v a maximizing vector of unit length. Clearly $\lambda \geq \lambda_1$. Decompose v as $\alpha v_i + \beta \overline{v_i}$, where $\overline{v_i}$ is the orthogonal complement of v_i, $|v_i|^2$ is taken to be 1, and v_i is the first eigenvector for which α is nonzero. (Since the collection of eigenvectors, including those corresponding to zero eigenvectors, spans $\mathcal{R}^p$, there must be such an i.) Clearly $\alpha^2 + \beta^2 = 1$. Now

$$
\lambda = v'Av = (\alpha v_i + \beta \overline{v_i})'A(\alpha v_i + \beta \overline{v_i})
$$
$$
= \alpha^2 \lambda_i + \beta^2 \overline{v_i}'A\overline{v_i}
$$
$$
\leq \alpha^2 \lambda_i + \beta^2 \lambda |\overline{v_i}|^2
$$
$$
\leq \alpha^2 \lambda_i + \beta^2 \lambda
$$
$$
= \alpha^2 \lambda_i + (1 - \alpha^2)\lambda.
$$

From this it follows that $\lambda \leq \lambda_i \leq \lambda_1$, which implies that $\lambda = \lambda_1$.

3.73. $(A - \lambda B)v = (AB^{-1} - \lambda I)Bv = B(B^{-1}A - \lambda I)v = B^{1/2}(B^{-1/2}AB^{-1/2} - \lambda I)B^{1/2}v$.

3.74. Use induction and the orthogonality of the v_i's.

3.75. Suppose that the eigenvalues corresponding to the columns of Γ are the diagonal elements of Λ. Then $A\Gamma = \Gamma\Lambda$, so that $A = \Gamma\Lambda\Gamma'$, since Γ is an orthogonal matrix. Thus, $A_1 = Q'AQ = Q'\Gamma\Lambda\Gamma'Q$, so that $A_1Q'\Gamma = Q'\Gamma\Lambda$, that is, the columns of $Q'\Gamma$ are eigenvectors of A_1 corresponding to the same set of eigenvalues Λ as A.

3.76. Since Q is a Householder matrix zeroing the last $p - 2$ positions of a vector, it leaves unchanged the first position of each vector to which it is applied. Thus, QA has the same first row as A does, and all positions in the first column are zero below the subdiagonal. Now we apply the same Householder transformation to the rows of QA (by post-multiplying by Q'). By definition of Q, the first row of QA must be zeroed to the right of the super-diagonal. What is more, since premultiplication by Q leaves the first row unchanged, postmultiplication by Q' must leave the first column unchanged; thus the earlier work of zeroing elements in the first column is untouched by the postmultiplication. Thus QAQ' has first row and column which are zero outside the tridiagonal area.

3.77. As in the previous exercise, zero the "off-tridiagonal" in the first row and column with a Householder transformation applied symmetrically. The same procedure can then be applied to the $(p - 1) \times (p - 1)$ lower right sub-block. It is easy to see that the previous step is unaffected by this transformation, at the end of which the first two rows and columns are zero in the off-tridiagonal. The process is continued for the first $p - 2$ rows and columns. It is important to take advantage of symmetry in this calculation. In particular, the QA matrix at each step need not be computed explicitly. Instead, the effect of each complete transformation on, say, the lower triangle of the current matrix can be computed, and only the lower triangle need be stored and updated. When this is done, the transformations can be accomplished quite efficiently; step k requiring only $2(p - k)^2$ flops.

3.80. Symmetry follows from writing $RQ = (Q'A)Q$. Let b_{ij} denote the i,j-th element of RQ. The matrix Q is simply a product of $p-1$ Givens transformations which convert the subdiagonal of A to zeros one at a time. What is more, the ith Givens rotation in this sequence can only affect two rows of A, namely, rows i and $i+1$. Thus, the Givens transformation which eliminates $a_{i+1,i}$ can introduce at most one new non-zero element, at the $(i, i + 2)$ position. (Thus, R must not only be upper triangular, but an upper band matrix with at most two off-diagonal bands, as well.) Now postmultiplying R by Q can again be interpreted as applying a sequence of $p - 1$ Givens rotations, each of which once again can affect just two adjacent columns at a time. By an analysis similar to the preceding one, at most one new non-zero element can be introduced at each step, and each of these is in the sub-diagonal position. From this we conclude that RQ has no non-zero entries below the subdiagonal, and by symmetry, the same must be true above the super-diagonal. Thus, RQ must be (at most) tridiagonal.

3.81. Denote the diagonal elements of A by $a_i \equiv a_{ii}$ and the subdiagonals by $b_i \equiv a_{i,i-1}$, $i = 2, \ldots, p$. The ith of the $p - 1$ Givens transformations required to triangularize $A - \mu I$ produces r_{ii}, which is of the form $\sqrt{x_i^2 + b_{i+1}^2}$, where x_i is the value of the ith diagonal element just before applying the ith Givens transformation, and b_{i+1} is the subdiagonal element immediately below, which has been unaffected by the preceding Givens rotations. Note that the first $p - 1$ diagonal elements of R must be nonzero, since $b_j \neq 0$, all j. However, $A - \mu I$ is rank deficient, so that one of the diagonal elements of R must be zero. Thus, it follows that $r_{pp} = 0$. The last row of RQ must then be zero, and translating back by adding μI, we see that $b_p \equiv a_{p-1,p} = 0$ and that $a_{pp} = \mu$.

3.82. (Golub and Van Loan, 1983). Denote the ith columns of Q and V by q_i and v_i, respectively. Let $W = V'Q$, with columns $w_1, \ldots, w_p$, so that $KW = WH$. From this it follows that, for $i \geq 2$,

$$h_{i,i-1}w_i = Kw_{i-1} - \sum_{j=1}^{i-1} h_{j,i-1}w_j.$$

Since the first columns of Q and V are the same, $w_1 = e_1$, where e_j denotes the vector that has 1 in the jth position, and zeros elsewhere. From this it follows that W must be upper triangular, and thus, $w_i = \pm e_i$ for $i \geq 2$. Now $w_i = V'q_i$, so that $v_i = \pm q_i$, and $h_{i,i-1} = w_i'Kw_{i-1}$, from which follows that $|h_{i,i-1}| = |k_{i,i-1}|$, for $i = 2, \ldots, p$.

3.83. We can write $\|A\| = \text{trace}(A'A) = \text{trace}(QQ'A'A) = \text{trace}(Q'A'AQ)$ $= \text{trace}(Q'A'QQ'AQ) = \|Q'AQ\|$.

3.84. Consider what happens in the ith and jth rows and columns when Q is applied. Let $\begin{pmatrix} c & s \\ -s & c \end{pmatrix}$ denote the i, j submatrix of Q. Then

$$\begin{pmatrix} b_{ii} & b_{ij} \\ b_{ji} & b_{jj} \end{pmatrix} = \begin{pmatrix} c & s \\ -s & c \end{pmatrix}' \begin{pmatrix} a_{ii} & a_{ij} \\ a_{ji} & a_{jj} \end{pmatrix} \begin{pmatrix} c & s \\ -s & c \end{pmatrix}.$$

Thus, $b_{ij} = (c^2 - s^2)a_{ij} + (a_{ii} - a_{jj})cs$. Trigonometry allows us to write $\text{ctn}(2\theta) = (a_{ii} - a_{jj})/2a_{ij}$, provided that the denominator is nonzero. The quantities $c = \cos(\theta)$ and $s = \sin(\theta)$ can be obtained with a little more trigonometric effort, which is left to the reader.

3.87. The solution to the WLS problem is equivalent to the OLS solution of a related problem, namely that of (3.10.4). Since (3.10.4) is just an OLS problem, the estimate for σ^2 is just the residual sum of squares, divided by degrees of freedom, that is,

$$\begin{aligned} (n - p)\hat{\sigma}^2 &= |Y^* - \hat{Y}^*|^2 \\ &= (Y^* - \hat{Y}^*)'(Y^* - \hat{Y}^*) \\ &= (Y^* - X^*\hat{\beta})'(Y^* - X^*\hat{\beta}) \\ &= (Y^* - X^*(X^{*'}X^*)^{-1}X^{*'}Y^*)'(Y^* - X^*(X^{*'}X^*)^{-1}X^{*'}Y^*) \\ &= (Y - X\hat{\beta}_{WLS})'V^{-1}(Y - X\hat{\beta}_{WLS}). \end{aligned}$$

3.88. $(Y - X\beta)'V^{-1}(Y - X\beta) = (Y^* - X^*\beta)'(Y^* - X^*\beta)$, which is minimized by $\beta = (X^{*'}X^*)^{-1}X^{*'}Y^* \equiv \hat{\beta}_{WLS}$.

3.89. Let $n^* = \sum w_k$, and let $\bar{X}_j^* = \sum w_k X_{kj}/n^*$. The latter is merely a weighted mean of column j. Then

$$\tilde{s}_{00} = -\frac{1}{s_{00}} = -\frac{1}{\sum w_k} = \frac{1}{n^*},$$

$$\tilde{s}_{0j} = \frac{s_{0j}}{s_{00}} = \frac{\sum w_k X_{kj}}{\sum w_k} = \bar{X}_j^*,$$

and

$$\tilde{s}_{ij} = s_{ij} - \frac{s_{i0}s_{0j}}{s_{00}}$$

$$= \sum w_k X_{ki} X_{kj} - \frac{1}{n^*} \left(\sum w_k X_{ki} \right) \left(\sum w_k X_{kj} \right)$$

$$= \sum w_k X_{ki} X_{kj} - n^* \bar{X}_i^* \bar{X}_j^*$$

$$= \sum w_k \left(X_{ki} - \bar{X}_i^* \right) \left(X_{kj} - \bar{X}_{kj}^* \right).$$

The last expression is a weighted sum of cross products of deviations from the weighted means.

3.91. (Golub and Van Loan, 1983). Let $r_k(\delta)$ be the kth component of $Y - X\hat{\beta}(\delta)$, where $\hat{\beta}(\delta)$ is the WLS estimator for $V^{-1} = W(\delta)$. Then

$$r_k(\delta) = \frac{r_k(0)}{1 + \delta w_k e_k' X(X'V^{-1}X)^{-1} X' e_k},$$

where $e_k = (0, 0, \ldots, 0, 1, 0, \ldots, 0)$ has a one only in the kth position. [*Hint.* First consider the case with all $w_i = 1$.]

3.93. $A\beta = c = (D+N)\beta$, the right-hand equality of which can be rearranged to give the result.

3.94. The symmetry of A implies that all eigenvalues of A are real, and that an orthonormal basis for $\mathcal{R}^p$ exists consisting of eigenvectors of A. Let $v_1, v_2, \ldots, v_p$ be such a basis, with $|\lambda_1| \geq \ldots \geq |\lambda_p|$ being the corresponding eigenvalues. First, assume that $|\lambda_1| \geq 1$. Then for $\epsilon^{(0)} = v_1$, we have that $|\epsilon^{(k)}| = |\lambda_1|^k$. The latter quantity is either identically one, or goes to infinity as $k \to \infty$, as k equals or exceeds unity. This establishes the necessity of the condition. Next assume that $|\lambda_1| < 1$. Then for any $\epsilon^{(0)}$, write

$$\epsilon^{(0)} = \sum_{i=1}^{p} w_i v_i,$$

so that

$$|\epsilon^{(k)}|^2 = \left| \sum_{i=1}^{p} w_i \lambda_i^k v_i \right|^2 \leq \sum_{i=1}^{p} w_i^2 |\lambda_i|^{2k} \to 0.$$

3.96. Under the conditions on $\epsilon^{(0)}$ and v, the result follows immediately from the power iteration described in section 3.9.1.1.

3.97. Suppose that both are strictly positive, and suppose that $\min(e_i^+, e_i^-) = c \geq 0$. By subtracting c from e_i^+ and e_i^-, the ith constraint is still satisfied, the inequality constraints remain satisfied, and the objective function to be minimized is reduced by $2c$. Thus, the objective function is minimized by taking $c = 0$, which implies the result.

3.98. Branch-and-bound techniques have been developed for ordinary least-squares regression and for least absolute-value regression. The similarity of M-estimation and iteratively reweighted least squares to the OLS case suggests that it is possible to modify the branch-and-bound techniques of the latter to apply to the former.

4.1. Using the factorization criterion, a statistic $T(x)$ is sufficient for a parameter θ if and only if the density $P(x \mid \theta)$ can be written in the factored form $P(x \mid \theta) = c(x)g(\theta)h(T(x), \theta)$. Setting $c(\cdot) = g(\cdot) \equiv 1$, and $h(L_x(\cdot), v) \equiv L_x(v)$ gives the result. Note that this is a *function-valued* sufficient statistic.

4.2. From equation (4.1.1), which applies in the *i.i.d.* case, $\ell_X(\theta) = \log L_X(\theta) = \log \prod P(X_i \mid \theta) = \sum \log P(X_i \mid \theta) = \sum \ell_{X_i}(\theta)$. Note that the last expression is the sum of the likelihoods for the individual random variables X_i.

4.3.

$$I(\theta) = -E_\theta \ddot{\ell}_X(\theta) = -E_\theta \frac{\partial^2}{\partial \theta^2} \ell_X(\theta) = -E_\theta \frac{\partial^2}{\partial \theta^2} \sum_{i=1}^{n} \ell_{X_i}(\theta) = n \cdot i(\theta).$$

4.4. $\dot{\ell}(\theta) = 2\sum_i (x_i - \theta)/[1 + (x_i - \theta)^2]$, $\ddot{\ell}(\theta) = 2\sum_i [(x_i - \theta)^2 - 1]/[1 + (x_i - \theta)^2]^2$, and $I(\theta) = n/2$.

4.5. Since $g(a) - a \geq 0$, and $g(b) - b \leq 0$, it follows that $g(x) - x$ has a zero in $[a, b]$, by the intermediate-value theorem of the calculus.

4.6. (Henrici, 1964). By the previous exercise, there is at least one solution, say s. Suppose that there is another solution $t \in I$. Then $|s - t| = |g(s) - g(t)| \leq L|s - t|$. Dividing both sides by $|s - t|$ we have $1 \leq L$, which is a contradiction.

4.7. (Henrici, 1964). Existence and uniqueness of a solution are guaranteed by the preceding two exercises. Denote this unique solution by s. Then $|x_i - s| = |g(x_{i-1}) - f(s)| \leq L|x_{i-1} - s|$. Thus, $|x_i - s| \leq L^i|x_0 - s|$, the right-hand side of which converges to zero, since $L < 1$.

4.8. Let $s, t \in I$, with $s < t$. By the mean-value theorem of the calculus, there exists a point $x \in [s, t]$ for which $g(s) - g(t) = g'(x)(s - t)$. Taking absolute values of both sides, and noting that $|g'(x)| \leq L$ by hypothesis, establishes the result.

4.9. Let $x \in (a, b]$. By the definition of derivative, $|g'(x)| = \lim_{s \uparrow x} |(g(x) - g(s))/(x - s)| \leq L$, by the Lipschitz condition. A similar argument applies to the derivative at a, except that the limit must be taken from the right.

4.10. Letting $M = \max_I |f'(x)|$ and $\alpha = \pm 1/M$, where the sign is chosen so that $\alpha \cdot f'(x) < 0$ will do it.

4.11. Suppose that $x_i = L$ and $x_{i-1} = U$, f is convex on the interval, and that $f(L) < 0 < f(U)$. Convexity asserts that the line segment connecting the points $(L, f(L))$ and $(U, f(U))$ lies above the graph of f on $[L, U]$. Since this line segment intersects the horizontal axis at x_{i+1}, we must have $f(x_{i+1}) < 0$. Thus, the new interval is $[x_{i+1}, U]$, and the conditions of the exercise remain in force. Thus, the right-hand endpoint will remain unchanged in all subsequent iterations. The other cases are handled similarly.

4.12. Let $f(x)$ be a continuously differentiable function for which $f(x) = x + 1$ for $x \geq 1/2$, $f(x) = x - 1$ for $x \leq -1/2$, $f(0) = 0$, and $f'(0) > 0$. If $x_i \geq 1$, then $x_{i+1} = -1$. Similarly, $x_i \leq 1 \Rightarrow x_{i+1} = +1$.

4.13. Always.

4.15. Hint: $E(m_i) = np_i$.

4.16. The Cauchy location problem is a thorny one numerically. The likelihood equation can—and for small n, does—have several roots. Plausible starting values are the median and a trimmed mean of the observations. See Reeds (1985) for a discussion of the number of roots of a Cauchy likelihood and their distribution.

4.17. The method of scoring uses the iteration $\theta_i = \theta_{i-1} + \dot{\ell}(\theta_{i-1})/\mathcal{I}(\theta_{i-1})$. In this case, the information is constant with value $n/2$.

4.18. (Ortega and Rheinboldt, 1970). Let $A = J_k^{-1}$, let $B = J_{k+1}^{-1}$, and drop the subscripts on d and g. For $x \neq 0$, write $x'Bx = x'Ax + (d'x)^2/d'g - (x'Ag)^2/g'Ag = [(x'Ax)(g'Ag) - (x'Ag)]/g'Ag + (d'x)^2/d'g$. The first term on the right-hand side is nonnegative by the Cauchy-Schwarz inequality, and the second term is nonnegative by virtue of $d'g > 0$. If the first is not strictly positive—that is, if it is zero—then we must have $x \propto g$, which implies that the second term is strictly positive.

4.19. Using the notation of Section 4.3.3.4, the BFGS inverse update is given by

$$J_{k+1}^{-1} = V_k J_k^{-1} V_k' + \frac{d_k d_k'}{d_k' g_k},$$

where the matrix

$$V_k = I - \frac{d_k g_k'}{d_k' g_k}.$$

4.20. Equation (4.2.7) can be considered as a log likelihood for the two-dimensional parameter (α, θ). First maximizing with respect to α, we find that $\ell(\alpha \mid \theta)$ is maximized at $\hat{\alpha} = \sum M_j/n = \overline{M}$, regardless of the value of θ. Thus, maximizing (4.2.7) with respect to θ with $\alpha = \overline{M}$ produces a global maximum. This is precisely what was done in Table 4.2.8.

4.21. Let X be a random variable which is zero with probability ξ and Poisson(λ) with probability $1 - \xi$. Then $Pr(X = 0) = \xi + (1 - \xi)e^{-\lambda}$, and

$Pr(X = i) = (1 - \xi)e^{-\lambda}\lambda^i/i!$. The likelihood function for a random sample of N observations from this population can then be written as

$$lik(\xi, \lambda) = [\xi + (1 - \xi)e^{-\lambda}]^{n_0}[(1 - \xi)e^{-\lambda}\lambda]^{n_1}[(1 - \xi)e^{-\lambda}\lambda^2/2]^{n_2}\cdots$$

$$= [\xi + (1 - \xi)e^{-\lambda}]^{n_0}(1 - \xi)^{N-n_0}e^{-\lambda(N-n_0)}\prod_{i=1}^{\infty}\left(\frac{\lambda^i}{i!}\right)^{n_i},$$

from which equation (4.3.8) follows after taking logarithms.

4.24. If the distribution of children were Poisson (λ), then $\overline{X}$ would be the maximum likelihood estimator for λ. In the example, however, there are extra, non-Poisson cases in category zero, making $\overline{X}$ smaller than it would be in the Poisson case. Similarly, if all widows in category zero were from population A (identically zero children), then n_0/N would estimate the binomial parameter. However, in this case, category zero also contains observations from population B (the Poisson population), thus inflating n_0 and making n_0/N larger than it would be in the pure binomial case.

4.25. After some simplification, one obtains

$$I_{\xi\xi} = N\left\{\frac{(1 - e^{-\lambda})^2}{\xi + (1 - \xi)e^{-\lambda}} + \frac{1 - e^{-\lambda}}{1 - \xi}\right\}$$

$$I_{\xi\lambda} = -N\left\{e^{-\lambda} + \frac{(1 - \xi)(1 - e^{-\lambda})e^{-\lambda}}{\xi + (1 - \xi)e^{-\lambda}}\right\}$$

$$I_{\lambda\lambda} = N\left\{-(1 - \xi)e^{-\lambda} + \frac{(1 - \xi)^2e^{-2\lambda}}{\xi + (1 - \xi)e^{-\lambda}} + \frac{1}{\lambda}\right\}.$$

4.27. The matrix of second derivatives of $\ell(\cdot)$ does not depend upon the observed data $\{y_j\}$, so that its expectation, and hence $I(\alpha, \beta)$ is a constant.

4.29. If β_0 is the only non-zero coefficient, then the linear logistic model asserts that $\text{logit}(p_i) = \text{logit}(p) = \log(p/(1 - p))$, which is approximately equal to $\log p$, when $p \approx 0$.

4.30. The estimated coefficient is $\hat{\beta}_1 = -0.431$, with a standard error of $0.408 = \sqrt{0.1664}$. The ratio of these quantities is $z = -1.057$; comparing this figure to tables of the normal distribution produces an observed significance level of 0.29.

4.31. Let $f(x) = ax^2 + bx + c$. Then the right-hand side of (4.4.1) is $a(x_i + x_{i-1}) + b$, so that if $a = 0$, the right-hand side of (4.4.1) equals the first derivative (b) of $f(x) = bx + c$, but not otherwise. On the other hand, the right-hand side of (4.4.2) is $[a((x_i + h)^2 - (x_i - h)^2) + b(2h)]/2h = (4ax_ih + 2bh)/2h = 2ax_i + b$, for any values of a and b.

4.32. $J_{ij}(x) \approx [f_i(x + \sum_1^j h_{in}e_n) - f(x + \sum_1^{j-1} h_{in}e_n - h_{ij}e_j)]/2h_{ij}$.

4.33. Write $f(x)$ in a Taylor series expansion about $x = x_i$. Then $f(x_i + h) = f(x_i) + hf'(x_i) + \frac{1}{2}h^2 f''(x_i) + \frac{1}{6}h^3 f'''(x^*)$, where $x^* = x_i + th$ for some $0 \leq t \leq 1$.

4.34. Suppose for instance that $M = I$, and that k_j points are required to evaluate x_j, the jth component of x, to a given resolution. Then the number of evaluation points required is $\Pi_{j=1}^{P} k_j$. Let $k = \min(k_j)$ and $K = \max(k_j)$. Then $k^P \leq \Pi k_j \leq K^P$.

4.35. Since S is bounded, $S \cap L$ has a finite number of elements, and so F achieves its minimum on $S \cap L$, say $F_{\min}$. Let F_j denote the minimum value of F observed at step j. Then $F_1 \geq F_2 \geq \ldots \geq F_{\min}$, so that the monotone sequence $\{F_j\}$ has a unique limit point.

4.36. Consider the four-point grid whose function values are $\left(\begin{smallmatrix} 0 & 2 \\ 2 & 1 \end{smallmatrix}\right)$. Starting at the lower-left grid point, if we first search in the left-right direction, the global minimum will never be found.

4.37. If g achieves its minimum at t^*, then $t^* \in [t_1, t_2]$.

4.38. Without loss of generality, take $[a, b]$ to be the unit interval. The first requirement has $t_1 = 1 - t_2$, with $t_1 < 1/2$. Suppose that the first step results in the shortened interval $[0, t_2]$. Then for t_1 to remain a trial point, it must bear the same ratio to the new length (t_2) as t_2 does to the original length (1). Thus, $t_1 = t_2^2$, and $t_1 = 1 - t_2$, which together determine that $t_1 = (3 - \sqrt{5})/2$.

4.39. The interval must be reduced to have length at most 0.02, so that $\xi^k < 0.02$, where $\xi = (\sqrt{5} - 1)/2$. The smallest such k is 9.

4.40. (Kiefer, 1953). Consider the case $M = 2$ evaluations and, without loss of generality, let the initial bracket be the unit interval. The two points $t_1 < t_2$ at which g is evaluated must be distinct, and any such choice is a strategy for $M = 2$. The final interval will either be $[0, t_2]$ or $[t_1, 1]$, and the figure of merit for such a strategy is $m(t_1, t_2) = \max(t_2, 1 - t_1) > \max(t_2, 1 - t_2) \geq 1/2$. The infimum is attained for $t_1 = t_2 = 1/2$, but this is not a search strategy, since the points must be distinct; hence no strategy for $M = 2$ attains the minimax value $1/2$, so that none is minimax. The case $M > 2$ is established by an inductive argument of the following form: if there were a minimax procedure for $M > 2$ function evaluations, then there must also be a minimax procedure with $M - 1$ function evaluations.

4.41. (Thisted, 1986c). The expected number of function evaluations is $(q^2 + 2q - 4)/2q$. The expected interval reduction per evaluation of g is $E(L^{(1/N)})$, where L and N are random variables depending upon the value of the random variable J. The conditional distribution of $L \mid J$ is constant, at $1/q$ for $J = 0$ or $J = q$, and $2/q$ otherwise. The conditional distribution of $N \mid J$ is also constant, at $J + 1$ for $J < q - 1$ and at $q - 1$ otherwise. Since $E(L^{1/N}) = E[E(L^{1/N} \mid J)]$, the left-hand side can be tabulated. The values for $q = 3$ through $q = 8$ are, respectively, 0.6961, 0.6836, 0.6788, 0.6781, 0.6798, and 0.6829. The values increase monotonically, with a limit of unity as $q \to \infty$.

4.42. Denote $F''(x)$ by A. Now an arbitrary direction d is a descent direction provided that $\partial/\partial t F(x + td) < 0$, that is, $d^t[F'(x)] < 0$. For the special form of d_{LM}, this means that $-[F'(x)]^t[A + \lambda I]^{-1}[F'(x)] < 0$, which will occur if the matrix $A + \lambda I$ is positive definite. Since the eigenvalues $ch(A + \lambda I)$ are simply $\lambda + ch(A)$, these will all be positive if and only if λ exceeds the smallest eigenvalue of A.

4.43. The ridge estimator is obtained from using the direction d_{LM} computed at $\beta = 0$.

4.44. Write $(X'X + kI)^{-1}X'Y = (X'X + kI)^{-1}(X'X)(X'X)^{-1}X'Y = (X'X + kI)^{-1}X'X\hat{\beta} \equiv A(k)\hat{\beta}$. Since $A(k)$ and $I - A(k)$ each have all eigenvalues less than unity, the ridge estimator can be written as a matrix convex combination of $\hat{\beta}$ and 0, namely $\hat{\beta}(k) = A(k)\hat{\beta} + (I - A(k))0$. As noted above, the ridge estimator is obtained from using the direction d_{LM} computed at $\beta = 0$. Except for convenience, there is no reason to start the iteration at $\beta = 0$; if the iteration is started at a point β_0 and taken for one step in the Levenberg-Marquardt direction, the resulting estimator is $\tilde{\beta} = \hat{\beta}(k) + k(X'X + kI)^{-1}\beta_0$, that is, $A(k)\hat{\beta} + (I - A(k))\beta_0$.

4.45. The ridge-like estimator $\tilde{\beta} = \hat{\beta}(k) + k(X'X + kI)^{-1}\beta_0$ is the Bayesian posterior mean for β using a prior distribution of $\beta \sim N(\beta_0, k^{-1}I)$ when ϵ is multivariate normal.

4.47. The problem of finding x to minimize $F(x)$ is the same as that of finding y to minimize $F(\Delta^{-1}\Delta x) = F(\Delta^{-1}y) \equiv G(y)$. The rescaled-Newton step produces $-\Delta^{-1}G''(y)^{-1}G'(y)$, which is equal to $-F''(x)^{-1}F'(x)$. [Hint: Express G' and G'' in terms of F.]

4.48. The steepest-descent direction in terms of the original scale is $-F'(x)$, while after rescaling, the descent direction becomes $-\Delta^{-1}G'(y) = -\Delta^{-1}[\Delta^{-1}F'(\Delta^{-1}y)] = -(\Delta^{-1})^2 F'(x)$. This is the same as the unscaled direction only if $\Delta = I$.

4.49. The only part of the log-likelihood function involving θ is simply $(4.5.2)$ multiplied by $-1/2\sigma^2$. Since the minimizer of $(4.5.2)$ maximizes the log likelihood for each fixed σ, it does so at the maximum-likelihood estimate of σ as well.

4.50. The orthogonality condition can be written as

$$
\begin{aligned}
0 &= \sum r_i(\theta)w_i(\theta)x_i \\
&= \sum [y_i - g(x_i^t\theta)]g'(x_i^t\theta)x_i \\
&\approx [y_i - g(x_i^t\theta_0) - (\theta - \theta_0)^t g'(x_i^t\theta_0)x_i]g'(x_i^t\theta_0)x_i \\
&= \sum [r_i - (\theta - \theta_0)^t w_i x_i]w_i x_i.
\end{aligned}
$$

The approximation step is obtained by taking a first-order Taylor approximation about θ_0 for the term in square brackets, and a zero-order (constant) Taylor approximation to the term $g'(x_i^t\theta)$. The last equality can be rearranged to have the form $(4.5.8)$.

4.51. Write $(4.5.8)$ as $\hat{\theta} = \theta_0 + [X^t W^2 X]^{-1}[X^t W R] = [X^t W^2 X]^{-1}X^t W^2 [X\theta_0 + W^{-1}R] \equiv [X^t W^2 X]^{-1}X^t W^2 Z^*$. This gives the parameter estimate from a weighted regression of the current linear predictor $x_i'\theta_0$, adjusted by the increment $[y_i - g(x_i'\theta_0)]/\sqrt{w_i^2}$, on the x_i's.

4.52. The constraint $c(\theta) = 0$ is satisfied if, and only if, $c(\theta) \geq 0$ and $-c(\theta) \geq 0$.

4.53. Since C is of full rank, its rows form a basis for a subspace of dimension m in $\mathcal{R}^p$. Thus, there exists a basis for the complementary space of dimension $p - m$. Let the rows of A be such a basis. Then every $1 \times p$ vector can be written uniquely as a weighted sum of the rows of A and the rows of C. In particular, for each row of X, say x_i', we may write $x_i' = z_i'A + w_i'C$. The matrices Z and W of (4.5.11) have the weight vectors z_i and w_i for rows.

4.54. As in section 4.5.7.2, we take C to have rank m. Consequently, we can apply the QR decomposition to the rows of C to obtain the factorization $C = R'F$, where R is an $m \times m$ nonsingular upper-triangular matrix, and $FF' = I_m$. Clearly $C\beta = 0$ if, and only if, $F\beta = 0$. Note also that $FA' = 0$, so that the matrix $Q \equiv \binom{A}{F}$ is orthogonal. Finally, since we must have $A\hat{\beta} = \hat{\gamma}$ and $F\hat{\beta} = 0$, we can write $Q\hat{\beta} = \binom{\hat{\gamma}}{0}$, which implies that $\hat{\beta} = Q'\binom{\hat{\gamma}}{0} = A'\hat{\gamma}$. (The reader should verify that the partitioned matrices are conformable.) Uniqueness follows from the fact that the rows of A are an orthonormal basis for the null space of C; if B is any other such basis, then $B = QA$, where Q is an orthogonal matrix of order $p - m$.

4.55. (Khuri, 1976). Let $c_j(\theta) \equiv c_j'\theta$, where c_j is a $p \times 1$ vector of coefficients, and denote by C the $m \times p$ matrix whose jth row is c_j'. Partition θ as $(\theta_1' \; \theta_2')$, where θ_1 is $m \times 1$ and θ_2 is $(p-m) \times 1$, partition $X = (X_1 \; X_2)$, where X_1 is $n \times m$, and use m Householder transformations on the left to write $C = Q(R\,G)$, where $QQ' = I_m$, and R is a nonsingular upper triangular matrix. Finally, set $\beta = \binom{C\theta}{\theta_2}$ and $W = (W_1\,W_2)$, where $W_1 = X_1R^{-1}Q'$ and $W_2 = X_2 - X_1R^{-1}G$. The original problem is equivalent to minimizing $|Y - W\beta|^2$ subject to $\beta_j = c_j(\theta) \geq 0$.

4.57. Part of the difficulty of this problem is to define a suitable notion of "average performance." Probabilistic notions of performance play an important role in the analysis of algorithms, but are not always easy to define in a natural or compelling manner. Some of these issues will be discussed in Chapter 22.

4.58. Consider the "obvious" code

for $i := 1$ to n do $x[i,j] := x[p[i,j],j]$;

applied to the column of X containing $(2, 3, 1)'$. [The corresponding column of P is $(3, 1, 2)'$.]

4.59. As defined in section 4.6.1, the jth column of P—denote it by P_j—contains the permutation of the original data required to order the contents of the jth column of X. Consider instead first sorting the rows of X (and the corresponding elements of Y) so that the elements of the first column of X are in increasing order. Then let P_2 contain the permutation (of this reorganized data) which would order the elements of the second column of X. Let P_3 contain the permutation which would, starting with the reordering specified by P_2, bring the elements of X_3 into ascending order. This process is repeated through P_p. Thus, $Z[i,1] = X[i,1]$, $Z[i,2] = X[P[i,2],2]$, $Z[i,3] = X[P[P[i,3],2],3]$, and so forth. Finally, P_1 is computed as that permutation, starting from the permutation of the last column, which restores the original order in the first column. The key to this incremental permutation method is that the Z matrix is computed once only, so that the implicitly nested array indices are never actually computed. At the end of a stage involving column j, the current residuals are stored in W. One need only apply P_{j+1} to this residual vector to re-order it for the next step.

4.60. The tractability of this problem, and to some extent its answer, may well depend upon the particular smoothing algorithm employed in the PSR algorithm. It may, for example, be possible to obtain analytical results for spline-based smoothing because of the linear structure of such estimated smooth functions. Significant progress on this question could be made through empirical investigation.

4.61. Take $k = p$ and $\alpha_j \equiv e_j$, where e_j is the jth column of the $p \times p$ identity matrix, to be fixed in advance.

4.63. Once again, there are two sets of iterations. The outer iteration alternates between estimating $\Theta = \sum \theta_k$ given fixed $\Phi = \sum \phi_j$, and estimating Φ given Θ. Each of these estimation steps in turn is comprised of an inner loop. The estimation of Θ, for example, cycles through the θ_k's, estimating each in turn while holding all of the other θ_k's and Φ fixed. Similar comments apply to estimating Φ.

4.64. Fowlkes and Kettenring (1985) suggest finding those transformations whose correlation matrix R^* has minimum determinant. Noting that $|R^*| = |R^*_{-j}|(1 - r_j^{*2})$, where R^*_{-j} is the correlation matrix of all but the jth variable, and r_j^{*2} is the squared correlation of the jth variable with all of the others, they suggest repeatedly fitting $\phi_j(x_j)$ as a function of the current values of the remaining ϕ's.

4.65. By analogy to GLIM, the "I" stands for *I*nteractive.

4.66. In exponential families we can write the log-likelihood function as

$$\log f(x \mid \theta) = \log a(\theta) + \log b(x) + \theta' t(x),$$

where $t(x)$ is a sufficient statistic for θ. The term $\log b(x)$ is a constant term (with respect to θ) and may be omitted; the conditional expected value of what remains is just a simple linear function in the conditional expectation of $t(x)$.

4.67. This is a standard exponential-family computation. Write the density of x as $L(\theta \mid x) = \exp(\theta' t(x) - a(\theta) - b(x))$. Since $\int L(\theta \mid x)dx = 1$, differentiating with respect to θ under the integral sign gives $0 = \int \{t(x) - a'(\theta)\}L(\theta \mid x)dx$, so that $E(t(x) \mid \theta) = a'(\theta)$. The likelihood equation for θ is obtained by setting the logarithmic derivative of $L(\theta \mid x)$ with respect to θ equal to zero; this produces the equation $t(x) = a'(\theta)$ which, taken with the previous calculation, gives the result.

4.68. The complete-data likelihood is multinomial, and the observed counts n_i for the cells $i = A, B, 1, \ldots, 6$, constitute the sufficient statistics. The conditional expectation in the E-step is $E(n_A, n_B, n_1, \ldots, n_6 \mid n_0, n_1, \ldots, n_6, \hat{\xi}, \hat{\lambda})$. The conditional expectations of n_1 through n_6 are simply those values. Writing $(n_A, n_B) = (n_A, n_0 - n_A)$, we see that the remainder of the conditional expectation depends only on the conditional expectation $E(n_A \mid n_0, \hat{\xi}, \hat{\lambda})$.

4.69. Conditional on $n_0 \equiv n_A + n_B$, the random variable n_A has a binomial distribution with parameter $\theta = p_A/(p_A + p_B)$, where $p_A = \xi$ and $p_B = (1 - \xi)e^{-\lambda}$ are respectively the probabilities under the full model of falling into categories A and B. The conditional expectation, then, is simply $\theta \cdot n_0$.

4.70. Wu (1983) relates certain theoretical properties of EM sequences to those of certain ascent algorithms; Wu's paper would be a good place from which to start an attack on this problem.

4.71. For the series to be stationary, the mean and variance must not depend on t, so let $E(X_t) = \mu$ and $\text{Var}(X_t) = V$. From (4.8.1a) we have that $\mu = \alpha_1 \mu$ and $(1 - \alpha_1^2)V = \sigma^2$. If $\alpha_1 = 1$ and $\sigma^2 = 0$ then the conditions are satisfied for any value of μ, and in this case X_t is a constant sequence. If $\sigma^2 > 0$ then we must have $\alpha_1^2 < 1$. Finally, $\text{Cov}(X_t, X_{t+h}) = \text{Cov}(X_t, \alpha_1^h X_t + \sum_0^{h-1} \alpha_1^j \epsilon_{t+h-j}) = \alpha_1^h V$, which depends on the time scale only through the increment h.

4.72. The first line follows from the definition of conditional densities, the second from the definition of an AR(2) process, and the third by induction.

4.73. From the definition of an AR(2) process, the conditional distribution of X_j given X_{j-1} and X_{j-2} is normal, with mean $\alpha_1 X_{j-1} + \alpha_2 X_{j-2}$ and variance σ^2. Thus, the jth term of the conditional likelihood is proportional to $\sigma^{-1} \exp(-(x_j - \alpha_1 x_{j-1} - \alpha_2 x_{j-2})^2/2\sigma)$. Since the unconditional mean and variance of the jth observation in the regression problem (4.8.4) have the same values as those just computed, the jth likelihood contribution is the same. Since the likelihood contributions in the regression problem multiply (since the ϵ's are independent), the total likelihood (computed from $j = 3$) is the same as that obtained by taking the product of L_C's in (4.8.3).

4.74. In regression the residual degrees of freedom are equal to the number of observations $(n - 3$ in this case) minus the number of parameters estimated in the mean function (2 in this case).

4.75. Let $\text{Var}(X_t) = V$ and $\text{Cov}(X_t, X_{t+1}) = C$. Then $C = \text{Cov}(X_t, \alpha_1 X_t + \alpha_2 X_{t-1}) = \alpha_1 V + \alpha_2 C$, so that $C = \alpha_1 V/(1 - \alpha_2)$. Note that $\delta_1 = C/V$, by definition. To obtain δ_2^2, compute

$$
\begin{aligned}
V &= \text{Var}(\alpha_1 X_{t-1} + \alpha_2 X_{t-2} + \epsilon_t) \\
&= \text{Var}(\alpha_1 X_{t-1} + \alpha_2 X_{t-2}) + \sigma^2 \\
&= \alpha_1^2 V + \alpha_2^2 V + 2\alpha_1 \alpha_2 C + \sigma^2 \\
&= V(\alpha_1^2 + \alpha_2^2 + 2\alpha_1 \alpha_2 \delta_1) + \sigma^2,
\end{aligned}
$$

from which the result follows.

4.76. The unconditional distribution of ϵ_0 is $\mathcal{N}(0, \sigma^2)$, and hence does not depend upon β.

4.77. The conditional distribution of X_j given Y_{j-1} is normal with mean $\hat{X}_j$ and variance σ^2. Thus, the conditional density depends on Y_{j-1} and ϵ_0 only through the linear predictor $\hat{X}_j$.

4.78. $f(x_j \mid \hat{x}_j, \sigma^2) = (2\pi\sigma^2)^{-1/2} \exp(-(x_j - \hat{x}_j)^2/(2\sigma^2))$.

4.79. Since the matrix K is lower triangular, its determinant is the product of its diagonal elements. As all of these are equal to one, $\det(K) = 1$.

4.80. Let θ denote the set of parameters $(\alpha, \beta, \sigma^2)$. We can write the joint density of the data $Y_j = (X_1, \ldots, X_j)$ through time j as $f(y_j \mid \theta) = f(x_j \mid y_{j-1}, \theta) \times f(y_{j-1} \mid \theta)$. The first term is a normal density whose mean is μ_j and whose variance is V_j, each of which is a function of y_{j-1} and θ. Thus, the conditional distribution of X_j depends on Y_{j-1} only through these two quantities, and we can then write $f(x_j \mid y_{j-1}, \theta) = f(x_j \mid \mu_j, V_j, \theta)$. Independence follows from the factorization theorem.

4.81. Differentiate (4.8.20) with respect to σ^2, set the result to zero, and solve for σ^2.

4.83. If (W, R, S) has a multivariate normal distribution, then in the obvious notation, $E(W \mid R, S) = E(W \mid S) + \Sigma_{ww.s}^{-1} \Sigma_{wr.s}(R - E(R \mid S))$. The result follows—after some further labor—by making the identification of W with the random variable $Z(t+1)$, R with X_{t+1}, and S with $Y_t = (X_1, \ldots, X_t)$.

5.1. $I'_R - I_R = h(f(x_n) - f(x_0)) = h(f(b) - f(a)) \geq 0$, since $f(x)$ is increasing.

5.2. Note that $I_R^{(n)} \leq I_R^{(2n)} \leq I_R^{(4n)} \leq \ldots$, and that this holds for any n. Since the limit of this sequence is just the integral I, we must have that $I_R^{(n)} \leq I$, for all n. Similarly, we have that $I \leq I_R'^{(n)}$, for all n.

5.3. Write f_i for $f(x_i)$. Then $I_T(a, b) = (h/2) \sum_{i=1}^{n}[f_{i-1} + f_i] = (h/2)[\{f_0 + \sum_{i=2}^{n} f_{i-1}\} + \{\sum_{i=1}^{n-1} f_i + f_n\}]$.

5.4. For convenience, write a and b for x_0 and x_1, respectively, and write $f(x)$ as $\alpha + \beta(x - a)$. Then $\int_a^b f(x)dx = [\alpha x + \beta(x-a)^2/2]_a^b = [\alpha(b-a) + \beta(b-a)^2/2] = (h/2)[2\alpha + \beta(b + a)] = I_T(a, b)$.

5.5. Using integration by parts we obtain $(x_0, x_1) = f(x)(x - \alpha)|_{x_0}^{x_1} - \int_{x_0}^{x_1}(x - \alpha)f'(x)\,dx$, for any α. Choosing $\alpha = m$ gives the result.

5.6. Write $-E_1 = (\int_{x_0}^{m} + \int_{m}^{x_1})(x-m)f'(x)\,dx$. In each of these two integrals, the term $(x - m)$ does not change sign, so that we may use the integral mean value theorem to write the sum as $f'(\xi_0)\int_{x_0}^{m}(x - m)\,dx + f'(\xi_1)\int_{m}^{x_1}(x - m)\,dx$. After a small amount of algebra, this reduces to $(h^2/8)[f'(\xi_1) - f'(\xi_0)] = (h^3/8)f''(\xi)$, where in the last step $\xi \in [\xi_0, \xi_1]$. [The last step uses the mean value theorem.]

5.7. The name of the rule comes from the fact that the basic trapezoidal rule simply gives the area between the line joining $(x_0, f(x_0))$ and $(x_1, f(x - 1))$ and the x-axis, and the shape of this region is trapezoidal.

5.8. Let $m = 2^{j+1}$, and let $f(x)$ be any nonnegative continuous function that is zero at the points i/m, $0 \le i \le m$ and that is not identically zero on $[0,1]$.

5.9. The intermediate-value theorem of the calculus asserts that, if f is continuous on $[a,b]$, then for any $0 \le \alpha \le 1$ the function value $\alpha f(a) + (1 - \alpha)f(b)$ must be attained at some point on the interval $[a,b]$. First, for $n = 1$ the desired result is trivially true. Suppose the result to hold for all $n < m$. Then $\sum_{i=1}^{m} f(y_i) = \sum_{i=1}^{m-1} f(y_i) + f(y_m) = (m-1)f(y_{m-1}^*) + f(y_m)$, by the induction assumption. Now apply the intermediate value theorem with $\alpha = (m-1)/m$.

5.10. Using the result of the previous exercise, and the fact that $h = O(n^{-1})$, we have that

$$I(a,b) = \sum_{i=1}^{n} I(x_{2(i-1)}, x_{2i})$$

$$= \sum_{i=1}^{n} [I_S(x_{2(i-1)}, x_{2i}) + O(h^5 f^{(4)}(x_{2i}^*))]$$

$$= I_S(a,b) + ch^5 \sum_{i=1}^{n} \{f^{(4)}(x_{2i}^*) + o(1)\}$$

$$= I_S(a,b) + cnh^5 f^{(4)}(x^*) + o(nh^5)$$

$$= I_S(a,b) + O(h^4 f^{(4)}(x^*)).$$

5.11. Writing $x_i = x_0 + ih$ and f_i for $f(x_i)$, we have $I_n = hf_0 + (2h)\sum_{i=1}^{n-1} f_{2i} + hf_{2n}$ and $I_{2n} = h[\frac{1}{2}f_0 + \sum_{i=1}^{n-1} f_{2i} + \sum_{i=1}^{n} f_{2i-1} + \frac{1}{2}f_{2n}]$.

5.12. I_{j+1} is based upon $2^{j+1} + 1 = 2 \cdots 2^j + 1 = 2n + 1$ points.

5.14. The sequence $\hat{k}_{n-1} \equiv -\log_2(|(I_{n-1} - I_n)/(I_{n-2} - I_n)|)$ for $n > 1$ gives a sequence of such estimates. Here, I_n is used as the best estimate for the true value of the integral; hence the estimate really gives an assessment for the order at I_{n-1}.

5.15. The estimate of the previous exercise gives, for instance, $\hat{k}_3 = 2.313$ and 3.871 for the trapezoidal estimate and Simpson's estimate, respectively. The actual values for k_3 (using the true value of the integral in place of I_n) are, respectively, 1.998 and 4.115.

5.16. The order estimates now are $\hat{k}_3 = 1.897$ (1.936), and $\hat{k}_9 = 1.939$ (1.955) for the trapezoidal (Simpson's) rule, with intermediate estimates approximately the same. The actual values are $k_3 = 1.454$ (1.499), and $k_9 = 1.496$ (1.505), suggesting that *both* Simpson's rule and the trapezoidal rule are actually of order 3/2 rather than of orders 4 and 2, respectively. As a result, Simpson's rule is little better than the trapezoidal rule for this integrand *on this interval*, and even the latter performs less well than the theory would imply.

5.17. Although both methods degrade due to the singularity in the derivative of the integrand at $t = 0$, Simpson's does so much more markedly because its error structure depends upon the magnitude of the fourth derivative of $\sqrt{t}$ near zero (proportional to $t^{-7/2}$) rather than the second derivative (proportional to $t^{-3/2}$). Consider, for instance, Simpson's estimate. From the second part of (5.1.6), we have that the error on each interval is of order $h^5 f^{(4)}(t_i^*)$, where t_i^* is in the ith interval, so that $t_i^* = O(i \cdot n^{-1})$. Hence each error term is $O(n^{-5}) \cdot O((i/n)^{-7/2})$. A sum of such terms for $1 \leq i \leq n$ is $O(n^{-3/2})$, since $\sum_{i=1}^{n} i^{-7/2} \approx \int_1^n t^{-7/2} \, dt = O(1)$. A similar calculation gives the same result for the trapezoidal rule.

5.18. Simpson requires 8 evaluations for exact relative error $< 10^{-5}$ and 16 evaluations (one more doubling of step size) to confirm this fact empirically. The trapezoidal rule needs 128 and 256 for the same levels of accuracy, sixteen times as much work. Some benchmark values: $I_S^{(8)} = 5.057072$, $I_T^{(128)} = 5.057091$, and $I = 5.057069$.

5.19. $\text{New}_j = (16 S_j - S_{j-1})/15$.

5.20. The weights are chosen so that the second-order terms in h vanish when the step size is tripled. (In obtaining Simpson's rule from the trapezoidal rule, the step size is only doubled.)

5.21. Let $n = 3^j$. The endpoints of the n intervals are of the form $a + (b-a)i/n$, so that the midpoints are of the form $a + (b-a)(2i-1)/(2n) = a + (b-a)[2(3i-1)-1]/(2 \cdot 3n)$.

5.22. It is sufficient to consider the basic midpoint and trapezoidal rules, and without loss of generality we may take $[a, b] = [0, 1]$. The trapezoidal inequality is an immediate consequence of Jensen's inequality. The midpoint inequality can be deduced by first writing the integrand in a Taylor's expansion using the integral form of the remainder: $f(x) = f(\frac{1}{2}) + (x - \frac{1}{2})f'(\frac{1}{2}) + \int_{1/2}^{x}(t - \frac{1}{2})f''(t) \, dt$. The inequalities are strict unless f is linear.

5.23. The actual transformation was $u(x) = e^{1-x}$. The only effect is to spread out the horizontal axis by a factor of e and to squash the vertical axis by the same factor. If $u(x) = e^{-x}$ were used, then the left-hand figure would be almost three times as tall and only one-third as wide as the figure on the right, making them harder to compare.

5.24. No. The resulting integrand near zero behaves like $(-\log(t))^{-1/2}$.

5.25. $[t^{\alpha-1}/t^{-1}(-\log(t))^{-3/2}] = t^{\alpha}(-\log(t))^{3/2} \to 0$ as $t \to 0$.

5.26. The 5-point rule has $n = 4$, and the coefficients are $(14, 64, 24, 64, 14)/45$. This formula is sometimes called *Bode's rule*.

5.27. Denote the rule by $a_1 f(x_1) + a_2 f(x_2)$. The four equations which must be satisfied are $\int_{-1}^{1} x^i \, dx = a_1 x_1^i + a_2 x_2^i$ for $0 \leq i \leq 3$.

5.28. Use equations for $i = 1$ and $i = 3$.

5.29. The first conclusion follows from the equation for $i = 0$, the second from the equations for $i = 1$ and $i = 2$.

5.30. Let $c = (b-a)/2$ and $m = (a+b)/2$. Then (5.3.2) can be written as $c \int_{-1}^{1} f(cy+m)w(cy+m)\,dy$.

5.31. (Outline) Let $r = \lfloor k/2 \rfloor$ denote the number of positive abscissæ $\{x_i\}$ in the k-point rule GL_k. Since $GL_k(f)$ is exact for any polynomial f of degree $2k-1$ or less, we have for odd $j \le (2r-1)$ that $GL_k(x^j) = 0 = \sum_{i=1}^{r}[a_i(x_i)^j + b_i(-x_i)^j] \equiv \sum_{1}^{r} c_i x_i^j$. Finally, let $v_i = (x_i, x_i^3, \ldots, x_i^{2r-1})$, so that this latter equation can be written as the vector equation $\sum c_i v_i = 0$. The v_i are linearly independent, since there are exactly r of them and the x_i's are distinct. Thus all $c_i = a_i - b_i$ must be zero.

5.32. Write the rule as $GL_3(f) = af(-x)+bf(0)+af(x)$. Integrating $f(x) = 1$ gives the relationship $b = 2(1-a)$; integrating $f(x) = x^2$ and $f(x) = x^4$ give the equations $3ax^2 = 1$ and $5ax^4 = 1$, from which it follows that $x = \sqrt{3/5}$, $a = 5/9$, and thus $b = 8/9$.

5.33. Because $w(x)$ is admissible, it has moments of all orders.

5.34. Compute $\langle f, p_i \rangle$ and use orthogonality.

5.35. The four-line MACSYMA program

```
p[n](x):=ratsimp(((2*n-1)*x*p[n-1](x)-(n-1)*p[n-2](x))/n)$
p[-1](x):=0$
p[0](x):=1$
for i:0 step 1 thru 8 do disp(p[i](x));
```

produces the first eight Legendre polynomials, of which the eighth is given by

$$\frac{6435x^8 - 12012x^6 + 6930x^4 - 1260x^2 + 35}{128}.$$

5.36. $T_8(x) = 256x^8 - 448x^6 + 240x^4 - 40x^2 + 1$.

5.37. $w^*(x) \propto w(2x-1)$.

5.38. $T_j^*(x) = T_j(2x-1)$, from which the first result follows using (5.3.4) applied to T_j. The subsequent computation produces $T_8^*(x) = 2048x^8 - 3072x^7 - 4864x^6 + 15232x^5 - 15232x^4 + 7744x^3 - 2128x^2 + 296x - 15$.

5.41. Some estimated values and their errors are:

Rule	Result	Error
Exact	0.308537538725987	
$GLT_{10}(\frac{1}{2},\infty)$	0.308540845665621	3.307×10^{-6}
$GLC_{10}(\frac{1}{2},3)$	0.308538309425469	7.707×10^{-7}
$GLC_{10}(\frac{1}{2},\frac{3}{2})$	0.308556426210155	1.889×10^{-5}

GLC_{10} is superior to GLT_{10} for roughly $1.9 < y < 3.3$.

5.42. $R_{2k}(x,y) = GLT_k(x,y) + GLT_k(y,\infty)$.

ANSWERS TO SELECTED EXERCISES

5.43. For $y < 1.9$, $GLT_{10}(\frac{1}{2}, \infty)$ is the winner. In this range, GLC and R are virtually indistinguishable, with R the winner by only a few percent over GLC. Here, the errors in GLC and R are coming almost exclusively from the integral on (y, ∞), where the two methods use the same rule. For $1.9 < y \le 2.0$, R maintains its slight edge over GLC. For $2.0 < y < 5.3$, R remains inferior to GLC. R is better than GLT only in the narrow range $(1.9, 2.2)$, and this only because its error curve is passing from positive errors to negative errors (and hence, must pass through zero *somewhere*). For $y > 3.3$, GLT is again the winner. GLC becomes worse that R in the extreme tail; as y gets large, the integrand begins its exponential asymptote on the right, so that the regular Gauss-Legendre rule on $(\frac{1}{2}, y)$ begins to fail.

5.44. Note that $\int_{-\infty}^{x} f_\lambda(x)\, dx = 0.5 + \arctan((x - \lambda)/\sigma)$. This makes it possible to obtain the exact value for $I(\lambda)$.

5.45. $\mathcal{P}_n$.

5.46. If $f(x)$ is a polynomial, it is a polynomial on every subinterval, which implies the first condition in the definition of a spline. Polynomials have infinitely many continuous derivatives, which implies the second condition.

5.47. If f is polynomial on an interval, it is polynomial on any subinterval.

5.48. If $s(x) \in \mathcal{S}_n(K)$ then $s(x) \in \mathcal{C}^{n-1}$ by definition. Moreover, the knots x_1 through x_m divide the real line into $m + 1$ intervals, on each of which $s(x) \in \mathcal{P}_n$. The nth derivative of such a polynomial is a constant; so that $s^{(n)}$ is constant except at the knots (which separate the intervals).

5.49. First, $s(x) \in \mathcal{C}^{n-1}$, since each of its components is in $\mathcal{C}^{n-1}$. Suppose that $x_1 < \cdots < x_m$. Now s_i is a polynomial of degree at most n on every interval not containing x_i. However the intervals (x_j, x_{j+1}) by construction contain none of the knots $x_1, \ldots, x_m$, so that $s(x)$ is a sum of nth degree polynomials on each such interval.

5.50. If $f \in \mathcal{P}_n$ then $f' \in \mathcal{P}_{n-1}$; similarly for $\mathcal{C}^n$.

5.51. $\mathcal{P}_n$.

5.52. To show that such a representation exists for an arbitrary $s(x) \in \mathcal{S}_n(K)$, consider the polynomial representation of $s(x)$ on two adjacent intervals, say $s(x) = p(x)$ on (x_{i-1}, x_i) and $s(x) = p(x) + q(x)$ on (x_i, x_{i+1}). Then show that x_i is a root of $q(x)$ of multiplicity n. To show that the representation is unique, suppose that two such representations exist for a single $s(x)$. For notational convenience, denote the ordered elements of K by $x_1 < \cdots < x_m$, and identify c_i with c_{x_i}. Suppose that, in addition to equation (5.5.1), we could also write $s(x) = q(x) + \sum_{k \in K} d_k \cdot (x - k)_+^n$, with $q \in \mathcal{P}_n$. For $x < x_1$, we can write $s(x) = p(x) = q(x)$, all of the terms in either sum being zero. Thus p and q are nth degree polynomials which agree at (more than) $n + 1$ points in $(-\infty, x_1]$. Therefore, p and q must coincide. Similarly, on $(x_1, x_2]$, we now have $s(x) = p(x) + c_1(x - x_1)^n = p(x) + d_1(x - x_1)^n$, so that $c_1 = d_1$. Equality of the other coefficients is established in the same fashion.

5.53. Write $p(x) = \sum_0^n a_i x^i$. Then $\int_0^t s(x)\,dx = \sum_0^n (a_i/(i+1))t^{i+1} + [\sum_k c_k(t-k)_+^{n+1}]/(n+1)$.

5.54. $b_{ij} = \sum_{l=j}^n a_{il}\binom{l}{j}x_i^{l-j}$.

5.55. $P(x) = \int p(x)\,dx = 3\alpha_3(x-\xi)^2 + 2\alpha_2(x-\xi) + \alpha_1$.

5.58. It helps to note that $dz = -dt$.

5.59. Let $\Delta = x_{j+1} - x_j$, $t = (x-x_{j+1})/\Delta$, and $z = 1-t$. Then $s(x) = zy_j + ty_{j+1} + (z^3 - z)s_j\Delta^2/6 + (t^3 - t)s_{j+1}\Delta^2/6$.

5.60. $3[y'(0) - (y_1 - y_0)/(x_1 - x_0)] = (x_1 - x_0)s_0$.

5.61. Let $\Delta = x_{j+1} - x_j$, $t = (x-x_{j+1})/\Delta$, and $z = 1-t$. Then

$$\int s(x)\,dx = -\frac{\Delta}{2}(z^2 y_j - t^2 y_{j+1}) - \frac{\Delta^3}{12}\left(\left[\frac{z^4}{2} - z^2\right] - \left[\frac{t^4}{2} - t^2\right]\right) + C.$$

5.63. The cubic natural spline gives $1 - \Phi(1.64485) \approx 0.05001143$. With adjustment for the derivatives, the approximate value is 0.05000039.

5.64. Note that subtracting a constant does not affect the order of the observations, so that $X_{(i)} - c = [X - c]_{(i)}$.

5.65. Since $E(\overline{X}_\alpha) = 0$, $\mathrm{Var}(\overline{X}_\alpha) = E\overline{X}_\alpha^2$.

5.66. $\sqrt{2} \approx 1.41$.

5.67. Write $E(Y \mid X) \approx \mu + \beta(X - \theta)$, where $\beta = \rho\sigma_y/\sigma_x$. This is exact when $X = \theta$, and for all X when (Y, X) is Gaussian. We are interested in the conditional expectation at $X = \theta$. Now we can fit the linear model $E(Y \mid X) = \alpha + \beta X$ by least-squares regression, and then take as our estimate $\hat{\mu} = \hat{\alpha} + \hat{\beta}\theta = \overline{Y} + \hat{\beta}(\theta - \overline{X})$. Thus, the regression estimator adjusts the raw estimate $\overline{Y}$ by an amount that depends on the deviation of $\overline{X}$ from its theoretical mean (θ). From the regression's residual mean square, one obtains an estimate of $\hat{\sigma}_{y \mid x}^2$, the *conditional* variance of Y. If ρ is at all large and positive, this estimate will be substantially smaller than $\hat{\sigma}_y^2$; indeed, this is where most of the gain in precision comes from. The standard deviation of the mean of Y at $X = \theta$ is given by the usual formula, $\sigma_{y \mid x}\sqrt{R^{-1} + (\theta - \overline{X})^2/\sum(X_i - \overline{X})^2}$.

5.69. $E[I(-\xi, 0, \xi)] = 2g(0) + 3\int_0^1 \frac{g(-\xi) - 2g(0) + g(\xi)}{3\xi^2}\xi^2\,d\xi$.

5.70. V and W have the same cdf, namely, $F_V(t) = t^3$ on $[0, 1]$. [Hint: $W < t$ if, and only if, U_1, U_2, and U_3 are *all* smaller than t.]

5.72. Let $\alpha_i = 1$ if o_i denotes "$\geq$" and $\alpha_i = -1$ otherwise, and let P be a diagonal matrix with entries α_i. Then write p in terms of Px.

5.73. Note that $P_k = 2^{-k}$. The problem becomes much more difficult when I is replaced by a general covariance matrix Σ. Why?

5.74. Work for $p = 4$ and $p = 5$ can be found in Gehrlein (1979) and in Dutt and Lin (1975).

386 *ANSWERS TO SELECTED EXERCISES*

5.75. Note that it is no longer sufficient just to be able to compute the positive orthant probability.

5.76. The first approximation simply replaces $L(\theta)$ by a second-order Taylor-series approximation at $\hat{\theta}$. (Why is the first-order term zero?) The second step just involves evaluating a normal integral after factoring out $\exp(L(\hat{\theta}))$.

5.77. Roughly, $p(t)dt = \int_{\{t>x_2\}} p(t, x_2)\, dx_2 + \int_{\{t>x_1\}} p(x_1, t)\, dx_1$.

5.78. Use integration by parts.

5.79. $p(y) = \phi(y - \mu)\Phi^{k-1}(y) + (k - 1)\phi(y)\Phi(y - \mu)\Phi^{k-2}(y)$.

5.80. Differentiate $x = F(F^{-1}(x))$.

5.81. Use the result of the previous exercise to show that a small perturbation in the input, say from p to $(1 + \epsilon)p$, perturbs the solution by a factor of at least $[F^{-1}(p)]^2$, which goes to $-\infty$ as p nears zero.

5.82. Following the text, we obtain the relationship $F(\phi(p)) - p \approx p - \phi^{-1}(x_p)$. Solve this for x_p.

5.83. Since $\phi(p) - F^{-1}(p) = 0$, it follows that $f(p) = F(\phi(p)) - p = 0$, where here we are using f as in Section 4.2.1 (f is *not* the density).

5.84. Convergence will occur if $0 < F'(\phi(y - x))\phi'(y - x) < 2$ for all $x, y \in (0, 1)$.

5.85. Hint: $d^k/dx^k \phi(x)$ can be expressed as $B_k(x)\phi(x)$, where $B_k(x)$ is a polynomial of degree k, and is even or odd depending on whether k is even or odd.

5.88. If we expand $\phi(t)$ about $t = 0$, then the leading term is the constant $\phi(0)$, which has an infinite integral on $(-\infty, 0)$.

5.89. It means that the result is good to approximately $22 \approx 6.6/\log_{10}(2)$ bits or $5.5 \approx 6.6/\log_{10}(16)$ hexadecimal digits.

5.90. For small values of t, the behavior is quite similar, with (5.9.4) converging very slightly faster for $t < 1.9$ and (5.9.3) converging somewhat faster for larger t. Almost 50 iterations are required to obtain $\Phi(5)$ to nine places using (5.9.4); "only" 39 iterations are required to obtain comparable accuracy with (5.9.3), which is good to almost fourteen places at the 47th iteration. Expression (5.9.4) does not remain below 1.0 until after the 40th iteration.

5.92. Hint: Define $S_n(w) = [A_n + A_{n-1}w]/[B_n + B_{n-1}w]$, then use induction.

5.93. $f_3 = 355/113$, which is accurate in the first seven digits. The well-known approximation of 22/7 is given by f_1.

5.96. $g(x) = -x$.

5.98. Section 5.9.4.3 contains some hints.

5.99. For computing tail probabilities, it is preferable to compute $(1 - p)$ and $(1 - \Phi(x_i))$ in (5.10.6). Why?

5.101. Note that $dz/dp = 1/\phi(z)$, which implies that absolute error grows at a faster rate than exponential, whereas $dz/z = [\Phi(z)/z\phi(z)]dp/p$. The latter expression is approximately $(-1/z^2)dp/p$ for p near zero. A similar computation holds for $(1-p)$ with p near one.

5.105. Let $a = (1 - 2/9n)$ and $b = (1 - 2/9m)$. Then

$$F^{1/3} = \frac{ab + Z\sqrt{a^2(1-b) + b^2(1-a) - (1-a)(1-b)Z^2}}{a^2 - (1-a)Z^2}.$$

Percent points derived in this way are not very accurate. If accuracy is important— as it often is—root finding methods using good approximations to the cdf must be used.

6.1. Hint: Compute $\lceil (x-a)/h \rceil$.

6.3. Look at $K'(x+)$ and $K'(x-)$ for $x = \pm 1$.

6.4. The most straightforward algorithm is $O(nk)$, where k is the predetermined number of bins. With some preprocessing this can be reduced to $O((n+k)\log k)$.

6.5. The first row is $(1/3, 1/3, 1/3, 0, 0, \ldots, 0)$.

6.6. As an example, suppose that near the left-hand portion of the data that $E[Y \mid x] \approx a + bx$, with $b > 0$. Then $\hat{y}_1$ is the average of $k+1$ values, k of which are *larger* than $E(Y_1)$.

6.7. Although both methods use equally-weighted means, the number of points on which $\hat{y}_i$ is based is a fixed function of i for the running-mean smoother, whereas the number depends on the spacings of the x_j's for the kernel smoother.

6.8. None in the middle of the data. Equation (6.2.6) shows that they are the same, since $\bar{x}_i = x_i$ when the data points are equally spaced and the neighborhoods are symmetric. Near the ends of the data, however, the running-line smoother is generally less biased. (It is unbiased if $M(x)$ is linear.)

6.9. Algorithm 6.2.1 can serve as an outline. At each step, both the mean of y and of x need to be updated, as well as the corrected sums of squares of x and cross-products of x and y. The cross-product updates can first be done relative to the old neighborhood means, the means can then be updated, and then the cross-products recentered using the centering formulæ of Section 2.3.

6.10. The regression computation at each of n points is $O(fn)$, where f is the span.

6.12. Hint: (6.2.8) can be derived directly from (6.2.7).

6.13. Use the binomial theorem and the normal approximation to the binomial distribution to show that after n applications, the weights are nearly proportional to $\phi([2k-n]/\sqrt{n})$, for $k = 0, \ldots, n$.

6.14. Yes, in a sense. The weights after n applications of the smoother are proportional to $\phi([k-np]/\sqrt{np(1-p)})$, with $k = 0, \ldots, n$. The convergence is less rapid than in the equally-weighted case.

6.15. Since M is assumed to be twice continuously differentiable, write $M(x) = a_0 + a_1 x + M_2(x)$ by expanding in a Maclaurin series. The penalty term depends only on M_2. As $\lambda \to \infty$, the penalty term goes to infinity as well for any $M_2 \neq 0$.

6.16. For each λ, the minimizer of (6.3.2) is a cubic spline with knots at the design points. The cubic natural interpolating spline is such a spline, and achieves the smallest possible value for the objective function, namely, zero. (The "natural" conditions make the spline interpolator linear to the left of x_1 and to the right of x_n.)

6.17. Regress y on 1, x, and $(x - a)_+$.

REFERENCES

Abramowitz, M., and Stegun, I. A., editors, (1964). *Handbook of Mathematical Functions,* National Bureau of Standards Applied Mathematics Series, Number 55. U. S. Government Printing Office: Washington.

Aitkin, Murray, and Tunnicliffe Wilson, Granville (1980). Mixture models, outliers, and the EM algorithm, *Technometrics,* **22,** 325–331.

Allen, David M. (1974). The relationship between variable selection and data augmentation and a method for prediction, *Technometrics,* **16,** 125–127.

Anderson, T. W. (1957). Maximum likelihood estimates for a multivariate normal distribution when some observations are missing, *Journal of the American Statistical Association,* **52,** 200–203.

Andrews, D. F. (1973). A general method for the approximation of tail areas, *Annals of Statistics,* **1,** 367–372.

Ansley, Craig F. (1979). An algorithm for the exact likelihood of a mixed autoregressive-moving average process, *Biometrika,* **66,** 59–65.

Ansley, Craig F. (1985). Quick proofs of some regression theorems via the QR Algorithm, *The American Statistician,* **39,** 55–59.

Ansley, Craig F., and Kohn, Robert (1983). Exact likelihood of vector autoregressive-moving average process with missing or aggregated data, *Biometrika,* **70,** 275–278.

Ansley, Craig F., and Kohn, Robert (1984). New algorithmic developments for estimation problems in time series, in *COMPSTAT 84: Proceedings in Computational Statistics,* T. Havránek, Z. Šidák, and M. Novák, editors. Physica-Verlag: Vienna, 23–34.

Ansley, Craig F., and Kohn, Robert (1985). Estimation, filtering, and smoothing in state space models with incompletely specified initial conditions, *Annals of Statistics,* **13,** 1286–1316.

Armstrong, R. D., Frome, E. L., and Kung, D. S. (1979). A revised simplex algorithm for the absolute deviation curve-fitting problem, *Communications in Statistics, Series B,* **8,** 175–190.

Armstrong, R. D. and Kung, M. T. (1982). An algorithm to select the best subset for a least absolute value regression problem, in *Optimization in Statistics*, S. H. Zanakis and J. S. Rustagi, editors. North-Holland: Amsterdam, 67–80.

Awad, Adnan M., and Shayib, Mohammed A. (1986). Tail area revisited, *Communications in Statistics, Series B*, **15**, 1215–1234.

Baker, R. J., and Nelder, J. A. (1978). *GLIM Manual, Release 3*. Numerical Algorithms Group and Royal Statistical Society: Oxford.

Barrodale, I., and Roberts, F. D. K. (1974). Algorithm 478: Solution of an overdetermined system of equations in the l_1 norm, *Communications of the ACM*, **17**, 319–320.

Barrodale, I., and Roberts, F. D. K. (1977). Algorithms for restricted LAV estimation, *Communications in Statistics, Series B*, **6**, 353–364.

Bartlett, M. (1951). An inverse matrix adjustment arising in discriminant analysis, *Annals of Mathematical Statistics*, **22**, 107–111.

Bates, Douglas M., Lindstrom, Mary J., Wahba, Grace, and Yandell, Brian S. (1987). GCVPACK—Routines for generalized cross validation, *Communications in Statistics, Series B*, **16**, 263–297.

Bates, Douglas M., and Watts, Donald G. (1984). A multi-response Gauss-Newton algorithm, *Communications in Statistics, Series B*, **13**, 705–715.

Beasley, J. D., and Springer, S. G. (1985). Algorithm AS 111: The percentage points of the normal distribution, in *Applied Statistics Algorithms*, P. Griffiths and I. D. Hill, editors. Ellis Horwood Limited: Chichester, 188–191.

Beaton, A. E. (1964). *The Use of Special Matrix Operators in Statistical Calculus*, Ed.D. thesis, Harvard University. [Reprinted as Research Bulletin 64-51, Educational Testing Service: Princeton, New Jersey.]

Beaton, A. E., Rubin, D. B., and Barone, J. L. (1976). The acceptability of regression solutions: Another look at computational accuracy, *Journal of the American Statistical Association*, **71**, 158–168.

Belsley, David A., Kuh, Edwin, and Welsch, Roy E. (1980). *Regression Diagnostics: Identifying Influential Data and Sources of Collinearity*. Wiley: New York.

Bentley, Donald L., and Cooke, Kenneth L. (1974). *Linear Algebra with Differential Equations*. Holt, Rinehart, and Winston: New York.

Berman, G. (1966). Minimization by successive approximation, *SIAM Journal on Numerical Analysis*, **3**, 123–133.

Best, D. J., and Roberts, D. E. (1985). Algorithm AS 91: The percentage points of the χ^2 distribution, in *Applied Statistics Algorithms*, P. Griffiths and I. D. Hill, editors. Ellis Horwood Limited: Chichester, 157–161.

Björck, Åke (1967). Solving linear least squares problems by Gram-Schmidt orthogonalization, *BIT*, **7**, 1–21.

Bloomfield, Peter (1976). *Fourier Analysis of Time Series: An Introduction.* Wiley: New York.

Bloomfield, P., and Steiger, W. (1980). Least absolute deviations curve-fitting, *SIAM Journal on Scientific and Statistical Computing*, **1**, 290–301.

Bohrer, R. E., and Schervish, M. J. (1981). An error-bounded algorithm for normal probabilities of rectangular regions, *Technometrics*, **23**, 297–300.

Box, George E. P., and Cox, David R. (1964). An analysis of transformations, *Journal of the Royal Statistical Society*, Series B, **26**, 211–252.

Box, George E. P., and Jenkins, Gwilym M. (1970). *Time Series Analysis: Forecasting and Control.* Holden-Day: San Francisco.

Bratley, Paul, Fox, Bennett L., and Schrage, Linus E. (1983). *A Guide to Simulation.* Springer-Verlag: New York.

Breiman, Leo, and Friedman, Jerome H. (1985). Estimating optimal transformations for multiple regression and correlation, (with Discussion), *Journal of the American Statistical Association*, **80**, 580–619.

Broyden, C. G. (1970). The convergence of a class of double-rank minimization algorithms, *Journal of the Institute of Mathematics and its Applications*, **6**, 76–90.

Buja, Andreas (1985). Theory of bivariate ACE, Technical Report Number 74, Department of Statistics, University of Washington.

Buja, Andreas, Hastie, Trevor, and Tibshirani, Robert (1987). Linear smoothers and additive models, Technical Report Number 10, Department of Statistics, University of Toronto.

Businger, P. A. (1970). Updating a singular value decomposition, *BIT*, **10**, 376–385.

Chambers, John M. (1968). Computing for statistics: Developments and issues, *The Logistics Review*, **4**, 27–40.

Chambers, John M. (1971). Regression updating, *Journal of the American Statistical Association*, **66**, 744–748.

Chambers, John M. (1973). Fitting nonlinear models: Numerical techniques, *Biometrika*, **60**, 1–13.

Chambers, John M. (1977). *Computational Methods for Data Analysis.* Wiley: New York.

Chan, Tony F. (1982a). An improved algorithm for computing the singular value decomposition, *ACM Transactions on Mathematical Software*, **8**, 72–83.

Chan, Tony F. (1982b). Algorithm 581: An improved algorithm for computing the singular value decomposition, *ACM Transactions on Mathematical Software*, **8**, 84–88.

Chan, Tony F., Golub, Gene H., and LeVeque, Randall J. (1983). Algorithms for computing the sample variance: Analysis and recommendations, *The American Statistician*, **37**, 242–247.

Chan, Tony F. C., and Lewis, J. G. (1978). Rounding error analysis of algorithms for computing means and standard deviations, Technical Report Number 284, The Johns Hopkins University, Department of Mathematical Sciences.

Chan, Tony F. C., and Lewis, J. G. (1979). Computing standard deviations: Accuracy, *Communications of the ACM*, **22**, 526–531.

Chernoff, Herman (1973). The use of faces to represent points in k-dimensional space graphically, *Journal of the American Statistical Association*, **68**, 361–368.

Chow, Y.-S., Geman, S., and Wu, L.-D. (1983). Consistent cross-validated density estimation, *Annals of Statistics*, **11**, 25–38.

Clasen, B. J. (1888). Sur une nouvelle méthode de résolution des équations linéaires et sur l'application de cette méthode au calcul des déterminants, *Ann. Soc. Sci. Bruxelles (2)*, **12**, 251–281.

Cleveland, William S. (1979). Robust locally weighted regression and smoothing scatter plots, *Journal of the American Statistical Association*, **74**, 829–836.

Cleveland, William S. (1981). LOWESS: A program for smoothing scatter plots by robust locally weighted regression, *The American Statistician*, **35**, 54.

Cody, W. J. (1981). Analysis of proposals for the floating-point standard, *Computer*, **14:3**, 63–68.

Coleman, D., Holland, P., Kaden, N., Klema, V., and Peters, S. C. (1980). A system of subroutines for iteratively reweighted least squares computations, Coleman, et al, *ACM Transactions on Mathematical Software*, **6**, 327–336.

Cook, R. Dennis, and Weisberg, Sanford (1982). *Residuals and Influence in Regression.* Chapman and Hall: New York.

Cook, R. Dennis, and Weisberg, Sanford (1983). Diagnostics for heteroscedasticity in regression, *Biometrika*, **70**, 1–10.

Coonen, Jerome T. (1981). Underflow and denormalized numbers, *Computer*, **14:3**, 75–87.

Cooper, B. E. (1968). Algorithm AS 2: The normal integral, *Applied Statistics*, **17**, 186–187.

Cooper, B. E. (1985). Algorithm AS 3: The integral of Student's t-distribution, in *Applied Statistics Algorithms*, P. Griffiths and I. D. Hill, editors. Ellis Horwood Limited: Chichester, 38–39.

Cotton, E. W. (1975). Remark on Stably updating mean and standard deviation of data, *Communications of the ACM*, **18**, 458. stable than textbook method

Cox, D. R., and Hinkley, D. V. (1974). *Theoretical Statistics*. Chapman & Hall: London.

Cramér, Harald (1955). *The Elements of Probability Theory and Some of its Applications*. Wiley: New York. (Reprinted by Robert E. Krieger, 1973).

Daniels, H. E. (1987). Tail probability approximations, *International Statistical Review*, **55**, 37–48.

Davidon, W. C. (1959). Variable metric methods for minimization, AEC Research and Development Report ANL-5990, Argonne National Laboratory, Illinois.

Davis, Philip J. and Rabinowitz, Philip (1984). *Methods of Numerical Integration*, second edition. Academic Press: New York.

de Boor, Carl (1978). *A Practical Guide to Splines*. Springer-Verlag: New York.

de Hoog, F. R., and Hutchinson, M. F. (1987). An efficient method for calculating smoothing splines using orthogonal transformations, *Numerische Mathematik*, **50**, 311–319.

Demmel, James (1984). Underflow and the reliability of numerical software, *SIAM Journal on Scientific and Statistical Computing*, **5**, 887–919.

Dempster, A. P (1969). *Continuous Multivariate Analysis*. Addison-Wesley: Reading, Massachusetts.

Dempster, A. P., Laird, N., and Rubin, D. B. (1977). Maximum likelihood from incomplete data via the EM algorithm, (with Discussion), *Journal of the Royal Statistical Society*, Series B, **39**, 1–38.

Dennis, John E., Jr., Gay, David M., and Welsch, Roy E. (1981). An adaptive nonlinear least-squares algorithm, *ACM Transactions on Mathematical Software*, **7**, 369–383.

Dennis, John E., Jr., and Moré, Jorge J. (1977). Quasi-Newton methods, motivation and theory, *SIAM Review*, **19**, 46–89.

Dennis, J. E., Jr., and Schnabel, Robert B. (1983). *Numerical Methods for Unconstrained Optimization and Nonlinear Equations*. Prentice-Hall: Englewood Cliffs.

Deuchler, Gustav Adolf (1914). Über die Methoden der Korrelationsrechnung in der Pädagogik und Psychologie, *Zeitschrift für Pädagogische Psychologie und Experimentelle Pädagogik*, **15**, 114–131; 145–159; 229–242.

Devroye, Luc (1986). *Non-uniform Random Variate Generation.* Springer-Verlag: New York.

Devroye, Luc, and Györfi, László (1985). *Nonparametric Density Estimation: The L_1 View.* Wiley: New York.

Diaconis, Persi, and Efron, Bradley (1983). Computer-intensive methods in statistics, *Scientific American*, **249**, (May), 116–130.

Diaconis, Persi, and Shahshahani, Mehrdad (1984). On nonlinear functions of linear combinations, *SIAM Journal on Scientific and Statistical Computing*, 5, 175–191.

DiDonato, A. R., and Hageman, R. K. (1982). A method for computing the integral of the bivariate normal distribution over an arbitrary region, *SIAM Journal on Scientific and Statistical Computing*, 3, 434–446.

DiDonato, A. R., Jarnagin, M. P., Jr., and Hageman, R. K. (1980). Computation of the integral of the bivariate normal distribution over convex polygons, *SIAM Journal on Scientific and Statistical Computing*, 1, 179–186.

Dielman, Terry, and Pfaffenberger, Roger (1982). LAV (Least Absolute Value) estimation in linear regression: A review, in *Optimization in Statistics*, S. H. Zanakis and J. S. Rustagi, editors. North-Holland: Amsterdam, 31–52.

Dierckx, P. (1984). Algorithms for smoothing data on the sphere with tensor product splines, *Computing*, **32**, 319–342.

Dongarra, J. J., Bunch, J. R., Moler, Cleve B., and Stewart, G. W. (1979). *Linpack Users Guide.* Society for Industrial and Applied Mathematics: Philadelphia.

Draper, Norman, and Smith, Harry (1981). *Applied Regression Analysis*, Second edition. Wiley: New York.

Durbin, J., and Watson, G. S. (1951). Testing for serial correlation in least squares regression, *Biometrika*, **37**, 409–428.

Dutt, J. E., and Lin, T. K. (1975). A short table for computing normal orthant probabilities of dimensions four and five, *Journal of Statistical Computation and Simulation*, 4, 95–120.

Dutter, R. (1975). Robust regression: Different approaches to numerical solutions and algorithms, Research Report Number 6, Fachgruppe für Statistik, Eidgenössische Technische Hochschule: Zürich.

Eddy, William F., and Gentle, James E. (1985). Statistical computing: What's past is prologue, in *A Celebration of Statistics*, Anthony C. Atkinson and Stephen E. Fienberg, editors. Springer-Verlag: New York, 233–249.

Efron, B. (1982). *The Jackknife, the Bootstrap, and Other Resampling Plans.* National Science Foundation-Conference Board of the Mathematical Sciences Monograph 38, Society for Industrial and Applied Mathematics: Philadelphia.

EISPACK (1976). *Matrix Eigensystem Routines—Eispack Guide*. Second edition. See B. T. Smith, et al.

EISPACK 2 (1972). *Matrix Eigensystem Routines: EISPACK Guide Extension*. See B. S. Garbow, et al.

Enslein, Kurt, Ralston, Anthony, and Wilf, Herbert S. (1977). *Statistical Methods for Digital Computers*. Wiley-Interscience: New York.

Evans, M., and Swartz, T. (1986). Monte Carlo computation of some multivariate normal probabilities, Technical Report Number 7, Department of Statistics, University of Toronto.

Evans, M., and Swartz, T. (1987). Sampling from Gauss rules, *SIAM Journal on Scientific and Statistical Computing*, **8**, in press.

Farebrother, R. W. (1986). Testing linear inequality constraints in the standard linear model, *Communications in Statistics, Series A*, **15**, 7–31

Fisher, R. A. (1925). *Statistical Methods for Research Workers*. Oliver and Boyd: Edinburgh.

Fisher, R. A. (1929). Moments and product moments of sampling distributions, *Proceedings of the London Mathematical Society, Series 2*, **30**, 199–238.

Fisherkeller, Mary Ann, Friedman, Jerome H., and Tukey, John W. (1974). PRIM-9: An interactive multidimensional data display and analysis system. *SLAC-PUB-1408*, Stanford Linear Accelerator Center, Stanford, CA.

Fix, G., and Heiberger, R. (1972). An algorithm for the ill-conditioned generalized eigenvalue problem, *SIAM Journal on Numerical Analysis*, **9**, 78–88.

Fletcher, R. (1970). A new approach to variable metric algorithms, *Computer Journal*, **13**, 317–322.

Fletcher, R. (1981). *Practical Methods of Optimization, Volume 2: Constrained Optimization*. John Wiley: Chichester.

Forsythe, George E., and Moler, Cleve B. (1967). *Computer Solution of Linear Algebraic Systems*. Prentice Hall: Englewood Cliffs, New Jersey.

Forsythe, George E., and Wasow, Wolfgang R. (1960). *Finite Difference Methods for Partial Differential Equations*. Wiley: New York.

Fowlkes, E. B., and Kettenring, J. R. (1985). Comment: The ACE method of optimal transformations, *Journal of the American Statistical Association*, **80**, 607–613. (Discussion of Breiman and Friedman, 1985).

Francis, J. G. F. (1961). The QR transformation: A unitary analog to the LR transformation, *Computer Journal*, **4**, 165–271, 332–334.

Friedman, Jerome H. (1984). A variable span smoother, Technical Report Number 5, Laboratory for Computational Statistics, Department of Statistics, Stanford University.

Friedman, J. H., Baskett, F., and Shustek, L. J. (1975). An algorithm for finding nearest neighbors, *IEEE Transactions on Computing*, **24**, 1000–1006.

Friedman, J. H., Bentley, J. L., and Finkel, R. A. (1977). An algorithm for finding best matches in logarithmic expected time, *ACM Transactions on Mathematical Software*, **3**, 209–226.

Friedman, Jerome H., Grosse, Eric, and Stuetzle, Werner (1983). Multidimensional additive spline approximation, *SIAM Journal on Scientific and Statistical Computing*, **4**, 291–301.

Friedman, Jerome H., and Stuetzle, Werner (1981). Projection pursuit regression, *Journal of the American Statistical Association*, **76**, 817–823.

Friedman, J H., Stuetzle, W., and Schroeder, A. (1984). Projection pursuit density estimation, *Journal of the American Statistical Association*, **79**, 599–608.

Friedman, Jerome H., and Tukey, John W. (1974). A projection pursuit algorithm for exploratory data analysis, *IEEE Transactions on Computers*, **C-23**, 881–889.

Friedman, J. H., and Wright, M. H. (1981a). A nested partitioning procedure for numerical multiple integration, *ACM Transactions on Mathematical Software*, **7**, 76–92.

Friedman, J. H., and Wright, M. H. (1981b). DIVONNE4, a program for multiple integration and adaptive importance sampling, CGTM Number 193, Computation Research Group, Stanford Linear Accelerator Center: Stanford.

Fuller, Wayne A. (1976). *Introduction to Statistical Time Series*. Wiley: New York.

Garbow, B. S., Boyle, J. M., Dongarra, J. J., and Moler, C. B. (1972). *Matrix Eigensystem Routines: EISPACK Guide Extension*. Springer-Verlag: New York. [This material has been incorporated into the second edition of the EISPACK Guide. See B. T. Smith (1976).]

Gautschi, Walter (1968). Construction of Gauss-Christoffel quadrature formulas, *Mathematics of Computation*, **22**, 251–270.

Gautschi, Walter (1970). On the construction of Gaussian quadrature rules from modified moments, *Mathematics of Computation*, **24**, 245–260.

Gautschi, Walter (1982). On generating orthogonal polynomials, *SIAM Journal on Scientific and Statistical Computing*, **3**, 289–317.

Gehrlein, W. V. (1979). A representation for quadrivariate normal positive orthant probabilities, *Communications in Statistics, Series B*, **8**, 349–358.

Geisser, Seymour (1975). The predictive sample reuse method with applications, *Journal of the American Statistical Association*, **70**, 320–328.

Gentle, James E. (1977). Least absolute values estimation: An introduction, *Communications in Statistics, Series B*, **6**, 313–328.

Gentleman, W. M., and Jenkins, M. A. (1968). An approximation for Student's t-distribution, *Biometrika*, **55**, 571–572.

Gill, P. E., Golub, Gene H., Murray, W., and Saunders, M. A. (1974). Methods for modifying matrix factorizations, *Mathematics of Computation*, **28**, 505–535.

Gill, Philip E., Murray, Walter, and Wright, Margaret (1981). *Practical Optimization*. Academic Press: London.

Godolphin, E. J., and Unwin, J. M. (1983). Evaluation of the covariance matrix for the maximum likelihood estimator of a Gaussian autoregressive-moving average process, *Biometrika*, **70**, 279–284.

Goldfarb, D. (1970). A family of variable metric methods derived by variational means, *Mathematics of Computation*, **24**, 23–26.

Golub, Gene H. (1969). Matrix decompositions and statistical calculations, in *Statistical Computation*, R. C. Milton and J. A. Nelder, eds. Academic Press: New York, 365–397.

Golub, Gene H., Heath, Michael, and Wahba, Grace (1979). Generalized cross validation as a method for choosing a good ridge parameter, *Technometrics*, **21**, 215–224.

Golub, Gene H., and Kahan, W. (1965). Calculating the singular values and pseudoinverse of a matrix, *SIAM Journal on Numerical Analysis*, **2**, 205–224.

Golub, Gene H., and Reinsch, C. (1970). Singular value decomposition and least squares solutions, *Numerische Mathematik*, **14**, 402–420.

Golub, Gene H., and Van Loan, Charles F. (1983). *Matrix Computations*. The Johns Hopkins University Press: Baltimore.

Golub, Gene H., and Welsch, J. H. (1969). Calculation of Gauss quadrature rules, *Mathematics of Computation*, **23**, 221-230. [*Microfiche Supplement*, **23**, Number 106, contains the associated ALGOL computer programs.]

Goodnight, James H. (1979). A tutorial on the SWEEP operator, *The American Statistician*, **33**, 149–158.

Gower, J. C. (1985). The development of statistical computing at Rothamsted, Rothamsted Annual Report, Part 2.

Green, P. J. (1984). Iteratively reweighted least squares for maximum likelihood estimation, and some robust and resistant alternatives, (with Discussion), *Journal of the Royal Statistical Society*, Series B, **46**, 149–192.

Greville, T. N. E., editor, (1969a). *Theory and Applications of Spline Functions*. Academic Press: New York.

Greville, T. N. E. (1969b). Introduction to spline functions, in *Theory and Applications of Spline Functions*, T. N. E. Greville, editor. Academic Press: New York, 1–35.

Griffiths, P., and Hill, I. D., editors, (1985). *Applied Statistics Algorithms*. Ellis Horwood Limited: Chichester.

Haber, S. (1970). Numerical evaluation of multiple integrals, *SIAM Review*, **12**, 481–526.

Hammersley, I. M., and Handscomb, D. C. (1964). *Monte Carlo Methods*. Chapman and Hall: London.

Hanson, R. J. (1975). Stably updating mean and standard deviation of data, *Communications of the ACM*, **18**, 57–58.

Harman, Harry H. (1967). *Modern Factor Analysis*, Second edition, revised. University of Chicago Press: Chicago.

Hastie, T., and Tibshirani, R. (1986). Generalized additive models, (with Discussion), *Statistical Science*, **1**, 297–318.

Henrici, Peter (1964). *Elements of Numerical Analysis*. Wiley: New York.

Hestenes, M. R., and Stiefel, E. (1952). Methods of conjugate gradients for solving linear systems, *J. Research of the National Bureau of Standards*, **49**, 409–436.

Hill, I. D. (1985). Algorithm AS 66: The normal integral, in *Applied Statistics Algorithms*, P. Griffiths and I. D. Hill, editors. Ellis Horwood Limited: Chichester, 126–129.

Hill, G. W. (1970a). Algorithm 395: Student's t-distribution, *Communications of the ACM*, **13**, 617–619.

Hill, G. W. (1970b). Algorithm 396: Student's t-quantiles, *Communications of the ACM*, **13**, 619–620.

Hoaglin, David C., and Welsch, Roy E. (1978). The hat matrix in regression and ANOVA, *The American Statistician*, **32**, 17–22. *Corrigendum*, **32**, 146.

Hocking, R. R. (1977). Selection of the best subset of regression variables, in *Statistical Methods for Digital Computers*, Kurt Enslein, Anthony Ralston, and Herbert S. Wilf, editors. Wiley-Interscience: New York, 39–57.

Holland, Paul, and Welsch, Roy E. (1977). Robust regression using iteratively reweighted least squares, *Communications in Statistics*, Series A, **6**, 813–827.

Householder, Alston S. (1953). *Principles of Numerical Analysis*. McGraw-Hill: New York. [Paper reprint, Dover: New York, 1974.]

Householder, Alston S. (1964). *The Theory of Matrices in Numerical Analysis*. Blaisdell: New York. [Paper reprint, Dover: New York, 1975.]

Huber, Peter J. (1981). *Robust Statistics*. Wiley: New York.

Huber, Peter J. (1985). Projection pursuit, (with Discussion), *Annals of Statistics*, **13**, 435–525.

Huh, Moon Y. (1986). Computation of percentage points, *Communications in Statistics, Series B*, **15**, 1191–1198.

Institute of Electrical and Electronics Engineers (1985). *IEEE standard for Binary Floating-Point Arithmetic*. (Standard 754-1985). IEEE: New York.

Jacobi, C. G. J. (1846). Über ein Leichtes Verfahren Die in der Theorie der Sacularstorungen Vorkommendern Gleichungen Numerisch Aufzulosen, *Crelle's Journal*, **30**, 51–94.

Jarratt, P. (1970). A review of methods for solving nonlinear algebraic equations in one variable, in *Numerical Methods for Nonlinear Algebraic Equations*, P. Rabinowitz, editor. Gordon and Breach: London.

Jennrich, Robert I. (1977). Stepwise regression, in *Statistical Methods for Digital Computers*, Kurt Enslein, Anthony Ralston, and Herbert S. Wilf, editors. Wiley-Interscience: New York, 58–75.

Johnson, Mark E. (1987). *Multivariate Statistical Simulation*. John Wiley: New York.

Johnstone, Iain M., and Velleman, Paul F. (1985). Efficient scores, variance decompositions, and Monte Carlo swindles, *Journal of the American Statistical Association*, **80**, 851–862.

Jones, R. H. (1980). Maximum likelihood fitting of ARMA models to time series with missing observations, *Technometrics*, **22**, 389–395.

Jones, William B., and Thron, W. J. (1980). *Continued Fractions: Analytic Theory and Applications*. Addison-Wesley: Reading. [Volume 11 of the *Encyclopedia of Mathematics and its Applications*.]

Jordan, W. (1904). *Handbuch der Vermessungskunde*, **1**, Fifth edition. J. B. Metzlerscher Verlag: Stuttgart.

Jørgensen, Bent (1983). Maximum likelihood estimation and large-sample inference for generalized linear and nonlinear regression models, *Biometrika*, **70**, 19–28.

Kahaner, David K., and Rechard, Ottis W. (1987). TWODQD an adaptive routine for two-dimensional integration, *Journal of Computational and Applied Mathematics*, **17**, 215–234.

Kendall, Sir Maurice, and Stuart, Alan (1977). *The Advanced Theory of Statistics, Volume 1*, Fourth edition. Macmillan: New York.

Kendall, Sir Maurice, and Stuart, Alan (1979). *The Advanced Theory of Statistics, Volume 2,* Fourth edition. Macmillan: New York.

Kennedy, William J., and Gentle, James E. (1980). *Statistical Computing.* Marcel Dekker: New York.

Khuri, A. I. (1976). A constrained least-squares problem, *Communications in Statistics, Series B,* **5**, 82–84.

Kiefer, J. (1953). Sequential minimax search for a maximum, *Proceedings of the American Mathematical Society,* **4**, 502–506.

Kiefer, J. (1957). Optimal sequential search and approximation methods under minimum regularity conditions, *Journal of the Society for Industrial and Applied Mathematics,* **5**, 105–136.

Knuth, Donald E. (1968). *The Art of Computer Programming, 1: Fundamental Algorithms,* First edition. Addison-Wesley: Reading, Massachusetts. [Second edition, 1973.]

Knuth, Donald E. (1981). *The Art of Computer Programming, 2: Seminumerical Algorithms,* Second edition. Addison-Wesley: Reading, Massachusetts.

Knuth, Donald E. (1973). *The Art of Computer Programming, 3: Sorting and Searching.* Addison-Wesley: Reading, Massachusetts.

Knuth, Donald E. (1984). *The TEXbook.* Addison-Wesley: Reading, Massachusetts.

Kohn, Robert, and Ansley, Craig (1982). A note on obtaining the theoretical autocovariances of an ARMA process, *Journal of Statistical Computation and Simulation,* **15**, 273–283.

Kohn, Robert, and Ansley, Craig F. (1987). A new algorithm for spline smoothing based on smoothing a stochastic process, *SIAM Journal on Scientific and Statistical Computing,* **8**, 33–48.

Kourouklis, S., and Paige, C. C. (1981). A constrained least-squares approach to the general Gauss-Markov linear model, *Journal of the American Statistical Association,* **76**, 620–625.

Kronrod, Aleksandr Semenovich (1964). *UZLY I VESA KVADRATURNYKH FORMUL.* Nauka Press of the Academy of Sciences of the USSR: Moscow.

Kronrod, A. S. (1965). *Nodes and Weights of Quadrature Formulas.* Consultants Bureau: New York. [English translation of Kronrod (1964)]

Kruskal, J. B. (1969). Toward a practical method which helps uncover the structure of a set of multivariate observations by finding a linear transformation which optimizes a new 'index of condensation'. in *Statistical Computation,* R. C. Milton and J. A. Nelder, editors. Academic Press: New York.

Kruskal, W. H. (1957). Historical notes on the Wilcoxon unpaired two-sample test, *Journal of the American Statistical Association*, **52**, 356–360.

Lawson, Charles L., and Hanson, Richard J. (1974). *Solving Least Squares Problems*. Prentice-Hall: Englewood Cliffs, New Jersey.

Lenth, R. V. (1977). Robust splines, *Communications in Statistics, Series A*, **6**, 847–854.

Levenberg, K. (1944). A method for the solution of certain problems in least squares, *Quarterly of Applied Mathematics*, **2**, 164–168.

LINPACK (1979). *Linpack Users Guide. See Dongarra, et al.*

Little, Roderick J. A., and Rubin, Donald B. (1983). On jointly estimating parameters and missing data by maximizing the complete-data likelihood, *The American Statistician*, **37**, 218–220.

Longley, J. W. (1967). An appraisal of least-squares for the electronic computer from the point of view of the user, *Journal of the American Statistical Association*, **62**, 819–841.

Louis, Thomas A. (1982). Finding the observed information matrix when using the EM algorithm, *Journal of the Royal Statistical Society, Series B*, **44**, 226–233.

Lyness, J. N. (1983). When not to use an automatic quadrature routine, *SIAM Review*, **25**, 63–87.

McCullagh, Peter (1983). Quasi-likelihood functions, *Annals of Statistics*, **11**, 59–67.

McCullagh, Peter (1987). *Tensor Methods in Statistics*. Chapman and Hall: London.

McCullagh, P., and Nelder, J. A. (1983). *Generalized Linear Models*. Chapman and Hall: London.

McDonald, John Alan (1986). Periodic smoothing of time series, *SIAM Journal on Scientific and Statistical Computing*, **7**, 665–688.

McDonald, John Alan, and Owen, Art B. (1986). Smoothing with split linear fits, *Technometrics*, **28**, 195–208.

McIntosh, Allen (1982). *Fitting Linear Models: An Application of Conjugate Gradient Algorithms*. Lecture Notes in Statistics **10**. Springer-Verlag: New York.

McLeod, A. I. (1975). Derivation of the theoretical autocovariance function of autoregressive moving average time series, *Applied Statistics*, **24**, 255–256.

Maindonald, J. H. (1984). *Statistical Computation*. Wiley: New York.

Mann, Nancy R. (1978). Everything you always wanted to know about the history of Computer Science and Statistics: Annual Symposia on the Interface—and more, in *Computer Science and Statistics: Eleventh*

Annual Symposium on the Interface, A. Ronald Gallant and Thomas M. Gerig, editors. Institute of Statistics, North Carolina State University: Raleigh, 2–5.

Marquardt, D. (1963). An algorithm for least-squares estimation of nonlinear parameters, *SIAM Journal of Applied Mathematics*, **11**, 431–441.

Marquardt, Donald W. (1970). Generalized inverses, ridge regression biased linear estimation, and nonlinear estimation, *Technometrics*, **12**, 591–612.

Marron, J. S. (1987). A comparison of cross-validation techniques in density estimation, *Annals of Statistics*, **15**, 152–162.

Martin, D. W., and Tee, G. J. (1961). Iterative methods for linear equations with symmetric positive definite matrix, *Computer Journal*, **3**, 242–254.

Martin, R. S., and Wilkinson, J. H. (1965). Symmetric decomposition of positive definite band matrices, *Numerische Mathematik*, **7**, 355–361.

Martin, R. S., and Wilkinson, J. H. (1968). Reduction of the symmetric eigenproblem $Ax = \lambda Bx$ and related problems to standard form, *Numerische Mathematik*, **12**, 349–368.

Mason, R. L., Gunst, Richard F., and Webster, J. T. (1975). Regression analysis and problems of multicollinearity, *Communications in Statistics*, **4**, 277–292.

Mélard, G. (1984a). Algorithm AS 197. A fast algorithm for the exact likelihood of autoregressive-moving average models, *Applied Statistics*, **33**, 104–114.

Mélard, G. (1984b). On fast algorithms for several problems in time series models, in *COMPSTAT 84: Proceedings in Computational Statistics*, T. Havránek, Z. Šidák, and M. Novák, editors. Physica-Verlag: Vienna, 41–45.

Milton, Roy C., and Nelder, John A., editors, (1969). *Statistical Computation*. Academic Press: New York.

Moler, Cleve B., and Stewart, G. W. (1973). An algorithm for generalized matrix eigenvalue problems, *SIAM Journal on Numerical Analysis*, **10**, 241–256.

Moolgavkar, Suresh H., Lustbader, Edward D., and Venzon, David J. (1984). A geometric approach to nonlinear regression diagnostics with application to matched case-control studies, *Annals of Statistics*, **12**, 816–826.

Morgan, Byron J. T. (1984). *Elements of Simulation*. Chapman and Hall: London.

Mosteller, Frederick, and Tukey, John W. (1977). *Data Analysis and Regression*. Addison-Wesley: Reading, Massachusetts.

Muller, D. E. (1956). A method for solving algebraic equations using an automatic computer, *Mathematical Tables and Aids Computation*, **10**,

208–215.

Murtagh, F. (1984). A review of fast techniques for nearest neighbour searching, in *COMPSTAT 84: Proceedings in Computational Statistics*, T. Havránek, Z. Šidák, and M. Novák, editors. Physica-Verlag: Vienna, 143–147.

Nash, J. C. (1975). A one-sided transformation method for the singular value decomposition and algebraic eigenproblem, *Computer Journal*, **18**,74–76.

Naylor, J. C., and Smith, A. F. M. (1982). Applications of a method for the efficient computation of posterior distributions, *Applied Statistics*, **31**, 214–225.

Neely, P. M. (1966). Comparison of several algorithms for computation of means and correlation coefficients, *Communications of the ACM*, **9**, 496–499.

Nelder, John A. (1984). Present position and potential developments: Some personal views, Statistical computing, *Journal of the Royal Statistical Society*, Series A, **147**, (Part 2), 151–160.

Nelder, John A. (1985). An alternative interpretation of the singular-value decomposition in regression, *The American Statistician*, **39**, 63–64.

Nelder, John A., and Wedderburn, R. W. M. (1972). Generalized linear models, *Journal of the Royal Statistical Society*, Series A, **135**, 370–384.

Neter, John, and Wasserman, Stanley (1974). *Applied Linear Statistical Models*. Richard D. Irwin: Homewood, Illinois.

Niederreiter, H. (1978). Quasi-Monte Carlo methods and pseudo-random numbers, *Bulletin of the American Mathematical Society*, **84**, 957–1042.

O'Sullivan, F., Yandell, B., and Raynor, W. (1986). Automatic smoothing of regression functions in generalized linear models, *Journal of the American Statistical Association*, **81**, 96–103.

Orchard, T., and Woodbury, M. A. (1972). A missing information principle: Theory and applications, *Sixth Berkeley Symposium on Mathematical Statistics and Probability*, L. M. LeCam, J. Neyman, and E. L. Scott, editors, **1**, 697–715.

Ortega, James M., and Rheinboldt, Werner C. (1970). *Iterative solution of nonlinear equations in several variables*. Academic Press: New York.

Paige, C. C. (1979a). Computer solution and perturbation analysis of generalized least squares problems, *Mathematics of Computation*, **33**, 171–184.

Paige, C. C. (1979b). Fast numerically stable computations for generalized linear least squares problems, *SIAM Journal on Numerical Analysis*, **16**, 165–171.

Paige, C. C., and Saunders, M. (1981). Towards a generalized singular value decomposition, *SIAM Journal on Numerical Analysis*, **18**, 398–405.

Patterson, T. N. L. (1968a). The optimum addition of points to quadrature formulae, *Mathematics of Computation*, **22**, 847–856. Errata, *Mathematics of Computation*, **23**, 892 (1969).

Patterson, T. N. L. (1968b). On some Gauss and Lobatto based integration formulae, *Mathematics of Computation*, **22**, 877–881.

Patterson, T. N. L. (1973). Algorithm 468. Algorithm for automatic numerical integration over a finite interval, *Communications of the ACM*, **16**, 694–699.

Paulson, E. (1942). An approximate normalization of the analysis of variance distribution, *Annals of Mathematical Statistics*, **13**, 233–235.

Peizer, David B., and Pratt, John W. (1968). A normal approximation for binomial, F, beta, and other common, related tail probabilities,I, *Journal of the American Statistical Association*, **63**,1416–1456.

Peters, Stephen C. (1980). Computational elements, Appendix 2B to Belsley, Kuh, and Welsch (1980), *Regression Diagnostics*. 69–84.

Piessens, R., de Doncker-Kapenga, E., Überhuber, C., and Kahaner, D. (1983). *QUADPACK, A Quadrature Subroutine Package*, Series in Computational Mathematics, **1**, Springer-Verlag: Berlin.

Plackett, R. L. (1950). Some theorems in least squares, *Biometrika*, **37**, 149–157.

Polasek, Wolfgang (1984). Regression diagnostics for general linear regression models, *Journal of the American Statistical Association*, **79**, 336–340.

Powell, M. J. D. (1981). *Approximation Theory and Methods*. Cambridge University Press: Cambridge.

Press, William H., Flannery, Brian P., Teukolsky, Saul A., and Vetterling, William T. (1986). *Numerical Recipes: The Art of Scientific Computing*. Cambridge University Press: Cambridge.

QUADPACK (1983). *QUADPACK: A Subroutine Package for Automatic Integration. See Piessens, et al.*

Rabinowitz, Philip, editor, (1971). *Numerical Methods for Nonlinear Algebraic Equations*. Gordon and Breach: London.

Redner, Richard A., and Walker, Homer F. (1984). Mixture densities, maximum likelihood and the EM algorithm, *SIAM Review*, **26**, 195–202.

Reeds, James A. (1985). Asymptotic number of roots of Cauchy location likelihood equations, *Annals of Statistics*, **13**, 775–784.

Rheinboldt, W. C. (1974). *Methods for Solving Systems of Nonlinear Equations*. CBMS Series in Applied Mathematics number 14. SIAM: Philadelphia.

Rice, John R. (1983). *Numerical methods, Software, and Analysis.* McGraw-Hill: New York.

Rosenbrock, H. H. (1960). An automatic method for finding the greatest or least value of a function, *Computer Journal*, **3**, 175–184.

ROSEPACK (1980).A system of subroutines for iteratively reweighted least squares computations, *see D. Coleman, et al.*

Rubinstein, Reuven Y. (1981). *Simulation and the Monte Carlo Method.* Wiley: New York.

Rutishauser, H. (1954). Der Quotienten-Differenzen-Algorithmus, *Zeitschrift für Angewandte Mathematik und Physik*, **5**, 233–251.

Sager, Thomas, and Thisted, Ronald A. (1982). Maximum likelihood estimation of isotonic modal regression, *Annals of Statistics*, **10**, 690–707.

SAS Institute, Inc. (1982a). *SAS User's Guide: Basics, 1982 Edition.* SAS Institute: Cary, North Carolina.

SAS Institute, Inc. (1982b). *SAS User's Guide: Statistics, 1982 Edition.* SAS Institute: Cary, North Carolina.

Schervish, M. J. (1984). Multivariate normal probabilities with error bound, *Applied Statistics*, **33**, 81–94. *Correction (1985)*, **34**, 103–104.

Schiffelbein, Paul (1986). A remark on algorithm AS 176. Kernel density estimation using the Fast Fourier Transform, *Applied Statistics*, **35**, 235–236.

Schnabel, Robert B., Koontz, John B., and Weiss, Barry E. (1985). A modular system of algorithms for unconstrained optimization, *ACM Transactions on Mathematical Software*, **11**, 419–440.

Scott, David W. (1979). On optimal and data-based histograms, *Biometrika*, **66**, 605–610.

Scott, David W. (1986). Choosing smoothing parameters for density estimation, in *Computer Science and Statistics: Eighteenth Symposium on the Interface, Proceedings*, Thomas J. Boardman, editor. American Statistical Association: Washington, 225–229.

Scott, David W., and Terrell, George R. (1986). Biased and unbiased cross-valiation in density estimation, Technical Report Number 23, Laboratory for Computational Statistics, Department of Statistics, Stanford University.

Shanno, David F. (1970). Conditioning of quasi-Newton methods for function minimization, *Mathematics of Computation*, **24**, 647–657.

Shanno, David F., and Phua, K. H. (1980). Remark on algorithm 500: Minimization of unconstrained multivariate functions, *ACM Transactions on Mathematical Software*, **6**, 618–622.

Shanno, David F., and Rocke, David M. (1986). Numerical methods for robust regression: Linear models, *SIAM Journal on Scientific and Statistical Computing*, **7**, 86–97.

Shaw, J. E. H. (1986). A quasirandom approach to integration in Bayesian statistics, Technical Report Number 14, Statistics Department, University of Warwick.

Sherman, J., and Morrison, W. J. (1949). Adjustment of an inverse matrix corresponding to changes in the elements of a given column or a given row of the original matrix, (Abstract). *Annals of Mathematical Statistics*, **20**, 621.

Siegel, Andrew F., and O'Brien, Fanny (1985). Unbiased Monte Carlo integration methods with exactness for low order polynomials, *SIAM Journal on Scientific and Statistical Computing*, **6**, 169–181.

Silverman, B. W. (1978). Choosing the window width when estimating a density, *Biometrika*, **65**, 1–11.

Silverman, B. W. (1982). Algorithm AS 176. Kernel density estimation using the Fast Fourier Transform, *Applied Statistics*, **31**, 93–99.

Silverman, B. W. (1985). Some aspects of the spline smoothing approach to non-parametric regression curve fitting, *Journal of the Royal Statistical Society*, Series B, **47**,1–52.

Silverman, B. W. (1986). *Density Estimation for Statistics and Data Analysis*. Chapman and Hall: London.

Silvey, S. D. (1969). Multicollinearity and imprecise estimation, *Journal of the Royal Statistical Society*, Series B, **31**, 539–552.

Singpurwalla, Nozer D., and Wong, Man-Yuen (1983). Kernel estimators of the failure-rate function and density estimation: An analogy, *Journal of the American Statistical Association*, **78**, 478–481.

Sint, P. (1984). Roots of computational statistics, in *COMPSTAT 84: Proceedings in Computational Statistics*, T. Havránek, Z. Šidák, and M. Novák, editors. Physica-Verlag: Vienna, 9–20.

Skates, Steven James (1987). *Laplacian and Uniform Expansions with Applications to Multidimensional Sampling*. Ph.D. thesis, Department of Statistics, The University of Chicago.

Smith, B. T., Boyle, J. M., Dongarra, J. J., Garbow, B. S., Ikebe, Y., Klema, Virginia C., and Moler, Cleve B. (1976) *Matrix Eigensystem Routines – EISPACK Guide*, Second edition. Lecture Notes in Computer Science, **6**, Springer-Verlag: Berlin.

Steen, N. M., Byrne, G. D., and Gelbard, E. M. (1969). Gaussian quadratures for the integrals $\int_0^\infty \exp(-x^2) f(x)\, dx$ and $\int_0^b \exp(-x^2) f(x)\, dx$, *Mathematics of Computation*, **23**, 661–671.

Stevenson, David (1981). A proposed standard for binary floating-point arithmetic. *Computer*, **14:3**, 51–62.

Stewart, G. W. (1973). *Introduction to Matrix Computations*. Academic Press: New York.

Stewart, G. W. (1974). Modifying pivot elements in Gaussian elimination, *Mathematics of Computation*, **28**, 527–542.

Stewart, G. W. (1979). The effects of rounding error on an algorithm for downdating a Cholesky factorization, *J. Inst. Math. Applic.*, **23**, 203–213.

Stewart, G. W. (1983). A method for computing the generalized singular value decomposition, in *Matrix Pencils*, B. Kagstrom and A. Ruhe, eds. Springer-Verlag: New York, 207–220.

Stewart, G. W. (1985). A Jacobi-like algorithm for computing the Schur decomposition of a nonhermitian matrix, *SIAM Journal on Scientific and Statistical Computing*, **6**, 853–864.

Stewart, G. W. (1986). Collinearity, scaling, and rounding error, in *Computer Science and Statistics: Seventeenth Symposium on the Interface, Proceedings*, David M. Allen, editor. North Holland: Amsterdam, 195–198.

Stewart, G. W. (1987). Collinearity and least squares regression, (with Discussion), *Statistical Science*, **2**, 68–100.

Stigler, Stephen M. (1980). Stigler's law of eponymy, *Transactions of the New York Academy of Sciences, Series II*, **39**, 147–158.

Stirling, W. Douglas (1984). Iteratively reweighted least-squares for models with a linear part, *Applied Statistics*, **31**, 7–17.

Stone, Charles J. (1985). Additive regression and other nonparametric models, *Annals of Statistics*, **13**, 689–705.]

Stone, Charles J., and Koo, Cha-Yong (1985). Additive splines in statistics, in *Proceedings of the Statistical Computing Section 1985*, American Statistical Association, 45–48.

Stone, M. (1974). Cross-validatory choice and assessment of statistical predictions, *Journal of the Royal Statistical Society*, Series B, **36**, 111-147.

Stroud, A. H. (1971). *Approximate Calculation of Multiple Integrals*. Prentice-Hall: Englewood Cliffs.

Stroud, A. H., and Secrest, D. H. (1966). *Gaussian Quadrature Formulas*. Prentice-Hall: Englewood Cliffs.

Switzer, P. (1970). Numerical classification. In *Geostatistics*, Plenum: New York.

Tanner, Martin A. (1984). Algorithm AS 202. Data-based non-parametric hazard estimation, *Applied Statistics*, **33**, 248–258.

Tanner, Martin, and Wong, Wing-Hung (1987). The calculation of posterior distributions by data augmentation, (with Discussion), *Journal of the American Statistical Association*, **82**, 528–550.

Terrell, George R., and Scott, David W. (1985). Oversmoothed nonparametric density estimates, *Journal of the American Statistical Association*, **80**, 209–214.

Thisted, Ronald A. (1976). *Ridge Regression, Minimax Estimation, and Empirical Bayes Methods.* PhD Thesis, Department of Statistics, Stanford University. [Also appeared as Division of Biostatistics Technical Report Number 27, now out of print. Available through University Microfilms: Ann Arbor, Michigan.]

Thisted, Ronald A. (1978). Multicollinearity, information, and ridge regression, Technical Report Number 66, Department of Statistics, The University of Chicago.

Thisted, Ronald A. (1981). The effect of personal computers on statistical practice, in *Computer Science and Statistics: Thirteenth Symposium on the Interface*, William F. Eddy, editor, 25–30.

Thisted, Ronald A. (1982). Decision-theoretic regression diagnostics, in *Statistical Decision Theory and Related Topics III*, **2**, S. S. Gupta and J. Berger, editors. Academic Press: New York, 363–382.

Thisted, Ronald A. (1984a). An appraisal of statistical graphics, in *Statistics: An Appraisal*, H. A. David and H. T. David, editors. Iowa State University Press: Ames, Iowa, 605–624.

Thisted, Ronald A. (1984b). Book review of *Statistical Software: A Comparative Review*, by Ivor Francis, *SIAM Review*, **26**, 294–297.

Thisted, Ronald A. (1986a). Computing environments for data analysis, (with Discussion), *Statistical Science*, **1**, 259–275.

Thisted, Ronald A. (1986b). Tools for data analysis management, in *Computer Science and Statistics: Eighteenth Symposium on the Interface, Proceedings*, Thomas J. Boardman, editor. American Statistical Association: Washington.

Thisted, Ronald A. (1986c). On minimization by successive approximation, Technical Report Number 190, Department of Statistics, The University of Chicago.

Thisted, Ronald A. (1987). Comment on Collinearity and least squares resgression, by G. W. Stewart, *Statistical Science*, **2**, 91–93.

Thisted, R. A., and Efron, B. (1987). Did Shakespeare write a newly-discovered poem? *Biometrika*, **74**, 445–455.

Tibshirani, Robert (1986). Estimating transformations for regression: A variation on ACE, Technical Report Number 1986-002, Biostatistics Group, University of Toronto.

Tierney, Luke, and Kadane, Joseph B. (1986). Accurate approximations for posterior moments and marginal densities, *Journal of the American Statistical Association*, **81**, 82–86.

Tishler, Asher, and Zang, Israel (1982). An absolute deviations curve-fitting algorithm for nonlinear models, in *Optimization in Statistics*, S. H. Zanakis and J. S. Rustagi, editors. North-Holland: Amsterdam, 81–103.

Titterington, D. M. (1985). Common structure of smoothing techniques in statistics, *International Statistical Review*, **53**, 141–170.

Traub, J. F. (1964). *Iterative Methods for the Solution of Equations*. Prentice-Hall: Englewood Cliffs.

Tufte, Edward R. (1983). *The Visual Display of Quantitative Information*. Graphics Press: Cheshire, Connecticut.

Tukey, John W. (1977). *Exploratory Data Analysis*. Addison-Wesley: Reading.

Tunnicliffe Wilson, G. (1979). Some efficient computational procedures for high order ARMA models, *Journal of Statistical Computation and Simulation*, **8**, 301–309.

Utreras, Florencio I. (1981). On computing robust splines and applications, *SIAM Journal on Scientific and Statistical Computing*, **2**, 153–163.

Vardi, Y., Shepp, L. A., and Kaufman, L. (1985). A statistical model for positron emission tomography, (with Discussion), *Journal of the American Statistical Association*, **80**, 8–37.

Wahba, Grace (1978). Improper priors, spline smoothing and the problem of guarding against model errors in regression, *Journal of the Royal Statistical Society*, Series B, **40**, 364–372.

Wahba, Grace (1981). Spline interpolation and smoothing on the sphere, *SIAM Journal on Scientific and Statistical Computing*, **2**, 5–16; *erratum* **3**, 385–386.

Wahba, Grace (1986). Partial and interaction splines for the semiparametric estimation of functions of several variables, in *Computer Science and Statistics: Eighteenth Symposium on the Interface, Proceedings*, Thomas J. Boardman, editor. American Statistical Association: Washington, in press.

Wahba, Grace (1987). Splines: An entry for the *Encyclopedia of Statistical Sciences*, Technical Report Number 806, Department of Statistics, University of Wisconsin.

Wallace, David L. (1959). Bounds on normal approximations to Student's and the Chi-square distributions, *Annals of Mathematical Statistics*, **30**,1121–1130.

Watkins, David S. (1982). Understanding the QR algorithm, *SIAM Review*, **24**, 427–440.

Webster, J. T., Gunst, Richard F., and Mason, R. L. (1974). Latent root regression analysis, *Technometrics*, **16**, 513–522.

Wecker, William E., and Ansley, Craig F. (1983). The signal extraction approach to nonlinear regression and spline smoothing, *Journal of the American Statistical Association*, **78**, 81–89.

Wedderburn, R. W. M. (1974). Quasi-likelihood functions, generalized linear models, and the Gauss-Newton method, *Biometrika*, **61**, 439–447.

Wegman, E. J. (1972a). Nonparametric probability density estimation I. A summary of available methods, *Technometrics*, **14**, 553–547.

Wegman, E. J. (1972b). Nonparametric probability density estimation II. A comparison of density estimation methods, *Journal of Statistical Computation and Simulation*, **1**, 225–245.

Wegman, Edward J., and Wright, Ian W. (1983). Splines in statistics, *Journal of the American Statistical Association*, **78**, 351–365.

Weisberg, Sanford (1980). *Applied Linear Regression*. Wiley: New York.

Welford, B. P. (1962). Note on a method for calculating corrected sums of squares and products, *Technometrics*, **4**, 419–420.

West, D. H. D. (1979). Updating mean and variance estimates: An improved method, *Communications of the ACM*, **22**, 532–535.

Wichura, Michael J. (1978). Unpublished tables.

Wichura, Michael J. (1987). An accurate algorithm for computing quantiles of the normal distribution, Department of Statistics, The University of Chicago.

Wilcoxon, Frank (1945). Individual comparisons by ranking methods, *Biometrics*, **1**, 80–83.

Wilkinson, J. H. (1963). *Rounding Errors in Algebraic Processes*. Prentice-Hall: Englewood Cliffs.

Wilkinson, J. H. (1965). *The Algebraic Eigenvalue Problem*. Clarendon Press: Oxford.

Wilson, E. B., and Hilferty, M. M. (1931). The distribution of chi-squares, *Proceedings of the National Academy of Sciences*, **17**, 684–688.

Woltring, H. J. (1986). A FORTRAN package for generalized, cross-validatory spline smoothing and differentiation, *Advances in Engineering Software*, **8**, 104–113.

Wong, Wing-Hung (1987). Personal communication.

Woodbury, M. (1950). Inverting modified matrices. Memorandum number 42, Statistical Research Group, Princeton University.

Wright, S. J., and Holt, J. N. (1985). Algorithms for nonlinear least squares with linear inequality constraints, *SIAM Journal on Scientific and Statistical Computing*, **6**, 1033–1048.

Wu, C. F. Jeff (1983). On the convergence properties of the EM algorithm, *Annals of Statistics*, **11**, 95–103.

Yip, E. L. (1986). A note on the stability of solving a rank-p modification of a linear system by the Sherman-Morrison-Woodbury formula, *SIAM Journal on Scientific and Statistical Computing*, **7**, 507–513.

Youngs, E. A., and Cramer, E. M. (1971). Some results relevant to choice of sum and sum-of-product algorithms, *Technometrics*, **13**, 657–665.

INDEX

for generalized linear models, 224.
for heteroscedasticity, 114.
for serial correlation, 114.
in nonlinear estimation, 224.
MLV (Moolgavkar, Lustbader, Venzon), 224.
multiple-row, 110.
variable, 110–111.
Diagnostics,, *See also* Regression diagnostics, 106–115.
Diagonal rational interpolant, 327.
DiDonato, A. R., 314, 394.
Dielman, Terry, 149, 394.
Dierckx, P., 360, 394.
Differences, central, 198.
Digital Equipment Corporation, 35.
Dirac delta function, 338.
Discrepancy, of random numbers, 306.
Discrete Newton methods, 183–184, 190, 197.
Dispersion function, 216.
Dispersion parameter, 217.
Display, *See* Graphics.
Distribution function, *See* Cumulative distribution function.
Divide-and-conquer, 46.
Divided differences, 296.
Dixon, Wilfrid Joseph, xx.
Dongarra, J. J., 394, 396, 406.
Double sweep, in Gauss-Seidel, 146.
Double-precision register model, 37.
Downdating, 116–119.
Draftsman's spline, 294.
Draper, Norman, 61, 394.
Dummy variable, 137.
Durbin, James, 114, 394.
Dutt, John E., 314, 385, 394.
Dutter, R., 151, 394.
Dynamic programming, 155.

Earthquakes, xvii.
ecdf: Empirical cumulative distribution function.
Eddy, William F., 29, 394.
Edgeworth series, 8.
Efron, Bradley, 18–19, 177, 192, 227, 394, 408.
Eigensystem, 119 (defined), 122.
Eigenvalue, 97 (defined), 112, 120.
Eigenvalue extraction, and underflow, 58.
Eigenvalue problem, symmetric, 123.
Eigenvector, 97 (defined), 113, 120.
EISPACK (linear algebra software), 122, 394.
EISPACK 2 (linear algebra software, too), 394.

Elbow, 354.
Elementary operations, 80.
Elimination methods, 80.
Elimination, solution of linear equality constraints by, 221.
Elliott-NRDC 401 (computer), 29.
ELS: Exact least squares (time series).
EM algorithm, 239–243, 248, 317.
E-step, 241.
M-step, 241.
Enslein, Kurt, 395.
Environment for data analysis, 17, 27.
Eponymy, *See* Stigler's Law.
Epsilon-minimax, defined, 204.
Equidistribution, 306.
Equivariance, location, 303, 307.
erf: Error function.
erfc: Complementary error function.
Error analysis.
backward, 41.
forward, 40–41, 55.
Error bounds, 40, 321.
Error bounds for quadrature.
differentiability assumption, 261.
smoothness condition, 261.
Error function (erf), 327, 330.
relation to normal cdf, 330.
Error structure.
autoregressive, 140.
intraclass, 140.
Error tolerances, relative, 265.
Euler, Leonhard, 324.
Euler summation formula, 269–270.
Evans, M., 313, 314, 395.
Even part, of continued fractions, 325.
Exact least squares (ELS), in time-series estimation, 251–252.
Exact likelihood, for stationary time series, 244.
Excess-128, 35.
Excess-64, 35.
Experimental design, 23.
sequential, 313.
Expert systems, 27.
Exponential distribution, 284.
Exponential family, 242, 243, 378.
Extended midpoint rule (quadrature), 274, 279.
Extended trapezoidal rule, 264–265, 268–269.
Extreme value distribution, 329.

F distribution, 333–334, 335 336.
Faces, *See* Chernoff's faces.
Factor analysis, 3, 95.
Farebrother, Richard W., 222, 314, 395.
Fast Fourier transform, 244, 343.